VOLUME SEVEN HUNDRED AND THREE

METHODS IN
ENZYMOLOGY

Mononuclear Non-heme Iron
Dependent Enzymes Part A

METHODS IN ENZYMOLOGY

Editors-in-Chief

ANNA MARIE PYLE

*Departments of Molecular, Cellular and Developmental
Biology and Department of Chemistry
Investigator, Howard Hughes Medical Institute
Yale University*

DAVID W. CHRISTIANSON

*Roy and Diana Vagelos Laboratories
Department of Chemistry
University of Pennsylvania
Philadelphia, PA*

Founding Editors

SIDNEY P. COLOWICK and NATHAN O. KAPLAN

VOLUME SEVEN HUNDRED AND THREE

METHODS IN
ENZYMOLOGY

Mononuclear Non-heme Iron Dependent Enzymes Part A

Edited by

JENNIFER BRIDWELL-RABB
Department of Chemistry, University of Michigan, Ann Arbor, MI, United States

Academic Press is an imprint of Elsevier
50 Hampshire Street, 5th Floor, Cambridge, MA 02139, United States
525 B Street, Suite 1650, San Diego, CA 92101, United States
125 London Wall, London, EC2Y 5AS, United Kingdom

First edition 2024

Copyright © 2024 Elsevier Inc. All rights are reserved, including those for text and data mining, AI training, and similar technologies.

Publisher's note: Elsevier takes a neutral position with respect to territorial disputes or jurisdictional claims in its published content, including in maps and institutional affiliations.

No part of this publication may be reproduced or transmitted in any form or by any means, electronic or mechanical, including photocopying, recording, or any information storage and retrieval system, without permission in writing from the publisher. Details on how to seek permission, further information about the Publisher's permissions policies and our arrangements with organizations such as the Copyright Clearance Center and the Copyright Licensing Agency, can be found at our website: www.elsevier.com/permissions.

This book and the individual contributions contained in it are protected under copyright by the Publisher (other than as may be noted herein).

Notices

Knowledge and best practice in this field are constantly changing. As new research and experience broaden our understanding, changes in research methods, professional practices, or medical treatment may become necessary.

Practitioners and researchers must always rely on their own experience and knowledge in evaluating and using any information, methods, compounds, or experiments described herein. In using such information or methods they should be mindful of their own safety and the safety of others, including parties for whom they have a professional responsibility.

To the fullest extent of the law, neither the Publisher nor the authors, contributors, or editors, assume any liability for any injury and/or damage to persons or property as a matter of products liability, negligence or otherwise, or from any use or operation of any methods, products, instructions, or ideas contained in the material herein.

ISBN: 978-0-443-31304-2
ISSN: 0076-6879

For information on all Academic Press publications
visit our website at https://www.elsevier.com/books-and-journals

Publisher: Zoe Kruze
Editorial Project Manager: Saloni Vohra
Production Project Manager: James Selvam
Cover Designer: Gopalakrishnan Venkatraman
Typeset by MPS Limited, India

Contents

Contributors   *xv*

Section 1 Methods for studying the catalytic mechanisms of mononuclear non-heme iron dependent enzymes

1. Characterization of O_2 uncoupling in biodegradation   **3**
reactions of nitroaromatic contaminants catalyzed by
rieske oxygenases

Charlotte E. Bopp, Nora M. Bernet, Sarah G. Pati, and
Thomas B. Hofstetter

1.	Introduction	4
2.	Mass Balances and Reaction Stoichiometries	6
	2.1 Quantification strategies and limitations	6
	2.2 Quantification of O_2 uncoupling	8
3.	Quantification of transient reactive oxygen species	16
	3.1 Survey of ROS quantification approaches	16
	3.2 Catalase-based assays	19
	3.3 Horseradish peroxidase-based assays	20
4.	Reaction kinetics	20
	4.1 Quantification of O_2 uncoupling from reaction kinetics	20
	4.2 Organic substrate kinetics	21
	4.3 Oxygen consumption kinetics	22
5.	Summary and Conclusions	24
	Acknowledgments	25
	Competing interests	25
	References	25

2. Spectroscopic definition of ferrous active sites in   **29**
non-heme iron enzymes

Edward I. Solomon and Robert R. Gipson

1.	d^6 Ligand field theory (LFT)	31
2.	LF spectroscopy = low temperature magnetic circular dichroism (LT MCD)	32
3.	Variable-temperature, variable-field (VTVH) MCD	36

v

4. LFT of spin-hamiltonian parameters from VTVH MCD	39
5. An early application of VTVH MCD on a non-heme Fe(II) enzyme	42
6. Perspective	45
Acknowledgments	46
References	46

3. Equilibrium dialysis with HPLC detection to measure substrate binding affinity of a non-heme iron halogenase   **51**

Elizabeth R. Smithwick, Ambika Bhagi-Damodaran, and
Anoop Rama Damodaran

1. Introduction	52
2. Materials	53
2.1 Quantitation of ligand and calibration curve	53
2.2 Equilibrium dialysis apparatus setup and initial measurements	54
2.3 Equilibrium dialysis for determination of substrate affinity in BesD	54
3. Methods	54
3.1 Quantitation of ligand and calibration curve	54
3.2 Equilibrium dialysis apparatus setup and initial measurements	55
3.3 Equilibrium dialysis for determination of substrate affinity in BesD	58
4. Conclusions	60
Acknowledgments	62
Author contributions	62
References	62

4. Preparation of reductases for multicomponent oxygenases   **65**

Megan E. Wolf and Lindsay D. Eltis

1. Introduction	66
2. General safety	69
3. Reductase production and activity	69
3.1 Overview	69
3.2 Expression vector cloning	69
3.3 Transformation of *E. coli* and RHA1 with expression vectors	70
3.4 Production of PbdB	71
3.5 Activity of lysates	72
3.6 Results	73
4. Codon optimization and protein purification	74
4.1 Overview	74
4.2 Codon optimization	74

Contents

4.3	Protein purification	76
4.4	Results	77
5.	Protein characterization	77
5.1	Overview	77
5.2	Cofactor analysis – labile sulfide	78
5.3	Cofactor analysis – non-heme iron	78
5.4	Cofactor analysis – FAD	79
5.5	Activity analysis – cytochrome c reduction	79
5.6	Results	80
6.	Summary and conclusions	81
	References	82

5. Development of a rapid mass spectrometric method for the analysis of ten-eleven translocation enzymes

87

Clara Graves and Kabirul Islam

1.	Introduction	88
2.	Preparation of the materials	90
2.1	Expression and purification of wild type TET2	90
2.2	Mutagenesis and expression of V1395A	95
2.3	Synthesis and characterization of oligonucleotides	99
3.	Biochemical assays and results	103
3.1	Development of a robust in-vitro assay	103
3.2	Measurement of IC_{50} of TET2 inhibitors NOG and 2HG	108
3.3	Validating the activity of wildtype TET2 and V1395A using BS-seq	111
4.	Notes	114
	Funding	118
	References	118

6. Non-standard amino acid incorporation into thiol dioxygenases

121

Zachary D. Bennett and Thomas C. Brunold

1.	Overview	122
2.	Eukaryotic thiol dioxygenases	123
2.1	CDO	123
2.2	ADO	126
3.	Genetic code expansion	128
3.1	Overview	128

3.2 Suppressor tRNA/aminoacyl-tRNA synthetase pairs and the pEVOL plasmid	129
3.3 Selenocysteine incorporation	131
4. Application to thiol dioxygenases	134
4.1 Fluorotyrosine incorporation into CDO and ADO using pEVOL F2Y	134
4.2 Sec incorporation into ADO	136
5. Conclusions	140
Acknowledgments	140
References	140
Further reading	145

7. Unveiling the mechanism of cysteamine dioxygenase: A combined HPLC-MS assay and metal-substitution approach — 147

Ran Duan, Jiasong Li, and Aimin Liu

1. Introduction	148
2. Protein expression, purification, and crystallization	151
2.1 Equipment	152
2.2 Reagents	153
2.3 Procedure	153
2.4 Note	155
3. Spectral characterization of Co-ADO	155
3.1 Equipment	156
3.2 Reagents	157
3.3 Optical spectral characterization	157
3.4 EPR spectral characterization	157
4. Cobalt reconstitution in ADO	157
4.1 Equipment	158
4.2 Reagents	158
4.3 Preparation of "apo-ADO" through 1,10-phenanthroline assay	158
4.4 Evaluation of the "apo-ADO" using ferrozine assay	159
4.5 Reconstitution of ADO enzyme by adding divalent metal ions	159
4.6 Note	159
5. HPLC-MS analysis of the hypotaurine formation by ADO	160
5.1 Equipment	161
5.2 Reagents	162
5.3 Procedure	162
5.4 Note	163

Contents ix

6. Summary and conclusions 164
Acknowledgments 164
References 164

8. *In vitro* analysis of the three-component Rieske oxygenase cumene dioxygenase from *Pseudomonas fluorescens* IP01   **167**

Niels A.W. de Kok, Hui Miao, and Sandy Schmidt

  1. Introduction 168
  2. *In vitro* analysis of Rieske oxygenases 171
  3. Expression of Cumene dioxygenase 173
    3.1 Materials 173
    3.2 Buffers and reagents 174
    3.3 Equipment 174
  4. Step-by-step method details 174
  5. General considerations 175
  6. Cell lysis and protein purification 177
    6.1 Materials and equipment 177
    6.2 Equipment 177
  7. Step-by-step method details 177
    7.1 Cell lysis by sonication 177
    7.2 Purification by immobilized metal ion chromatography (IMAC) 178
    7.3 Desalting by size exclusion chromatography (SEC) 179
    7.4 Concentration by centrifugal ultrafiltration 180
  8. General considerations 180
  9. Enzymatic activity assay 182
    9.1 Materials and equipment 182
    9.2 Equipment 183
 10. Step-by-step method details 183
    10.1 Enzymatic reaction 183
    10.2 Sample extraction 183
    10.3 Non-chiral GC-MS and chiral GC-FID analysis 184
 11. General considerations 185
 12. Summary and conclusions 187
Acknowledgments 187
References 187

Section 2 Leveraging mononuclear non-heme iron enzymes for biocatalysis

9. Radical-relay C(sp³)–H azidation catalyzed by an engineered nonheme iron enzyme
195

Qun Zhao, Jinyan Rui, and Xiongyi Huang

1. Introduction	196
2. Materials	198
2.1 Cloning	198
2.2 Enzyme expression in *E. coli*	201
2.3 Whole-cell reaction	201
2.4 GCMS (gas chromatography–mass spectrometry) and normal phase HPLC (high performance liquid chromatography) analysis	201
3. Protocols	202
3.1 Cloning for a site-saturated mutagenesis screening library	202
3.2 High-throughput experimentation in 96-well plates	205
3.3 Analytical scale reactions to validate the screening hits	207
3.4 Preparative-scale reactions	209
4. Summary	211
Acknowledgments	211
References	211

10. Purification and characterization of a Rieske oxygenase and its NADH-regenerating partner proteins
215

Gage T. Barroso, Alejandro Arcadio Garcia, Madison Knapp, David G. Boggs, and Jennifer Bridwell-Rabb

1. Introduction	216
2. Considerations for assembling a Rieske oxygenase pathway *in vitro*	219
3. Protein constructs for recombinant expression and purification	220
3.1 Assembly of needed constructs for protein isolation	220
3.2 Transformation protocol for the TsaMBCD pathway encoding genes	221
4. Recombinant expression and purification of the TsaM, TsaC, TsaD, and VanB	221
4.1 Recombinant expression of the TsaM, TsaC, TsaD, and VanB encoding genes	221
4.2 Purification of TsaM, VanB, and TsaC	223
4.3 Purification of the NAD⁺-dependent aldehyde dehydrogenase TsaD	225

5.	Methods for assessing the quality of the purified TsaMBCD pathway proteins	226
	5.1 Biochemical analysis of purified proteins	226
	5.2 Quantification of the iron content in TsaM and VanB	227
6.	Enzymatic assays for the TsaMBCD pathway	228
	6.1 Liquid chromatography mass spectrometry (LC-MS) methods for activity assays	228
	6.2 Separation of TsaMBCD pathway intermediates using LC-MS	230
	6.3 Identification of the optimal conditions for measuring the activity of TsaM	231
	6.4 Total turnover number (TTN) determination using LC-MS	232
	6.5 Spectroscopic assay for analysis of NAD(P)H consumption and production	234
7.	Crystallization of the short-chain dehydrogenase/reductase (SDR) enzyme TsaC	236
8.	Conclusions	238
	Acknowledgments	239
	References	239

11. Whole-cell Rieske non-heme iron biocatalysts 243

Meredith B. Mock, Shuyuan Zhang, and Ryan M. Summers

1.	Introduction	244
2.	Before you begin timing: 4–5 days	246
3.	Key resources table	248
4.	Materials and equipment	249
	4.1 Equipment	249
	4.2 Materials and reagents	249
5.	Step-by-step method details	249
	5.1 Cell growth and gene expression	250
	5.2 Resting cell reactions	251
	5.3 Reaction sampling and sample preparation	252
6.	High-performance liquid chromatography (HPLC) analysis	253
7.	Establishing a calibration curve	254
8.	Expected outcomes	255
9.	Quantification and statistical analysis	256
10.	Advantages	258
11.	Limitations	258
12.	Optimization and troubleshooting	258

12.1	No or low activity detected	258
12.2	Potential solutions to optimize the procedure	259
13.	Safety considerations and standards	259
14.	Alternative methods/procedures	260
References		260

12. Photo-reduction facilitated stachydrine oxidative N-demethylation reaction: A case study of Rieske non-heme iron oxygenase Stc2 from *Sinorhizobium meliloti*   **263**

Tao Zhang, Kelin Li, Yuk Hei Cheung, Mark W. Grinstaff, and Pinghua Liu

1.	Introduction	265
2.	Heterologous expression and purification of Stc2	267
3.	Materials and equipment	269
	3.1 Equipment	269
	3.2 Solutions and consumables	269
4.	Step-by-step procedure details	270
	4.1 *Escherichia coli* starter culture	270
	4.2 Stc2 protein overexpression	270
	4.3 Anaerobic Stc2 protein purification	271
	4.4 Concentrating the eluted Stc2 in the COY chamber	275
5.	Expected outcomes, advantages, and disadvantages	276
6.	Optimization and troubleshooting	276
7.	Safety considerations and standards	277
8.	Stc2 characterization	277
9.	Materials and equipment	278
	9.1 Equipment	278
	9.2 Solutions and consumables	278
10.	Step-by-step details	279
	10.1 Stc2 protein analysis	279
	10.2 Stc2 iron content quantification	280
	10.3 Quantification of labile sulfur in Stc2 based on the reaction shown in Fig. 6	282
	10.4 Optimization and troubleshooting	285
11.	Stc2 photo-reduction using eosin Y and Na_2SO_3	285
12.	Materials and equipment	286
	12.1 Equipment	286
	12.2 Solutions and consumables	286

13.	Step-by-step method details	286
13.1	Light-driving Stc2 Fe-S cluster reduction using eosin Y and Na_2SO_3 as photosensitizer/sacrificial reagent pair	287
13.2	Light-driving demethylation of stachydrine using eosin Y and Na_2SO_3 under multiple turnover condition	288
13.3	Stc2-catalysis in a flow-setting	290
13.4	Optimization and troubleshooting	292
	References	294

13. Functional and spectroscopic approaches to determining thermal limitations of Rieske oxygenases 299

Jessica Lusty Beech, Julia Ann Fecko, Neela Yennawar, and Jennifer L. DuBois

1.	Introduction	300
1.1	Multimeric mononuclear iron oxygenases and their engineering: Historical perspectives	301
1.2	Rieske iron oxygenases and their engineering: Efforts towards substrate expansion and functional improvement	302
1.3	Multimeric Rieske iron oxygenases and their engineering: Outline of unmet potential	304
1.4	Overview and key parameters derived from each method	305
2.	Expression and purification of a RO system	306
2.1	Equipment	306
2.2	Expression and purification of TPA_{DO}	307
2.3	Expression and purification of TPA_{RED}	309
2.4	Notes	309
3.	Iron cofactor lability	311
3.1	Equipment	311
3.2	Procedure	311
3.3	Data analysis	312
3.4	Notes	313
4.	Temperature dependent kinetics	313
4.1	Equipment	314
4.2	Procedure	314
4.3	Data analysis	314
4.4	Notes	316
5.	Lifetime	317
5.1	Equipment	317
5.2	Procedure	318

5.3 Data analysis	318
5.4 Notes	320
6. Differential scanning calorimetry (DSC)	320
6.1 Equipment	320
6.2 Procedure	321
6.3 Data analysis	322
6.4 Notes	322
References	328

Contributors

Gage T. Barroso
Department of Chemistry, University of Michigan, Ann Arbor, MI, United States

Jessica Lusty Beech
Department of Chemistry and Biochemistry, Montana State University, Bozeman, MT, United States

Zachary D. Bennett
Department of Chemistry, University of Wisconsin-Madison, Madison, WI, United States

Nora M. Bernet
Eawag, Swiss Federal Institute of Aquatic Science and Technology, Dübendorf; Institute of Biogeochemistry and Pollutant Dynamics (IBP), ETH Zurich, Zurich, Switzerland

Ambika Bhagi-Damodaran
Department of Chemistry, University of Minnesota – Twin Cities, Minneapolis, MN, United States

David G. Boggs
Department of Chemistry, University of Michigan, Ann Arbor, MI, United States

Charlotte E. Bopp
Eawag, Swiss Federal Institute of Aquatic Science and Technology, Dübendorf; Institute of Biogeochemistry and Pollutant Dynamics (IBP), ETH Zurich, Zurich, Switzerland

Jennifer Bridwell-Rabb
Department of Chemistry, University of Michigan, Ann Arbor, MI, United States

Thomas C. Brunold
Department of Chemistry, University of Wisconsin-Madison, Madison, WI, United States

Yuk Hei Cheung
Department of Chemistry, Boston University, Boston, MA, United States

Anoop Rama Damodaran
Department of Chemistry, University of Minnesota – Twin Cities, Minneapolis, MN, United States

Jennifer L. DuBois
Department of Chemistry and Biochemistry, Montana State University, Bozeman, MT, United States

Ran Duan
Department of Chemistry, University of Texas at San Antonio, San Antonio, TX, United States

Lindsay D. Eltis
Microbiology and Immunology, The University of British Columbia, Vancouver, BC, Canada

Julia Ann Fecko
The Huck Institutes of the Life Sciences, The Pennsylvania State University, University Park, PA, United States

Alejandro Arcadio Garcia
Department of Chemistry, University of Michigan, Ann Arbor, MI, United States

Robert R. Gipson
Department of Chemistry, Stanford University, Stanford, CA, United States

Clara Graves
Department of Chemistry, University of Pittsburgh, Pittsburgh, PA, United States

Mark W. Grinstaff
Department of Chemistry, Boston University, Boston, MA, United States

Thomas B. Hofstetter
Eawag, Swiss Federal Institute of Aquatic Science and Technology, Dübendorf; Institute of Biogeochemistry and Pollutant Dynamics (IBP), ETH Zurich, Zurich, Switzerland

Xiongyi Huang
Department of Chemistry, Johns Hopkins University, Baltimore, MD, United States

Kabirul Islam
Department of Chemistry, University of Pittsburgh, Pittsburgh, PA, United States

Madison Knapp
Department of Chemistry, University of Michigan, Ann Arbor, MI, United States

Jiasong Li
Department of Chemistry, University of Texas at San Antonio, San Antonio, TX, United States

Kelin Li
Department of Chemistry, Boston University, Boston, MA, United States

Aimin Liu
Department of Chemistry, University of Texas at San Antonio, San Antonio, TX, United States

Pinghua Liu
Department of Chemistry, Boston University, Boston, MA, United States

Hui Miao
Department of Chemical and Pharmaceutical Biology, Groningen Research Institute of Pharmacy, University of Groningen, Groningen, The Netherlands

Meredith B. Mock
Department of Chemical and Biological Engineering, The University of Alabama, Tuscaloosa, AL, United States

Sarah G. Pati
Department of Environmental Geosciences, Centre for Microbiology and Environmental Systems Science, University of Vienna, Vienna, Austria

Jinyan Rui
Department of Chemistry, Johns Hopkins University, Baltimore, MD, United States

Sandy Schmidt
Department of Chemical and Pharmaceutical Biology, Groningen Research Institute of Pharmacy, University of Groningen, Groningen, The Netherlands

Elizabeth R. Smithwick
Department of Chemistry, University of Minnesota – Twin Cities, Minneapolis, MN, United States

Edward I. Solomon
Department of Chemistry, Stanford University, Stanford; Stanford Synchrotron Radiation Lightsource, SLAC National Acceleration Laboratory, Stanford University, Menlo Park, CA, United States

Ryan M. Summers
Department of Chemical and Biological Engineering, The University of Alabama, Tuscaloosa, AL, United States

Megan E. Wolf
Microbiology and Immunology, The University of British Columbia, Vancouver, BC, Canada

Neela Yennawar
The Huck Institutes of the Life Sciences, The Pennsylvania State University, University Park, PA, United States

Shuyuan Zhang
Department of Chemical and Biological Engineering, The University of Alabama, Tuscaloosa, AL, United States

Tao Zhang
Department of Chemistry, Boston University, Boston, MA, United States

Qun Zhao
School of Biotechnology and Key Laboratory of Industrial Biotechnology of Ministry of Education, Jiangnan University, Wuxi, P.R. China

Niels A.W. de Kok
Department of Chemical and Pharmaceutical Biology, Groningen Research Institute of Pharmacy, University of Groningen, Groningen, The Netherlands

SECTION 1

Methods for studying the catalytic mechanisms of mononuclear non-heme iron dependent enzymes

CHAPTER ONE

Characterization of O_2 uncoupling in biodegradation reactions of nitroaromatic contaminants catalyzed by rieske oxygenases

Charlotte E. Bopp[a,b], Nora M. Bernet[a,b], Sarah G. Pati[c], and Thomas B. Hofstetter[a,b,*]

[a]Eawag, Swiss Federal Institute of Aquatic Science and Technology, Dübendorf, Switzerland
[b]Institute of Biogeochemistry and Pollutant Dynamics (IBP), ETH Zürich, Zürich, Switzerland
[c]Department of Environmental Geosciences, Centre for Microbiology and Environmental Systems Science, University of Vienna, Vienna, Austria
*Corresponding author. e-mail address: thomas.hofstetter@eawag.ch

Contents

1. Introduction	4
2. Mass Balances and Reaction Stoichiometries	6
2.1 Quantification strategies and limitations	6
2.2 Quantification of O_2 uncoupling	8
3. Quantification of transient reactive oxygen species	16
3.1 Survey of ROS quantification approaches	16
3.2 Catalase-based assays	19
3.3 Horseradish peroxidase-based assays	20
4. Reaction kinetics	20
4.1 Quantification of O_2 uncoupling from reaction kinetics	20
4.2 Organic substrate kinetics	21
4.3 Oxygen consumption kinetics	22
5. Summary and Conclusions	24
Acknowledgments	25
Competing interests	25
References	25

Abstract

Rieske oxygenases are known as catalysts that enable the cleavage of aromatic and aliphatic C–H bonds in structurally diverse biomolecules and recalcitrant organic environmental pollutants through substrate oxygenations and oxidative heteroatom dealkylations. Yet, the unproductive O_2 activation, which is concomitant with the release of reactive oxygen species (ROS), is typically not taken into account when

Methods in Enzymology, Volume 703
ISSN 0076-6879, https://doi.org/10.1016/bs.mie.2024.05.010
Copyright © 2024 Elsevier Inc. All rights are reserved, including those for text and data mining, AI training, and similar technologies

characterizing Rieske oxygenase function. Even if considered an undesired side reaction, this O_2 uncoupling allows for studying active site perturbations, enzyme mechanisms, and how enzymes evolve as environmental microorganisms adapt their substrates to alternative carbon and energy sources. Here, we report on complementary methods for quantifying O_2 uncoupling based on mass balance or kinetic approaches that relate successful oxygenations to total O_2 activation and ROS formation. These approaches are exemplified with data for two nitroarene dioxygenases (nitrobenzene and 2-nitrotoluene dioxygenase) which have been shown to mono- and dioxygenate substituted nitroaromatic compounds to substituted nitrobenzylalcohols and catechols, respectively.

1. Introduction

Rieske oxygenases play important roles in many catabolic and biosynthetic processes owing to the ability of this class of non-heme Fe enzymes to activate molecular O_2 for the oxygenations and oxidative heteroatom dealkylations of numerous aliphatic and (poly)aromatic substrates (Barry & Challis, 2013; Dunham & Arnold, 2020; Jiang, Wilson, & Weeks, 2013; Knapp, Mendoza, & Bridwell-Rabb, 2021; Lukowski et al., 2018; Münch, Püllmann, Zhang, & Weissenborn, 2021; Perry, de Los Santos, Alkhalaf, & Challis, 2018; Tian, Garcia, Donnan, & Bridwell-Rabb, 2023). Even though the role of Rieske oxygenases for biocatalysis is well-known, the factors that lead to successful substrate oxygenation remain elusive. On the one hand, Rieske oxygenases require the transfer of two electrons from the substrate and retrieve two additional reduction equivalents for O_2 activation. This process occurs through long-range electron transfer from an NAD(P)H reductase to the Rieske cluster of the oxygenase component often involving a Rieske [2Fe–2S] ferredoxin as third component of the enzyme system (Ashikawa et al., 2012; Friemann et al., 2005; Furusawa et al., 2004; Karlsson et al., 2003; Runda, De Kok, & Schmidt, 2023a). On the other hand, the catalytic outcome is also defined by structural elements of the oxygenase which are determined especially by residues of substrate and O_2 tunnels, flexible loops, and in the active site (Aukema et al., 2017; Brimberry, Garcia, Liu, Tian, & Bridwell-Rabb, 2023; Csizi, Eckert, Brunken, Hofstetter, & Reiher, 2022; Heinemann, Armbruster, & Hauer, 2021; Knapp et al., 2021; Liu, Knapp, Jo, Dill, & Bridwell-Rabb, 2022; Lukowski et al., 2018; Lukowski, Liu, Bridwell-Rabb, & Narayan, 2020; Tian et al., 2023). However, Rieske oxygenase substrates do not coordinate at the non-heme Fe^{II} center. It is thus especially difficult to pinpoint the molecular events that initiate structural and electronic changes in Rieske oxygenases which lead to O_2 activation and successful substrate oxygenation. Moreover, as we have recently illustrated for nitroarene Rieske oxygenases exhibiting a broad substrate spectrum, these enzymes activate O_2 quite inefficiently concomitant with the generation and release of reactive oxygen species (ROS) (Bopp, Bernet, Kohler, & Hofstetter, 2022; Bopp et al., 2024; Pati, Bopp, Kohler, & Hofstetter, 2022). This phenomenon of so-called O_2 uncoupling further complicates the study of Rieske oxygenase activity. Under these circumstances, the lack of reactivity towards a substrate is no longer due to a lack of enzyme activity but the consequence of unproductive O_2 activation to ROS and potential enzyme inactivation through self-hydroxylation.

The extent of O_2 uncoupling can indeed be considerable even if Rieske oxygenases react with their eponymous (sometimes referred to as "native") substrates. Naphthalene dioxygenase (NDO), for example, catalyses the reaction of naphthalene with O_2 to *cis*-naphthalene dihydrodiol stoichiometrically (Lee & Gibson, 1996). Conversely, only 30% of the available O_2 is used by nitrobenzene dioxygenase (NBDO) to oxygenate nitrobenzene (Bopp, Kohler, & Hofstetter, 2020; Pati et al., 2022). Measures for Rieske oxygenase activity such as k_{cat}, k_{cat}/K_m of the organic substrate, and turnover numbers are thus smaller if O_2 uncoupling occurs and silently include the inefficient use of reduction equivalents (i.e., NAD(P)H) for wasteful O_2 activation. While knowledge of O_2 uncoupling may or may not be of interest for the characterization of Rieske oxygenase activity, quantification of O_2 uncoupling allows for insights into important aspects of Rieske oxygenase function. In fact, considerations of O_2 uncoupling are not limited to Rieske oxygenases but apply to all O_2-activating enzymes (Huang & Groves, 2018). Examples include the evaluation of O_2 uncoupling as proxy for active-site perturbations affecting substrate positioning (McCusker & Klinman, 2009, 2010), the evaluation of enzyme mechanisms (Pati et al., 2022), and, most recently, the consequences of adaptations to ROS formation (Bopp et al., 2024). For protection against oxidative damage, oxidases and oxygenases can transport potentially damaging oxidizing equivalents away from active sites towards the protein surface through redox-active tyrosine/tryptophan chains (Gray & Winkler, 2015, 2018; Ravanfar, Sheng, Gray, & Winkler, 2023; Teo et al., 2019). As a consequence of this hole hopping process and the scavenging of oxidation equivalents by cellular reductants, reconfigurations of metabolic fluxes with faster turnover of reduction equivalents have been observed (Nikel et al., 2021). Such phenomena affect microorganisms that are able to adapt to alternative growth substrates, for example, in soil and aqueous environments where microbes may evolve novel metabolic capabilities due to the presence of organic contaminants (Bopp et al., 2024; Pérez-Pantoja, Nikel, Chavarria, & de Lorenzo, 2013).

Despite the relevance of O_2 uncoupling for studying functions of Rieske oxygenase, procedures for its systematic quantification have just recently been developed (Bopp et al., 2022; Pati et al., 2022; Runda, Miao, De Kok, & Schmidt, 2024). From a practical perspective, this omission is likely due to the tedious work associated with the quantification of many different species in laboratory enzyme assays. Species like O_2 and ROS are of transient nature due to the various reductants in the biological matrices required for studying Rieske oxygenases such as ferrous iron and NAD(P)H. Here, we introduce our approach to the quantification of O_2 uncoupling which we have established for one class of Rieske oxygenases, namely nitroarene dioxygenases. As we will describe in the following, reliable estimates for O_2 uncoupling are ideally based on a combination of approaches. Those include (i) mass balances and determination of operational stoichiometric coefficients (Section 2). (ii) For quantification of ROS (Section 3), we discuss options to quantify selected ROS in assay matrices used for the study of Rieske oxygenases. Several approaches exist for the detection of ROS, especially hydrogen peroxide, H_2O_2, and superoxide, $O_2^{\bullet-}$, but their quantitative interpretation may be limited. Finally, (iii) we discuss the complementary evaluation of enzyme kinetics (Section 4) as independent means of supporting the mass balance approach.

2. Mass Balances and Reaction Stoichiometries
2.1 Quantification strategies and limitations

Quantification of reaction mass balances for oxygenations catalyzed by nitroarene dioxygenases are based on the generalized reaction scheme of Fig. 1. Nitroarene dioxygenases for different substrates have been described and include enzymes for nitrobenzene (NBDO), nitrotoluene isomers (2NTDO, 3NTDO), dinotrotoluene isomers (DNTDO), and dichloronitrobenzene (34DCNBDO) (Friemann et al., 2005; Gao et al., 2021; Singh, Kumari, Ramaswamy, & Ramanathan, 2014). Oxygenations of nitroarenes generally lead to dioxygenations (reactions **1 → 2**) whereas monooxygenations of aliphatic substituents to alkyl alcohols (**1 → 3**) are

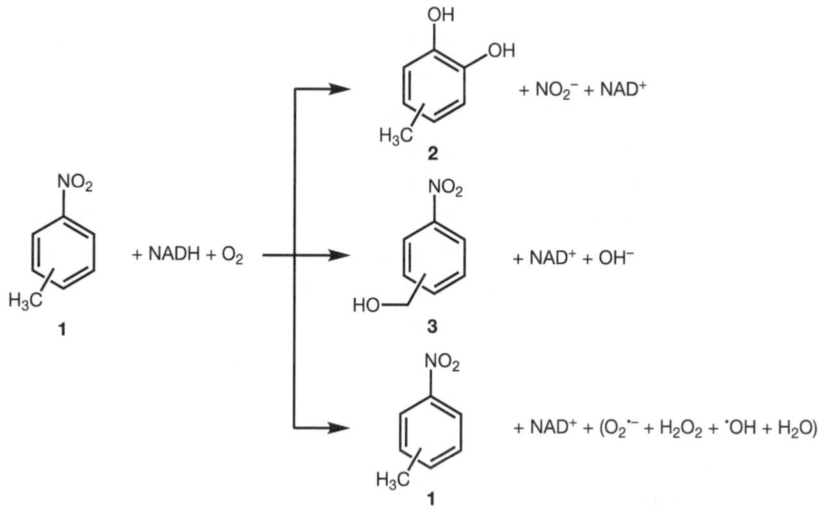

Fig. 1 Generalized reaction scheme for the activity of nitroarene dioxygenases with some of the most commonly studied substrates, nitrobenzene and nitrotoluene isomers. The three reactions include dioxygenation (top), monooxygenation (middle), and O_2 uncoupling (bottom). The O_2 uncoupling pathway lists all partially and fully reduced oxygen species in parenthesis. Reaction stoichiometries are not balanced for the sake of simplicity.

considered metabolically unproductive side reactions. Fig. 1 exemplifies the possible reaction pathways of nitroarene dioxygenases that allow for delineating the principal experimental approaches for the quantification of enzymatic turnover. Note that we use the term "substrate" to refer to nitroaromatic compounds whereas the second substrate, O_2, is addressed separately.

Enzymatic turnover is typically quantified from substrate and product mass balances. Most often, this approach implies quantification of substrate and hydroxylated aromatic products. Nitroarene oxygenations, however, are quantified through measurement of the second reaction product, nitrite (NO_2^-). NO_2^- is spontaneously eliminated from the *cis*-dihydrodiol intermediates formed in the initial reaction step leading to a re-aromatization of the intermediates as (substituted) catechols (Nishino & Spain, 1995). This feature of Rieske oxygenase reactivity is unique to nitroarene dioxygenases and simplifies quantification of its activity. This avenue, however, exclusively applies to the dioxygenation pathway.

Mass balance considerations less frequently include the quantification of oxidized and reduced forms of oxygen. The quantification of some of these species is technically straightforward. Standard approaches include optical and electrochemical O_2 sensors for dissolved O_2, or liquid chromatography UV/Vis or mass spectrometry detectors for hydroxylated aromatic compounds. However, the accurate quantification of both O_2 and hydroxylated aromatic compounds strongly hinges on the control of loss processes (e.g., O_2 consumption by non-enzymatic processes) and difficulties of analyte preservation (e.g., complexation and oxidation of catechols). To that end, systematic evaluation of unintentional analyte losses are required. In addition to these challenges, partially reduced oxygen species are very reactive and of transient nature. Superoxide, H_2O_2, and hydroxyl radicals can be detected and quantified with specific assays. The functioning of such assays, however, can be compromised by matrix components of Rieske oxygenase assays such as the high concentrations of Fe^{II} complexes.

A second option for turnover quantification is the measurement of NAD(P)H oxidation to $NAD(P)^+$. Oxidation of NAD(P)H is an indirect measure of Rieske oxygenase activity because this process is coupled to generation of the resting state with a reduced non-heme Fe and reduced Rieske cluster. Consumption of NAD(P)H is a measure for long-range electron transfer in a Rieske oxygenase system and implies stoichiometric transfer of reduction equivalents. These details are rarely elaborated on for enzyme assays (Runda, Kremser, Özgen, & Schmidt, 2023b). Comparison of NAD(P)H consumption and O_2 consumption is used as a measure of O_2 uncoupling in cytochromes (Ravanfar et al., 2023). However, $NAD(P)H/NAD(P)^+$ quantification in matrices of Rieske oxygenase assays are not reported. Nevertheless, information from NAD (P)H is very useful because the amount of reduction equivalents in enzyme assays allow for controlling the extent of enzymatic activity and substrate turnover.

2.2 Quantification of O_2 uncoupling

The general approach for quantifying O_2 uncoupling consists of determining species concentrations in enzyme assays where the extent of substrate conversion is deliberately restricted. To that end, the enzymatic turnover is limited systematically by carrying out experiments with different initial concentrations of reduction equivalents, that is NADH in work with the nitroarene dioxygenases NBDO and 2NTDO portrayed in this work (Fig. 2) (Bopp et al., 2022; Pati et al., 2022). The quantification of stoichiometric coefficients, v_j, through linear regressions with

Oxygen uncoupling in biodegradation reactions

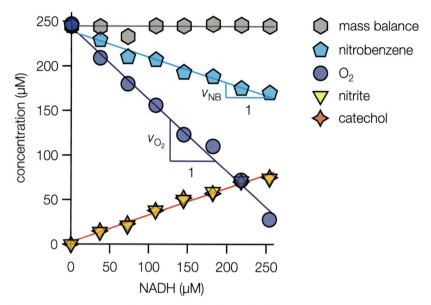

Fig. 2 Species concentration in a limited enzymatic turnover experiment with NBDO and nitrobenzene as substrate. Stoichiometric coefficients, v_j, are calculated from linear regressions vs. nominal NADH concentrations. *Data reproduced from Pati, S. G., Bopp, C. E., Kohler, H.-P. E., & Hofstetter, T. B. (2022). Substrate-specific coupling of O_2 activation to hydroxylations of aromatic compounds by Rieske non-heme iron dioxygenases. ACS Catalysis, 12(11), 6444–6456. https://doi.org/10.1021/acscatal.2c00383.*

Eq. (1) are critical to confirm Rieske oxygenase function, specifically the stoichiometry of long range electron transfer from the reductase to the non-heme Fe^{II} center.

$$[j] = v_j \cdot [\text{NADH}] + q \qquad (1)$$

where $[j]$ is the experimentally determined molar concentration of substrates, O_2, and reaction products after conclusion of enzymatic activity, [NADH] is the nominal concentration of NADH, and q is the y-intercept. O_2 uncoupling, denoted here as the fraction of unproductively consumed O_2, $f_{O_2\text{-uc}}$, can be determined by several complementary approaches. Typically, one compares substrate disappearance (ΔS) or product formation (ΔP) to O_2 consumption (i.e., change in O_2 concentration, ΔO_2) as in Eq. (2). Alternatively, $f_{O_2\text{-uc}}$ can be approached with stoichiometric coefficients for O_2, substrates, and products (v_{O_2}, v_S, and v_P, respectively) as in Eq. (3). As we outline in the following, control of O_2 background consumption and control experiments are critical to allocate disappearance of O_2 to O_2 uncoupling.

$$f_{O_2-uc} = 1 - \frac{\Delta S}{\Delta O_2} \approx 1 - \frac{-\Delta P}{\Delta O_2} \tag{2}$$

$$= \frac{v_{O_2} - v_S}{v_{O_2}} \approx \frac{v_{O_2} + v_P}{v_{O_2}} \tag{3}$$

2.2.1 Experimental design

The assay composition is optimized for activity by choice of the buffer type and pH, optimal enzyme ratios, and the concentration of ferrous ammonium sulfate, $(NH_4)_2Fe(SO_4)_2$, for stabilization of the Rieske cluster (Pati, Kohler, & Hofstetter, 2017). Enzyme concentrations are chosen to balance experiment time with available amounts of purified enzyme. The volume of the assays is determined by the requirements of the type of analyses to be carried out subsequently.

A minimum of two distinct limited substrate turnover assays are required to distinguish O_2 uncoupling from O_2 background consumption (Section 2.2.5). The dissolved O_2 concentration in the buffer, typically 250 μM, limits the enzymatic turnover that is targeted by NADH dosing. Due to the stoichiometric ratio of NADH and O_2, the maximum NADH concentration should be less than equimolar to avoid complete O_2 consumption unless the NADH is used very inefficiently by the enzyme. The minimum NADH concentration targets a small turnover that will allow detection of O_2 consumption and product formation above the limit of quantification. A preliminary experiment used to estimate the extent of O_2 uncoupling is recommended to identify approximate target concentrations of the analytes.

2.2.2 Equipment

- Clear glass crimp-top vials (12 mL) with minimal variation in content volume.
- Septum lids for air-tight sealing of the vials (i.e., butyl rubber stoppers).
- A set of identical magnetic stirrers and magnetic stirrer plate with multiple places for simultaneous measurement.
- Gas-tight glass syringe for NADH addition to sealed reactors.
- Needle-type oxygen microsensor covering concentration ranges from low concentrations (a few μM) up to O_2 saturation (220–280 μM) with online O_2 recording and display (PreSens Precision Sensing GmbH, Germany).

- High-performance liquid chromatography (HPLC) coupled to UV–vis detection (Dionex UltiMate 3000 System, Thermo Scientific).

2.2.3 Procedure for controlled substrate turnover experiments

1. Purify all components of the Rieske oxygenase, namely reductase, ferredoxin, and oxygenase, separately to >95% purity in the assay buffer (50 mM MES, pH 6.8). The oxygenase compound is oxygen-sensitive and may require addition of dithiothreitol (Runda et al., 2023b). Despite the complexing nature of a His-tag, recent studies have demonstrated successful purification of Rieske oxygenases with affinity chromatography (Halder, Nestl, & Hauer, 2018; Runda et al., 2023b). Once thawed, the oxygenase compound looses activity within a few hours. Reductase and ferredoxin remain functional even after several days if cooled.

2. At the start of the experiment, saturate the buffer with O_2 at reaction temperature by equilibration with the laboratory atmosphere.

3. Prepare the NADH stock solution freshly and determine the exact concentration photometrically ($6.220 \, M^{-1} \, cm^{-1}$ at 340 nm) (Bergmeyer, 1975). Adjust the stock solution concentration (50 mM in 10 mM NaOH) to minimize the volume of the added solution (10–50 μL).

4. Before filling the vials with enzyme assay, weight the empty reactor with the magnetic stirrer and lid for determining the actual assay volume and NADH concentration.

5. Carry out each limited turnover experiment with a predefined NADH concentration in a separate reactor. First, mix the MES buffer (pH 6.8, 50 mM), 100 μM $(NH_4)_2Fe(SO_4)_2$ (100 mM in 20 mM HCl), and organic substrate (50 mM methanolic stock), followed by the individual enzyme compounds (0.15 μM reductase, 1.8 μM ferredoxin, 0.15 μM oxygenase). Fill the reactor to the top with buffer and close with a lid and crimp cap in an air-tight fashion. Prepare at least one vial without enzymes, $(NH_4)_2Fe(SO_4)_2$, or NADH to determine the initial substrate concentration. Additional blank experiments are detailed below (Section 2.2.4).

6. Insert the needle-type oxygen microsensor into the reactor to measure the initial O_2 concentration. Secure the reactor on a magnetic stir plate throughout the entire experiment. With all enzyme components present, the O_2 concentration may decrease slightly. Initiate the reaction by injection of 10–50 μL NADH stock solution with a pointed needle through the septum. The onset of the reaction is visible in a marked decrease in measured O_2 concentration.

7. Remove the oxygen microsensor when termination of the enzymatic reaction is evident in constant O_2 concentrations. Weight the filled reactor to determine the assay volume and derive the NADH concentration from the difference to the empty reactor. Analyze for reactant, products, and transient oxygen species (Section 3) immediately after the reaction. One can do so by opening the vessel or removing part of the assay solution while replacing it with N_2 to create a headspace for the oxygen isotope analysis (Pati et al., 2017). Plan the analysis of substrates and products according to their stability in the assay.

8. Analyze aromatic compounds by reversed-phase liquid chromatography coupled to UV–vis detection using a cooled autosampler. A suite of different chromatographic columns and eluent mixtures are necessary to separate the various possible mixtures of nitroarene substrates and aromatic products. Separation of isomers of nitrotoluene, methylcatechol, and nitrobenzyl alcohol is achieved with phosphate buffers as eluents and by applying a temperature gradient. Details are compiled in Pati et al. (2016, 2017). Quantify nitrite using a spectrophotometric method at 540 nm with the reagents sulfanilamide and N-(1-naphthyl)ethylene diamine (Griess test) (An, Gibson, & Spain, 1994).

2.2.4 Blank experiments

Accurate quantification of O_2 uncoupling requires prevention of O_2 consuming side-reactions in the assay. While a comprehensive list of such reactions is so far not available, impurities from the purified enzyme batches are the likely sources of O_2 reductants. Impurities (e.g., aromatic amino acids) can also act as substrates for Rieske dioxygenases. Therefore, we recommend a series of blank experiments as part of the enzyme assay characterization for each batch of purified enzyme to exclude and minimize "fortuitous" O_2 consumption.

Table 1 gives an overview of the recommended control experiments in which O_2 consumption is monitored continuously with an oxygen microsensor. The experiments are carried out in the same manner as described above in Section 2.2.3. except for the omission of selected enzyme assay components, substrate, or NADH. We note that $(NH_4)_2Fe(SO_4)_2$ typically has no effect on O_2 concentrations in the complete assay (i.e., assay 7 in Table 1) but ferrous iron can become oxidized by O_2 in control experiments with low enzyme contents. Significant rates of O_2 consumption ($>3\,\mu M\ min^{-1}$) in assays 1–5 indicate O_2 consuming

Table 1 Control experiments for Rieske oxygenase-catalyzed reactions performed for the assessment of O_2 consumption in controlled substrate turnover experiments.

Assay no.	Description	Reductase	Ferredoxin	Oxygenase	Substrate	NADH
1	Reductase only	✓	–	–	✓	✓
2	Ferredoxin only	–	✓	–	✓	✓
3	Oxygenase only	–	–	✓	✓	✓
4	No NADH	✓	✓	✓	✓	–
5	No enzyme	–	–	–	✓	✓
6	No substrate	✓	✓	✓	–	✓
7	Full assay	✓	✓	✓	✓	✓

Checkmarks (✓) indicate presence of the specified component, hyphens (-) their absence.

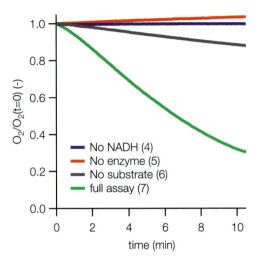

Fig. 3 Normalized O$_2$ consumption determined in control experiments with assays for 2NTDO in the absence of enzymes, NADH, or substrate. For comparison, the O$_2$ consumption in the fully functional enzyme system is shown. Numbers in legend refer to entries in Table 1. *Data reproduced from Bopp, C. E., Bernet, N. M., Kohler, H.-P. E., & Hofstetter, T. B. (2022). Elucidating the role of O$_2$ uncoupling in the oxidative biodegradation of organic contaminants by Rieske non-heme iron dioxygenases. ACS Environmental Au, 2(5), 428–440. https://doi.org/10.1021/acsenvironau.2c00023.*

impurities that must be eliminated in the respective component before performing the full assay. Assay 6 without the substrate can show a certain O$_2$ consumption as the full enzyme system is capable of activating oxygen (Fig. 3). The initial O$_2$ consumption in substrate-free assays (assay 6 in Table 1) should be ≥6-times smaller than in the full assay (7) (Bopp et al., 2022).

2.2.5 Quantification of O$_2$ uncoupling and background O$_2$ consumption

Quantification of O$_2$ uncoupling by Eqs. (2) and (3) do not account for background O$_2$ consumption thereby resulting in a possible overestimation of the extent of O$_2$ uncoupling. O$_2$ uncoupling is more accurately determined using a linear correlation with data from at least two enzyme assays in which the loss of substrate (ΔS) or formation of products (ΔP) is compared to changes in O$_2$ concentration (ΔO$_2$). Note that product formation must include both monooxygenation and dioxygenation reactions.

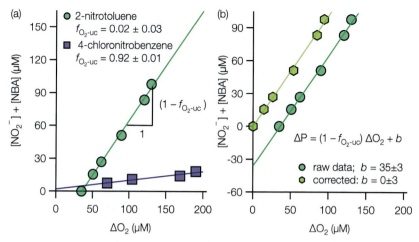

Fig. 4 Quantification of O₂ uncoupling from the formation of oxygenation products vs. the consumption of O₂ in assays of 2NTDO. Lines are linear fits to Eq. (4). (a) Examples of two 2NTDO substrates, 2-nitrotoluene and 4-chloronitrobenzene. (b) Corrections for O₂ background consumption, $[O_2]^{bg}$, by adjusting parameter b. $[O_2]^{bg}$ is obtained from Eq. (5). Data reproduced form Bopp, C. E., Bernet, N. M., Kohler, H.-P. E., & Hofstetter, T. B. (2022). Elucidating the role of O₂ uncoupling in the oxidative biodegradation of organic contaminants by Rieske non-heme iron dioxygenases. ACS Environmental Au, 2(5), 428–440. https://doi.org/10.1021/acsenvironau.2c00023.

Fig. 4 shows such linear correlations of ΔP vs. ΔO₂ for data from individual assays. The correlation slopes correspond to $1 - f_{O_2-uc}$ from Eq. (4).

$$\Delta S = |\Delta P| = \left(1 - f_{O_2-uc}\right) \cdot \Delta O_2 + b \qquad (4)$$

A slope close to unity implies negligible O₂ uncoupling and stoichiometric transfer of activated O₂ species to the substrate forming NO_2^- from dioxygenation and nitrobenzylalcohol (NBA) from monooxygenation. Two boundary cases are shown in Fig. 4a. The steep regression slope for 2-nitrotoluene stands for a small f_{O_2-uc} of 0.02 ± 0.03 whereas the shallow slope points to substantial O₂ uncoupling for 4-chloronitrobenzene (f_{O_2-uc} of 0.92 ± 0.01).

The fit for 2-nitrotoluene in Fig. 4a and b illustrates a typical case where the slope does not go through the origin and Eq. (4) exhibits a negative intercept b. Negative y-intercepts indicate that a constant amount of O₂ is consumed in all individual enzyme assays from one set of experiments. This background consumption of O₂ is independent of the enzymatic substrate turnover. The O_2 background consumption, $[O_2]^{bg}$, can be quantified with Eq. (5) as

$$[O_2]^{bg} = -\frac{b}{\left(1 - f_{O_2-uc}\right)} \quad (5)$$

According to our observations (Bopp et al., 2024; Bopp, Bernet, Kohler, & Hofstetter, 2022; Pati, Bopp, Kohler, & Hofstetter, 2022), selected batches of nitroarene dioxygenases can show $[O_2]^{bg}$-values of up to 60 μM. Further elaboration of possible limitations of the enzyme purification procedure are necessary to elucidate this phenomenon. O_2 background consumption is taken into account during data evaluation as follows. A good quality of the linear fit of ΔS or ΔP vs. ΔO_2 as in Fig. 4a indicates that the observed O_2 consumption is indeed due to O_2 activation from enzymatic activity as well as a constant contribution of $[O_2]^{bg}$ from unknown side reactions. The observed, overall O_2 concentration $[O_2]^{raw}$ can then be corrected by the value of $[O_2]^{bg}$ to arrive at $[O_2]^{corr}$ as in Eq. (6) and Fig. 4b. Note that correlations of ΔS or ΔP with ΔO_2 that include samples without added NADH, where neither O_2 is consumed nor NO_2^- is formed, provide inaccurate estimates of f_{O_2-uc} in that they distort the linear fit while not accounting for $[O_2]^{bg}$.

$$\Delta[O_2]^{corr} = \Delta[O_2]^{raw} + [O_2]^{bg} \quad (6)$$

2.2.6 Procedure for quantification of O_2 uncoupling
1. Plot the sum of the reaction products (mono- and dioxygenation) versus the O_2 consumed in each individual assay.
2. Apply a linear correlation to all data according to Eq. (4).
3. Determine the O_2 background consumption according to Eq. (5).
4. Correct the ΔO_2 concentration according to Eq. (6).
5. Include one data point that goes through the origin (0|0).
6. Determine the slope of the corrected data to calculate f_{O_2-uc} with the standard deviation.

3. Quantification of transient reactive oxygen species
3.1 Survey of ROS quantification approaches

Establishing oxygen mass balances were key for the discovery of Rieske oxygenase activity and the use of ^{18}O-labeled O_2 and H_2O_2 enabled tracing the transfer of oxygen to hydroxylated reaction products (Wolfe & Lipscomb, 2003). Keeping track of partially reduced oxygen species, namely the potential formation of $O_2^{\bullet-}$, H_2O_2, and $^{\bullet}OH$ in O_2 uncoupling pathways (Fig. 1), however, is difficult owing to their transient nature. The reactivity of the ROS and up to milimolar concentrations of reductants and electron donors in typical Rieske oxygenase assays often only allows for qualitative evidence of their formation. Typically studied ROS from Rieske oxygenase activity, namely $O_2^{\bullet-}$ and H_2O_2, are quantified indirectly using probe compounds that enable chemical and enzymatic conversion of ROS to products for spectroscopic and photometric detection (Messner & Imlay, 2004). The release of activated oxygen as hydroxyl radicals from the active site, by contrast, is not documented and would be difficult to prove in Rieske oxygenase assays given the possibility of formation from Fenton-type reactions of H_2O_2. Trapping of $O_2^{\bullet-}$ in reactions catalyzed by Rieske oxygenase was documented using nitroblue tetraazolium and hydroethidine (Pérez-Pantoja et al., 2013; Vaillancourt, Labbe, Drouin, Fortin, & Eltis, 2002). In agreement with their earlier characterization (Messner & Imlay, 2004), rapid "side" reactions with the enzyme's substrate and $O_2^{\bullet-}$ disproportionation interfered with $O_2^{\bullet-}$ detection by the probes and made its use non-quantitative. Because of these experimental circumstances, ROS quantification from Rieske oxygenase activity focused exclusively on H_2O_2.

While incomplete as an approach for total ROS, H_2O_2 concentrations can provide a lower boundary for the extent of O_2 uncoupling. The difference between O_2 uncoupling derived from reaction stoichiometries and H_2O_2 is a qualitative measure for other ROS even though both parameters can exhibit uncertainty that make detection of other ROS than H_2O_2 difficult. We observed that assays relying on the quantitative conversion of H_2O_2 by horseradish peroxidase (HRP) and concomitant quantification of the conversion of aromatic amines (aniline, 4-methoxyaniline) (Bopp et al., 2024; Bopp, Bernet, Kohler, & Hofstetter, 2022; Pati, Bopp, Kohler, & Hofstetter, 2022) and phenoxazin (Ampliflu, Amplex Red) (Messner & Imlay, 2004; Morlock, Böttcher, & Bornscheuer, 2018) provide the most accurate outcomes. Fig. 5 illustrates the necessity for calibration of the HRP-based assay in the Rieske oxygenase assay matrix. Given that the matrix of enzyme assays provides several reductants for H_2O_2 in HRP-catalyzed as well as non-biological

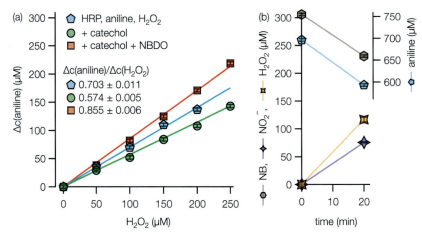

Fig. 5 (a) Calibration rows for the HRP-catalyzed oxidation of aniline with H_2O_2 used for H_2O_2 quantification in assays of NBDO. (b) Quantification of H_2O_2 formation upon O_2 uncoupling associated with the dioxygenation of nitrobenzene with NBDO. The disappearance of 103 μM of aniline corresponded to production of 120 μM of H_2O_2. The extent of O_2 uncoupling can be derived here from the ratio of $H_2O_2/(H_2O_2 + NO_2^-)$ and amounts to 60%. This number is identical within uncertainty to $f_{O_2\text{-uc}}$ quantified through NO_2^- formation and O_2 consumption and suggests that no additional ROS were formed in this experiment. *Data reproduced from Pati, S. G., Bopp, C. E., Kohler, H.-P. E., & Hofstetter, T. B. (2022). Substrate-specific coupling of O_2 activation to hydroxylations of aromatic compounds by Rieske non-heme iron dioxygenases. ACS Catalysis, 12(11), 6444–6456. https://doi.org/10.1021/acscatal.2c00383.*

reactions, the HRP-substrate does not necessarily react with H_2O_2 according to the expected reaction stoichiometry in Eq. (7).

$$\text{Ar–NH}_2 + \frac{1}{2}H_2O_2 + H^+ \xrightarrow{\text{HRP}} \text{Ar–NH}_2^{\bullet+} + H_2O \quad (7)$$

where Ar–NH$_2$ and Ar–NH$_2^{\bullet+}$ stand for aniline and the aniline radical cation, respectively. The presence of an oxidation product of nitroarene dioxygenase (catechol), for example, decreases this stoichiometry. Serendipitously for work with nitroarene dioxygenases, this yield is substantially higher (85%) in the presence of the Rieske oxygenase. Note that the quantification of O_2 consumption in the combined Rieske oxygenase-HRP assays is no longer possible and is obtained from separate experiments.

Qualitative evidence for H_2O_2 from O_2 uncoupling can be obtained more rapidly through the addition of catalase to Rieske oxygenase assay solution prior to or shortly after reaction completion and monitoring of O_2

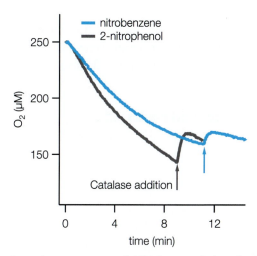

Fig. 6 Addition of catalase to assays of NBDO amended with nitrobenzene and 2-nitrophenol. The increase of O_2 stands for the disproportionation reaction $2H_2O_2 \rightarrow O_2 + 2H_2O$ and implies a H_2O_2 concentration that is twice as large as the increase in O_2. *Data reproduced from Pati, S. G., Bopp, C. E., Kohler, H.-P. E., & Hofstetter, T. B. (2022). Substrate-specific coupling of O_2 activation to hydroxylations of aromatic compounds by Rieske non-heme iron dioxygenases. ACS Catalysis, 12(11), 6444–6456. https://doi.org/10.1021/acscatal.2c00383.*

concentrations (Fig. 6) (Pati et al., 2022). This assay is especially useful for a screening of H_2O_2 formation prior to deploying the HRP assay. Assuming a stoichiometric disproportionation of H_2O_2 to O_2 and H_2O, an approximate estimate of H_2O_2 concentration corresponds to twice the measured increase of dissolved O_2.

3.2 Catalase-based assays
3.2.1 Equipment
- Optical oxygen microsensor (PreSens Precision Sensing GmbH) and equipment for experiments as described in Section 2.2.2.

3.2.2 Procedure
- Prepare completely filled nitroarene dioxygenase assay in 11-mL reactor with O_2 microsensor introduced into closed crimp vials through a stainless-steel needle (Section 2.2.3).
- Initiate reaction through addition of 300–500 μL of NADH stock solution (10–100 mM in 0.01 M NaOH) to closed reaction vessels with a gas-tight glass syringe.

- Monitor O₂ consumption continuously thereafter.
- After approximately 10 min reaction time (corresponding to <40% of the predetermined O₂ turnover), add 3.5 mg of catalase (100 μL of a 35 mg/mL stock solution).

3.3 Horseradish peroxidase-based assays
The following procedure applies for separate HRP experiments.

3.3.1 Equipment
- Clear glass crimp-top vials (2 mL).
- High-performance liquid chromatography (HPLC) coupled to UV–vis detection (Dionex UltiMate 3000 System, Thermo Scientific).
- Plate reader (Synergy Mx, Biotek Instruments).

3.3.2 Procedure
1. Prepare 2-mL filled crimp vials containing 200–300 μM of nitroarene substrate, 100 μM of $(NH_4)_2Fe(SO_4)_2$, 600 μM of aniline, 4-methoxy-aniline or Ampliflu (HRP substrates), and 10 mg/L of HRP in MES buffer. Add all three enzyme components at the optimal ratio (Section 2.2.3).
2. Determine initial concentrations of nitrobenzene, HRP substrates, and NO_2^- in one assay without NADH addition.
3. Add NADH (100–500 μM final concentration) through the rubber septum while continuously stirring the solution and monitoring dissolved O₂ concentrations until constant.
4. After 20 min, determine species concentrations from calibration rows in the respective matrices and concentration ranges.

4. Reaction kinetics
4.1 Quantification of O₂ uncoupling from reaction kinetics

Evaluation of Rieske oxygenase kinetics offers an additional approach for determining O₂ uncoupling. To that end, one uses the catalytic turnover numbers of the organic substrate, k_{cat}^S, and O₂, $k_{cat}^{O_2}$, as in Eq. (8).

$$f_{O_2-uc} \approx 1 - \frac{k_{cat}^S}{k_{cat}^{O_2}} \tag{8}$$

Data for k_{cat}^S and $k_{cat}^{O_2}$ are obtained from assessing catalysis on the basis of Michaelis-Menten kinetics. In contrast to procedures following the mass balance approach, the need to establish steady-state conditions in kinetic experiments requires that all samples are withdrawn from the same reactor at different time points. Both organic substrate consumption and product formation are meaningful endpoints for this kinetic evaluation. O_2 consumption, however, needs to be assessed separately from organic substrate turnover to avoid assay contamination with ambient air. For consistency, all kinetic experiments should be performed with the same purification batch and/or the reaction kinetics of a reference compound should be tested for each batch.

We note that the quantification of substrate oxygenation rates from colorimetric quantification of NO_2^- (Griess test, Section 2.2.3), which is widely used to assess nitroarene oxygenase kinetics (An et al., 1994; Ju & Parales, 2006; Mahan et al., 2015) is prone to interferences from residual NADH. Excess NADH used in kinetic assays interferes with the colorimetric assay and results in an underestimation of NO_2^- concentrations and thus enzyme activity described by k_{cat}^S. Conversely, enzyme-substrate combinations with large O_2 uncoupling activity will cause NADH to be depleted in enzyme assays more quickly than at low O_2 uncoupling. Decreased NADH concentrations in the assay will counteract the interferences for colorimetric NO_2^- determination thus making it difficult to anticipate the nonlinear interferences of NADH on the quantification of $f_{O_2\text{-uc}}$ from kinetic data. Note that interferences of NADH on the colorimetric determination of NO_2^- have no consequence for NO_2^- measurements as part of the mass balance approach (Section 2.2.3) where NADH is completely consumed once enzymatic activity ceases.

4.2 Organic substrate kinetics

Initial reaction rates, ν_0^i, of different enzyme-substrate combinations are obtained from repeated sampling during the first 60 s after substrate addition. The reaction is quenched in the withdrawn sample by lowering the pH by diluting with a strong acid (e.g., HCl). For nitroarene dioxygenases, the turnover through dioxygenation is typically assessed from increasing NO_2^- concentrations determined with the Griess test (Section 2.2.3). Alternatively, concentrations of nitroaromatic substrates or oxygenated products are determined by HPLC/UV–vis.

Maximum rates, ν_{max}^i, and Michaelis constants, K_{m}^i, of nitrite formation in the presence of different substrates i can be determined through nonlinear least square regression according to Eq. (9).

$$\nu_{0,\text{NO}_2^-}^i = \frac{\nu_{\text{max}}^i \cdot c_0^i}{K_{\text{m}}^i + c_0^i} = \frac{k_{\text{cat}}^i \cdot \text{E}_0 \cdot c_0^i}{K_{\text{m}}^i + c_0^i} \tag{9}$$

where $\nu_{0,\text{NO}_2^-}^i$ is the initial rate of NO_2^- formation from substrate i, c_0^i is the nominal initial substrate concentration, k_{cat}^i is the observable first-order rate constant, and E_0 is the nominal concentration of active sites in the nitroarene dioxygenase trimer, corresponding to 3 mol per mol of oxygenase.

4.2.1 Procedure

1. Test at least 6 different initial substrate concentrations in triplicate.
2. Fill 1.5 mL plastic tubes with equilibrated 0.5 mL MES buffer (50 mM, pH 6.8) containing reductase (0.3 μM), ferrodoxin (3.6 μM), oxygenase (0.15 μM), 500 μM $(\text{NH}_4)_2\text{Fe}(\text{SO}_4)_2$, and the respective substrate concentration (10 to 300 μM). Compared to the assays for O_2 uncoupling quantification (Section 2.2.3), the concentrations are elevated to ensure that only substrate concentrations limit the reaction.
3. Initiate the reaction by addition of 500 μM NADH (20 mM in 10 mM NaOH) and immediate mixing.
4. At predefined times between 10 and 120 s, withdraw 100 μL of sample and quench the reaction by mixing the sample with 200 μL sulfanilamide (10 g L^{-1} in 1.5 M HCl) in a measurement cuvette.
5. Add 200 μL N-(1-naphthyl)ethylenediamine dihydrochloride (1 g^{-1} in 1.5 M HCl) to each cuvette.
6. Determine NO_2^- concentrations photometrically at 540 nm.
7. Evaluate kinetic parameters according to standard Michaelis-Menten procedures and Eq. (9).

4.3 Oxygen consumption kinetics

ν_{max}^i and K_{m}^i for O_2 consumption are obtained from the continuous measurement of the O_2 concentration, c_{O_2}, over time in a single assay provided that product inhibition can be ruled out (Pati et al., 2022). A typical O_2 concentration time course is shown in Fig. 7. The rate of O_2 consumption at each time-point, $\nu_{\text{O}_2}^i$, is then calculated as the derivative of c_{O_2} vs. time t, that is, the pairwise differences of c_{O_2} and t for each

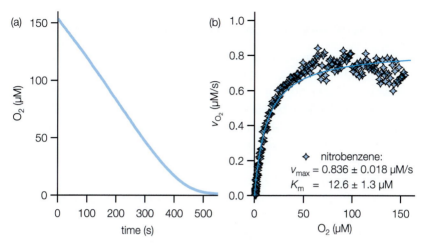

Fig. 7 Evaluation of the kinetics of O_2 activation by NBDO with nitrobenzene as substrate. (A) Continuous measurements of O_2 in 3 second intervals (c_0(nitrobenzene) = 200 µM). (B) Rates of O_2 consumption, $v^i_{O_2}$, obtained from derivative of O_2 concentration ($\Delta c_{O_2}/\Delta t$) vs. O_2 concentrations for derivation of v^i_{max} and K^i_m. Data reproduced from Pati, S. G., Bopp, C. E., Kohler, H.-P. E., & Hofstetter, T. B. (2022). Substrate-specific coupling of O_2 activation to hydroxylations of aromatic compounds by Rieske non-heme iron dioxygenases. ACS Catalysis, 12(11), 6444–6456. https://doi.org/10.1021/acscatal.2c00383.

measurement as $\Delta c_{O_2}/\Delta t$. $v^i_{O_2}$ and measured $c^i_{O_2}$-values are used to estimate v^i_{max} and K^i_m with non-linear least square regression according to Eq. (10).

$$v^i_{O_2} = \frac{v^i_{max} \cdot c^i_{O_2}}{K^i_m + c^i_{O_2}} = \frac{k^i_{cat} \cdot E_0 \cdot c_{O_2}}{K^i_m + c^i_{O_2}} \qquad (10)$$

4.3.1 Procedure

1. Prepare equilibrated MES buffer (50 mM, pH 6.8) containing reductase (0.3 µM), ferrodoxin (3.6 µM), oxygenase (0.15 µM), 500 µM $(NH_4)_2Fe(SO_4)_2$, and the organic substrate concentration (1 mM) to completely fill 2-mL vials. Close the crimp vials and introduce the O_2 microsensor through a stainless-steel needle.
2. Initiate the reaction by adding NADH (100 mM in 10 mM NaOH) with a gas tight syringe to reach final concentrations of 1 mM.
3. Monitor the O_2 consumption throughout the experiment (every 3 seconds) until all O_2 is consumed (5–30 min).
4. Determine the derivative ($\Delta O_2/\Delta t$) for each point and plot against O_2 concentrations. Apply a nonlinear least-square regression fit according to Eq. (10) for obtaining kinetic parameters.

5. Summary and Conclusions

O$_2$ uncoupling is a well-known phenomenon pertinent to the activity of O$_2$-activating enzymes such as Rieske oxygenases. Yet, the systematic consideration of O$_2$ uncoupling as the primary outcome of interactions of Rieske oxygenases with their many substrates is still missing. A likely cause is the lack of established methodologies for its quantification. A compilation of our observations of O$_2$ uncoupling for nitroarene oxygenases and some of its variants in Fig. 8 reveals more than 40 O$_2$ uncoupling instances for the various enzyme-substrate combination studies. In the majority of these many observations, O$_2$ uncoupling even makes up for more than 60% of the overall Rieske oxygenase activity. A more comprehensive consideration of O$_2$ uncoupling in the activity assessment of Rieske oxygenases is thus warranted.

O$_2$ uncoupling is causally related to the formation of ROS in the enzyme's active site and, because of the risk of hydroxylation of amino acid side chains, represents an inherent limitation to enzyme function. Even if damages from reaction of ROS can be avoided through hole hopping

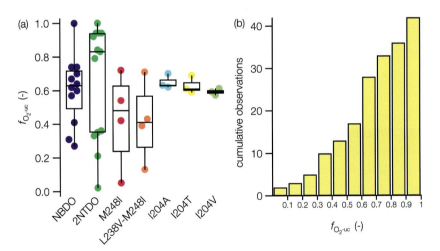

Fig. 8 Survey of O$_2$ uncoupling observations for all enzyme-substrate combinations involving the nitroarene dioxygenases NBDO and 2NTDO as well as 2NTDO variants from laboratory evolution experiments (M248I, L238V-M248I, I204A, I204T, I204V) (Bopp et al., 2024; Bopp, Bernet, Kohler, & Hofstetter, 2022; Pati, Bopp, Kohler, & Hofstetter, 2022). (A) Enzyme-specific distribution of $f_{O_2\text{-uc}}$-values. Box plots indicate median values for each enzyme as well as 25th and 75th quantiles, and bar lengths stand for one standard deviation. (B) Cumulative number of O$_2$ uncoupling observations for nitroarene dioxygenases and enzyme variants in 0.1-increments of $f_{O_2\text{-uc}}$.

processes, the additional need for cellular reductants imposes some kind of metabolic burden on the affected (micro)organism. Further work is required to identify the enzymatic and cellular consequences of O_2 uncoupling activity of Rieske oxygenases and whether/how ROS are effectively transported to the exterior of the enzyme. Quantification of O_2 uncoupling is a first and essential step to address such fundamental questions of Rieske oxygenase reactivity.

Acknowledgments

This work was supported by the Swiss National Science Foundation grant no. 200021_172950-1.

Competing interests

The authors declare no competing interests.

References

An, D., Gibson, D. T., & Spain, J. C. (1994). Oxidative release of nitrite from 2-nitrotoluene by a three-component enzyme system from *Pseudomonas* sp. strain JS42. *Journal of Bacteriology, 176*(24), 7462–7467. https://doi.org/10.1128/jb.176.24.7462-7467.1994.

Ashikawa, Y., Fujimoto, Z., Usami, Y., Inoue, K., Noguchi, H., Yamane, H., & Nojiri, H. (2012). Structural insight into the substrate- and dioxygen-binding manner in the catalytic cycle of Rieske nonheme iron oxygenase system, carbazole 1,9a-dioxygenase. *BMC Structural Biology, 12*(1), 1472.

Aukema, K. G., Escalante, D. E., Maltby, M. M., Bera, A. K., Aksan, A., & Wackett, L. P. (2017). *In silico* identification of bioremediation potential: Carbamazepine and other recalcitrant personal care products. *Environmental Science & Technology, 51*(2), 880–888. https://doi.org/10.1021/acs.est.6b04345.

Barry, S. M., & Challis, G. L. (2013). Mechanism and catalytic diversity of Rieske nonheme iron-dependent oxygenases. *ACS Catalysis, 3*(10), 2362–2370. https://doi.org/10.1021/cs400008/p.

Bergmeyer, H. (1975). New values for the molar extinction coefficients of nadh and nadph for the use in routine laboratories (author's transl). *Zeitschrift fur Klinische Chemie und Klinische Biochemie, 13*(11), 507–508. http://europepmc.org/abstract/MED/3038.

Bopp, C. E., Bernet, N. M., Kohler, H.-P. E., & Hofstetter, T. B. (2022). Elucidating the role of O_2 uncoupling in the oxidative biodegradation of organic contaminants by rieske non-heme iron dioxygenases. *ACS Environmental Au, 2*(5), 428–440. https://doi.org/10.1021/acsenvironau.2c00023.

Bopp, C. E., Bernet, N. M., Meyer, F., Khan, R., Robinson, S. L., Kohler, H.-P. E., & Hofstetter, T. B. (2024). Elucidating the role of O_2 uncoupling for the adaptation of bacterial biodegradation reactions catalyzed by Rieske oxygenases. *ACS Environmental Au, 4*(4), https://doi.org/10.1021/acsenvironau.4c00016.

Bopp, C. E., Kohler, H.-P. E., & Hofstetter, T. B. (2020). Enzyme kinetics of organic contaminant oxygenations. *Chimia, 74*(3), 108–114. https://doi.org/10.2533/chimia.2020.108.

Brimberry, M., Garcia, A. A., Liu, J., Tian, J., & Bridwell-Rabb, J. (2023). Engineering Rieske oxygenase activity one piece at a time. *Current Opinion in Chemical Biology, 72*, 102227. https://doi.org/10.1016/j.cbpa.2022.102227.

Csizi, K.-S., Eckert, L., Brunken, C., Hofstetter, T. B., & Reiher, M. (2022). The apparently unreactive substrate facilitates the electron transfer for dioxygen activation in rieske dioxygenases. *Chemistry—A European Journal, 28*(16), e202103937. https://doi.org/10.1002/chem.202103937.

Dunham, N. P., & Arnold, F. H. (2020). Nature's machinery, repurposed: Expanding the repertoire of iron-dependent oxygenases. *ACS Catalysis, 10*(20), 12239–12255. https://doi.org/10.1021/acscatal.0c03606.

Friemann, R., Ivkovic-Jensen, M. M., Lessner, D. J., Yu, C. L., Gibson, D. T., Parales, R. E., & Ramaswamy, S. (2005). Structural insight into the dioxygenation of nitroarene compounds: The crystal structure of nitrobenzene dioxygenase. *Journal of Molecular Biology, 348*(5), 1139–1151. https://doi.org/10.1016/j.jmb.2005.03.052.

Furusawa, Y., Nagarajan, V., Tanokura, M., Masai, E., Fukuda, M., & Senda, T. (2004). Crystal structure of the terminal oxygenase component of biphenyl dioxygenase derived from *Rhodococcus* sp. Strain RHA1. *Journal of Molecular Biology, 342*(3), 1041–1052.

Gao, Y., Palatucci, M. L., Waidner, L. A., Li, T., Guo, Y., Spain, J. C., & Zhou, N. (2021). A nag-like dioxygenase initiates 3,4-dichloronitrobenzene degradation via 4,5-dichlorocatechol in *Diaphorobacter* sp. strain JS3050. *Environmental Microbiology, 23*(2), 1053–1065. https://doi.org/10.1111/1462-2920.15295.

Gray, H. B., & Winkler, J. R. (2015). Hole hopping through tyrosine/tryptophan chains protects proteins from oxidative damage. In: *Proceedings of the National Academy of Sciences of the United States of America, 112*(35), 10920–10925. https://doi.org/10.1073/pnas.1512704112.

Gray, H. B., & Winkler, J. R. (2018). Living with oxygen. *Accounts of Chemical Research, 51*(8), 1850–1857. https://doi.org/10.1021/acs.accounts.8b00245.

Halder, J. M., Nestl, B. M., & Hauer, B. (2018). Semirational engineering of the naphthalene dioxygenase from *Pseudomonas* sp. NCIB 9816-4 towards selective asymmetric dihydroxylation. *ChemCatChem, 10*(1), 178–182. https://doi.org/10.1002/cctc.201701262.

Heinemann, P. M., Armbruster, D., & Hauer, B. (2021). Active-site loop variations adjust activity and selectivity of the cumene dioxygenase. *Nature Communications, 12*(1), 1095. https://doi.org/10.1038/s41467-021-21328-8.

Huang, X., & Groves, J. T. (2018). Oxygen activation and radical transformations in heme proteins and metalloporphyrins. *Chemical Reviews, 118*(5), 2491–2553. https://doi.org/10.1021/acs.chemrev.7b00373.

Jiang, W., Wilson, M. A., & Weeks, D. P. (2013). O-Demethylations catalyzed by Rieske nonheme iron monooxygenases involve the difficult oxidation of a saturated C–H bond. *Chemistry & Biology, 8*, 1687–1691. https://doi.org/10.1021/cb400154a.

Ju, K. S., & Parales, R. E. (2006). Control of substrate specificity by active-site residues in nitrobenzene dioxygenase. *Applied and Environmental Microbiology, 72*(3), 1817–1824. https://doi.org/10.1128/AEM.72.3.1817-1824.2006.

Karlsson, A., Parales, J. V., Parales, R. E., Gibson, D. T., Eklund, H., & Ramaswamy, S. (2003). Crystal structure of naphthalene dioxygenase: Side-on binding of dioxygen to iron. *Science, 299*(5609), 1039–1042. https://doi.org/10.1126/science.1078020.

Knapp, M., Mendoza, J., & Bridwell-Rabb, J. (2021). Enzymes—An aerobic route for C–H bond functionalization: The Rieske non-heme iron oxygenases. In J. Jez (Ed.). *Encyclopedia of biological chemistry III enzymes—An aerobic route for C–H bond functionalization: The Rieske non-heme iron oxygenases* (pp. 413–424)(3rd ed.). Oxford: Elsevier. https://doi.org/10.1016/B978-0-12-819460-7.00140-7.

Lee, K.-S., & Gibson, D. T. (1996). Toluene and ethylbenzene oxidation by purified naphthalene dioxygenase from *Pseudomonas* sp. strain ncib 9816-4. *Applied and Environmental Microbiology*, 3101–3106. https://doi.org/10.1128/aem.62.9.3101-3106.1996.

Liu, J., Knapp, M., Jo, M., Dill, Z., & Bridwell-Rabb, J. (2022). Rieske oxygenase catalyzed c–h bond functionalization reactions in Chlorophyll b biosynthesis. *ACS Central Science, 8*(10), 1393–1403. https://doi.org/10.1021/acscentsci.2c00058.

Lukowski, A., Ellinwood, D., Hinze, M., DeLuca, R., Du Bois, J., Hall, S., & Narayan, A. (2018). C–H Hydroxylation in paralytic shellfish toxin biosynthesis. *Journal of the American Chemical Society, 140*(37), 11863–11869. https://doi.org/10.1021/jacs.8b08901.

Lukowski, A. L., Liu, J., Bridwell-Rabb, J., & Narayan, A. R. H. (2020). Structural basis for divergent C—H hydroxylation selectivity in two rieske oxygenases. *Nature Communications, 11*(1), 2991. https://doi.org/10.1038/s41467-020-16729-0.

Mahan, K. M., Penrod, J. T., Ju, K.-S., AlKass, N., Tan, W. A., Truong, R., Parales, J. V., & Parales, R. E. (2015). Selection for growth on 3-nitrotoluene by 2-nitrotoluene-utilizing *Acidovorax* sp. strain JS42 identifies nitroarene dioxygenases with altered specificities. *Applied and Environmental Microbiology, 81*(1), 309–319. https://doi.org/10.1128/aem.02772-14.

McCusker, K. P., & Klinman, J. P. (2009). Modular behavior of tauD provides insight into the origin of specificity in α-ketoglutarate-dependent nonheme iron oxygenases. In: *Proceedings of the National Academy of Sciences, 106*(47), 19791–19795. https://doi.org/10.1073/pnas.0910660106.

McCusker, K. P., & Klinman, J. P. (2010). An active-site phenylalanine directs substrate binding and c-h cleavage in the α-ketoglutarate-dependent dioxygenase TauD. *Journal of the American Chemical Society, 132*(14), 5114–5220. https://doi.org/10.1021/ja909416z.

Messner, K. R., & Imlay, J. A. (2004). In vitro quantitation of biological superoxide and hydrogen peroxide generation. In L. Packer (Ed.). *Superoxide dismutase, volume 349 of Methods in enzymlology* (pp. 354–361). Academic Press. https://doi.org/10.1016/S0076-6879(02)49351-2.

Morlock, L., Böttcher, D., & Bornscheuer, U. (2018). Simultaneous detection of NADPH consumption and H_2O_2 production using the Ampliflu™ Red assay for screening of P450 activities and uncoupling. *Applied Microbiology and Biotechnology, 102*(2), 985–994. https://doi.org/10.1007/s00253-017-8636-3.

Münch, J., Püllmann, P., Zhang, W., & Weissenborn, M. J. (2021). Enzymatic hydro-xylations of sp^3-carbons. *ACS Catalysis, 11*(15), 9168–9203. https://doi.org/10.1021/acscatal.1c00759.

Nikel, P. I., Fuhrer, T., Chavarría, M., Sánchez-Pascuala, A., Sauer, U., & de Lorenzo, V. (2021). Reconfiguration of metabolic fluxes in *Pseudomonas putida* as a response to sub-lethal oxidative stress. *The ISME Journal, 15*, 1751–1766. https://doi.org/10.1038/s41396-020-00884-9.

Nishino, S., & Spain, J. (1995). Oxidative pathway for the biodegradation of nitrobenzene by *Comamonas* sp. Strain JS765. *Applied and Environmental Microbiology, 61*(6), 2308–2313. https://doi.org/10.1128/AEM.61.6.2308-2313.1995.

Pati, S. G., Bopp, C. E., Kohler, H.-P. E., & Hofstetter, T. B. (2022). Substrate-specific coupling of O_2 activation to hydroxylations of aromatic compounds by Rieske non-heme iron dioxygenases. *ACS Catalysis, 12*(11), 6444–6456. https://doi.org/10.1021/acscatal.2c00383.

Pati, S. G., Kohler, H.-P. E., & Hofstetter, T. B. (2017). Characterization of substrate, co-substrate, and product isotope effects associated with enzymatic oxygenations of organic compounds based on compound-specific isotope analysis. In M. E. Harris, & V. E. Anderson (Eds.). *Measurement and analysis of kinetic isotope effects, volume 596 of Methods in enzymlology* (pp. 292–329). Academic Press. https://doi.org/10.1016/bs.mie.2017.06.044.

Pati, S. G., Kohler, H.-P. E., Pabis, A., Paneth, P., Parales, R. E., & Hofstetter, T. B. (2016). Substrate and enzyme specificity of the kinetic isotope effects associated with the dioxygenation of nitroaromatic contaminants. *Environmental Science & Technology, 50*(13), 6708–6716. https://doi.org/10.1021/acs.est.5b05084.

Pérez-Pantoja, D., Nikel, P. I., Chavarria, M., & de Lorenzo, V. (2013). Endogenous stress caused by faulty oxidation reactions fosters evolution of 2,4-dinitrotoluene-degrading bacteria. *PLoS Genetics, 9*(8), e1003764–11. https://doi.org/10.1371/journal.pgen.1003764.

Perry, C., de Los Santos, E., Alkhalaf, L., & Challis, G. (2018). Rieske non-heme iron-dependent oxygenases catalyse diverse reactions in natural product biosynthesis. *Natural Product Reports, 35*(7), 622–632. https://doi.org/10.1039/C8NP00004B.

Ravanfar, R., Sheng, Y., Gray, H. B., & Winkler, J. R. (2023). Tryptophan extends the life of cytochrome P450. In: *Proceedings of the National Academy of Sciences of the United States of America, 120*(50), e2317372120. https://doi.org/10.1073/pnas.2317372120.

Runda, M. E., De Kok, N. A. W., & Schmidt, S. (2023a). Rieske oxygenases and other ferredoxin-dependent enzymes: Electron transfer principles and catalytic capabilities. *ChemBioChem, 24*(15), e202300078. https://doi.org/10.1002/cbic.202300078.

Runda, M. E., Kremser, B., Özgen, F. F., & Schmidt, S. (2023b). An optimized system for the study of Rieske oxygenase-catalyzed hydroxylation reactions in vitro. *ChemCatChem, 15*(16), e202300371. https://doi.org/10.1002/cctc.202300371.

Runda, M. E., Miao, H., De Kok, N. A. W., & Schmidt, S. (2024). Developing hybrid systems to address O₂ uncoupling in multi-component Rieske oxygenases. *BioRxiv.* https://doi.org/10.1101/2024.02.16.580709.

Singh, D., Kumari, A., Ramaswamy, S., & Ramanathan, G. (2014). Expression, purification and substrate specificities of 3-nitrotoluene dioxygenase from *Diaphorobacter* sp strain DS2. *Biochemical and Biophysical Research Communications, 445*(1), 36–42.

Teo, R. D., Wang, R., Smithwick, E. R., Migliore, A., Therien, M. J., & Beratan, D. N. (2019). Mapping hole hopping escape routes in proteins. In: *Proceedings of the National Academy of Sciences of the United States of America, 116*(32), 15811–15816. https://doi.org/10.1073/pnas.1906394116.

Tian, J., Garcia, A. A., Donnan, P. H., & Bridwell-Rabb, J. (2023). Leveraging a structural blueprint to rationally engineer the Rieske oxygenase TsaM. *Biochemistry, 62*(11), 1807–1822. https://doi.org/10.1021/acs.biochem.3c00150.

Vaillancourt, F. H., Labbe, G., Drouin, N. M., Fortin, P. D., & Eltis, L. D. (2002). The mechanism-based inactivation of 2, 3-dihydroxybiphenyl 1, 2-dioxygenase by catecholic substrates. *Journal of Biological Chemistry, 277*(3), 2019–2027. https://doi.org/10.1074/jbc.M106890200.

Wolfe, M. D., & Lipscomb, J. D. (2003). Hydrogen peroxide-coupled *cis*-diol formation catalyzed by naphthalene 1,2-dioxygenase. *Journal of Biological Chemistry, 278*(2), 829–835. https://doi.org/10.1074/jbc.M209604200.

CHAPTER TWO

Spectroscopic definition of ferrous active sites in non-heme iron enzymes

Edward I. Solomon[a,b,*] and Robert R. Gipson[a]

[a]Department of Chemistry, Stanford University, Stanford, CA, United States
[b]Stanford Synchrotron Radiation Lightsource, SLAC National Acceleration Laboratory, Stanford University, Menlo Park, CA, United States
*Corresponding author. e-mail address: edward.solomon@stanford.edu

Contents

1. d^6 Ligand field theory (LFT)	31
2. LF spectroscopy = low temperature magnetic circular dichroism (LT MCD)	32
3. Variable-temperature, variable-field (VTVH) MCD	36
4. LFT of spin-hamiltonian parameters from VTVH MCD	39
5. An early application of VTVH MCD on a non-heme Fe(II) enzyme	42
6. Perspective	45
Acknowledgments	46
References	46

Abstract

Non-heme iron enzymes play key roles in antibiotic, neurotransmitter, and natural product biosynthesis, DNA repair, hypoxia regulation, and disease states. These enzymes had been refractory to traditional bioinorganic spectroscopic methods. Thus, we developed variable-temperature variable-field magnetic circular dichroism (VTVH MCD) spectroscopy to experimentally define the excited and ground ligand field states of non-heme ferrous enzymes (Solomon et al., 1995). This method provides detailed geometric and electronic structure insight and thus enables a molecular level understanding of catalytic mechanisms. Application of this method across the five classes of non-heme ferrous enzymes has defined that a general mechanistic strategy is utilized where O_2 activation is controlled to occur only in the presence of all cosubstrates.

The five major classes of mononuclear non–heme iron (NHFe) enzymes, presented in Table 1, use a high spin ferrous site to activate dioxygen (Solomon et al., 2000). The non–cofactor dependent enzymes on the left, the cysteine oxidases and oxygenases (Baldwin & Abraham, 1988; Fernandez, Juntunen, & Brunold, 2022), the Rieske dioxygenases

Methods in Enzymology, Volume 703
ISSN 0076-6879, https://doi.org/10.1016/bs.mie.2024.05.019
Copyright © 2024 Elsevier Inc. All rights reserved, including those for text and data mining, AI training, and similar technologies.

Edward I. Solomon and Robert R. Gipson

Table 1 Major classes of non-heme ferrous enzymes.

Mononuclear non-heme iron enzymes

Non-cofactor Dependent Enzymes	Cofactor Dependent Enzymes

Non-cofactor Dependent Enzymes:

C-S Bond Formation — Cysteine Oxidase IPNS: $Cys-S^-$ → Cysteine Oxygenase CDO: $Cys-SO_2^-$

Rieske Dioxygenase → (structure with OH, OH)

Intradiol Dioxygenase (CO_2^-, CO_2^-) ← (structure with OH, OH) → Extradiol Dioxygenase (structure with OH, CO_2^-, O)

Cofactor Dependent Enzymes:

αKG Dependent: $Cl/HO-$ R', R ← $H-$ R', R → R', R (alkene)

Pterin Dependent: (benzene) → (phenol OH)

$Fe(II) + cof. (2e^-) \xrightarrow{O_2} Fe(IV)=O \xrightarrow{sub.} HAA (\alpha KG)$ / EAS (pterin)

Source: Adapted with permission from Solomon, E.I., Deweese, D.E., Babicz, J.T. (2021). Mechanisms of O_2 activation by mononuclear non-heme iron enzymes. *Biochemistry, 60*(46), 3497–3506. https://doi.org/10.1021/acs.biochem.1c00370 (Solomon, Deweese, Babicz, 2021). Copyright 2021 American Chemical Society.

(Barry & Challis, 2013), and the diol dioxygenases (Lipscomb, 2008; Vaillancourt, Bolin, & Eltis, 2006) have all been determined to react through high-spin ferric-superoxo intermediates that electrophilically attack their bound substrate (Babicz et al., 2023; Goudarzi, Babicz, Kabil, Banerjee, & Solomon, 2018; Sutherlin, Rivard, et al., 2018; Sutherlin, Wasada-Tsutsui, et al., 2018). In the cofactor dependent enzymes (cofactor = α-ketoglutarate (αkg) and pterin), the cofactor and the Fe(II) active site together transfer 4 electrons to O_2 to generate Fe(IV)=O S=2 intermediates (Krebs, Galonić Fujimori, Walsh, & Bollinger, 2007). For the αkg dependent enzymes, this intermediate H-atom abstracts (HAA) from a substrate and then rebound hydroxylates, halogenates, or desaturates the subsequent substrate radical (Blasiak, Vaillancourt, Walsh, & Drennan, 2006; Purpero & Moran, 2007). In the pterin dependent NHFe enzymes, the Fe(IV)=O S=2 intermediate instead performs electrophilic aromatic substitution (EAS) on an aromatic amino acid substrate (Fitzpatrick, 2023; Kappock & Caradonna, 1996).

Non-heme ferrous metalloenzymes are much more challenging to spectroscopically study than heme enzymes as they do not exhibit the intense $\pi \rightarrow \pi$ * transitions of the porphyrin. As shown in Fig. 1, these NHFe enzymes generally have a non-chromophoric facial triad set of two histidine and one carboxylate ligand (Kal & Que, 2017), or for the αkg dependent halogenases the carboxylate is replaced by the halide (Cl^-, Br^-) (Blasiak et al., 2006), with the remaining Fe(II) coordination positions occupied by H_2O's from solvent. Also, these ferrous active sites do not exhibit the low-energy, intense ligand to metal charge transfer features

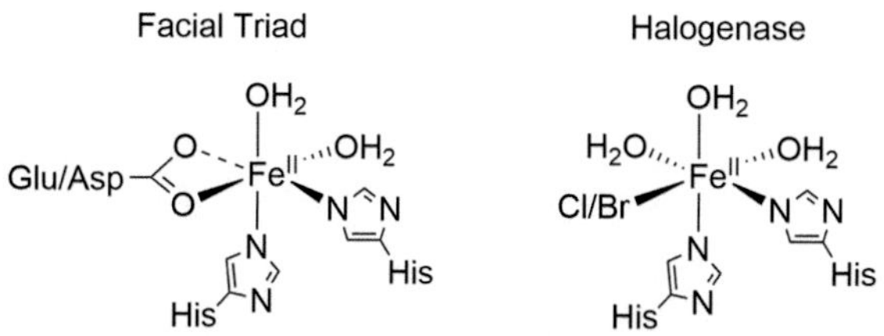

Facial Triad / Halogenase

Fig. 1 Schematic structures of non-heme ferrous enzyme active sites.

exhibited by ferric sites, and as these are d^6 even electron ions, they have integer spin non-Kramers' ground states that are generally not accessible by X-band EPR (as described below). However, high spin ferrous sites do have d→d transitions that are well described by ligand field theory.

1. d^6 Ligand field theory (LFT)

The Tanabe–Sugano (T&S) diagram in Fig. 2 gives the energies of all the ligand field (LF) states of an octahedral d^6 ion that are dependent on the combined effects of the ligand field splittings of the d orbitals (10 Dq) and the repulsions between the electrons within these orbitals (quantified by the Racah parameter B) (Solomon & Hanson, 2006). For a high spin ferrous octahedral site, the left side of this T&S diagram is appropriate, which shows that the ground state is a $^5T_{2g}$ with the extra electron in one of the three-fold degenerate t_{2g} orbitals. There is one spin allowed ligand field transition from this ground state to the ^5E_g excited state that corresponds to the extra electron promoted into the two-fold degenerate e_g set of d orbitals. As shown in this T&S diagram (Fig. 2) and on the left in Fig. 3, this transition is at an energy of 10 Dq (the LF splitting energy of the two-fold degenerate e_g above the three-fold degenerate t_{2g} orbitals) that for the facial triad ligation is approximately $10{,}000 \text{ cm}^{-1}$.

As shown in Fig. 3, the two-fold orbital degeneracy of the ^5E_g excited state is predicted by LFT to split in energy in a characteristic way dependent on the active site coordination number and structure (Pavel, Kitajima, & Solomon, 1998). For distorted 6 coordinate (6C) high spin ferrous protein sites, the ^5E excited state can split by up to approximately 2000 cm^{-1}, predicting 2 spin-allowed LF transitions, from the lowest energy component of the ground state to these ^5E derived excited states, at approximately $10{,}000 \text{ cm}^{-1}$ split by up to 2000 cm^{-1}. Removing one

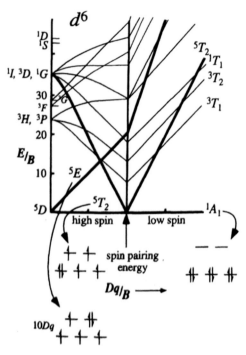

Fig. 2 Elaborated Tanabe-Sugano Diagram quantifying the energies of the states of a d^6 transition metal ion in an octahedral ligand field in terms of Dq and B. *Used with permission from John Wiley & Sons from (Solomon & Hanson, 2006).*

ligand to generate a square pyramidal, 5C structure predicts a large splitting of the ^5E state with spin allowed LF transitions at > 10,000 and at approximately 5000 cm^{-1}. Distortion of the 5C structure to trigonal bipyramidal changes the LF and somewhat lowers the energy of both transitions, still predicting a splitting of ~5000 cm^{-1}. Removal of another ligand generates a 4C distorted tetrahedral Fe(II) site. From LFT, 10 Dq (tetrahedron) = (−4/9) 10 Dq (octahedron), thus predicting only low energy spin allowed d→d transitions in the 5000–7000 cm^{-1} energy region.

2. LF spectroscopy = low temperature magnetic circular dichroism (LT MCD)

From the summary above, electronic spectroscopy in the ligand field region is a very sensitive probe of high spin ferrous active site structure.

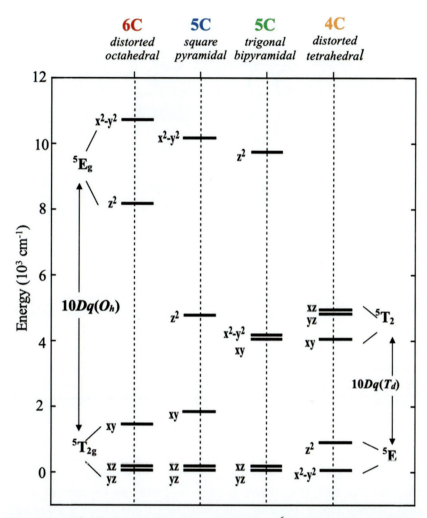

Fig. 3 Ligand field splitting of the high spin quintet d⁶ LF states with coordination number and structure. *Adapted with permission from Solomon, E.I., Brunold, T.C., Davis, M.I., Kemsley, J.N., Lee, S.K., Lehnert, N., ... Zhou, J. (2000). Geometric and electronic structure/function correlations in non-heme iron enzymes. Chemical Reviews, 100(1), 235–349. https://doi.org/10.1021/cr9900275. Copyright 2000 American Chemical Society.*

However, these spin-allowed transitions (Fig. 3) are still parity forbidden (d→d), thus very weak in absorption ($\varepsilon \sim 10\,\text{M}^{-1}\,\text{cm}^{-1}$). Additionally, they are in the near infrared (NIR) 12,000–4000 cm⁻¹ spectral region that is obscured by intense overlapping vibrational transitions of the protein and solvent. Thus, the ferrous enzyme LF transitions are not accessible by absorption spectroscopy. However, the high spin Fe(II) ground state is an

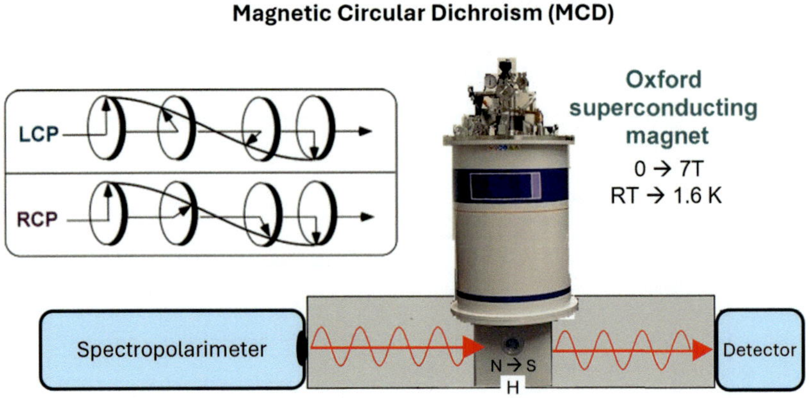

Magnetic Circular Dichroism (MCD)

Fig. 4 Experimental configuration for VTVH MCD spectroscopy.

$S = 2$, quintet that is paramagnetic. The way to study paramagnetic metal ion active sites is by MCD at low temperatures, due to the C-term intensity mechanism (Buckingham & Stephens, 1966) explained below, where at low temperatures, paramagnetic MCD signals are two to three orders of magnitude more intense than those of the diamagnetic background. As shown in Fig. 4, MCD spectroscopy involves the differential absorption of left (LCP) versus right (RCP) circularly polarized light induced by a longitudinal magnetic field of up to 7T at temperatures down to 1.6 K, where LCP light has its $\overrightarrow{E}$ vector rotating counterclockwise as it propagates to the detector and RCP rotates clockwise.

We have measured the low temp LF MCD spectra of more than 20 structurally defined mononuclear high spin Fe(II) complexes, and the predictions of LFT (Fig. 3) are strongly supported by experiment (Pavel et al., 1998). Fig. 5 presents representative LT MCD spectra for each structural type. 6C shows two LF transitions in the $\sim 10,000\ \mathrm{cm}^{-1}$ region, split by approximately $2000\ \mathrm{cm}^{-1}$; 5C square pyramidal sites show transitions at $> 10,000\ \mathrm{cm}^{-1}$ and $\sim 5000\ \mathrm{cm}^{-1}$; for trigonal bipyramidal, these shift to lower energies still with an $\sim 5000\ \mathrm{cm}^{-1}$ splitting; and 4C distorted tetrahedral Fe(II) sites show only low energy LF transitions in the $5000–7000\ \mathrm{cm}^{-1}$ region. Note that while MCD transitions can be either positively or negatively signed, most LF transitions of high spin ferrous active sites exhibit transitions of the same sign (positive bands in Fig. 5) in MCD due to the presence of low lying LF excited states that spin-orbit couple with the ground state leading to this "deviation from the sum rule" (Gerstman & Brill, 1985; Solomon & Hanson, 2006).

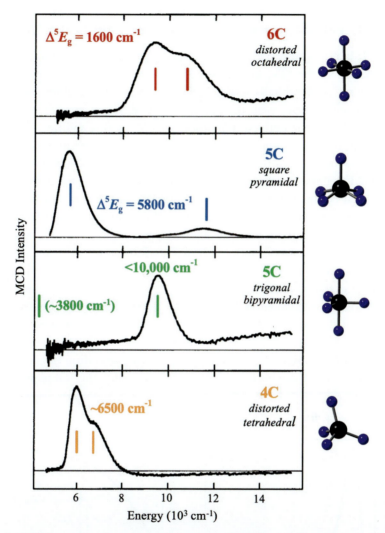

Fig. 5 LT MCD spectra of representative structurally defined Fe(II) model complexes. 6C, octahedral [Fe(H$_2$O)$_6$](SiF$_6$). 5C, square-pyramidal [Fe(HB(3,5-^{i}Pr$_2$pz)$_3$)(OAc)]. 5C, trigonal-bipyramidal [Fe(tris(2-(dimethylamino)ethyl)amine)Br]$^+$. 4C, tetrahedral [Fe(HB(3,5-^{i}Pr$_2$pz)$_3$)(Cl)]. *Adapted with permission from Pavel, E.G., Kitajima, N., & Solomon, E.I. (1998). Magnetic circular dichroism spectroscopic studies of mononuclear non-heme ferrous model complexes. Correlation of excited- and ground-state electronic structure with geometry. Journal of the American Chemical Society, 120(16), 3949–3962. https://doi.org/10.1021/JA973735L. Copyright 1998 American Chemical Society.*

3. Variable-temperature, variable-field (VTVH) MCD

It is important to also observe in Fig. 3, that the three-fold orbital degeneracy of the $^5T_{2g}$ ground state (deriving from the extra electron in the t_{2g} set of d orbitals) is also split in energy due to differences in π bonding interactions with the ligands in the different coordination geometries. We have derived a method to obtain this splitting from the temperature (T) and field (H) dependence of the MCD intensity (VTVH MCD) of the LF transitions (Whittaker & Solomon, 1988). Fig. 6A shows the field dependence of the MCD intensity at 4.2 K for a 5C square pyramidal complex as in Fig. 5. The intensity of the bands increases in a nonlinear way with increasing field (and decreasing temperature) and levels off at high fields and low temperatures. This saturation magnetization behavior is shown in Fig. 6B (left to right), where the intensity first increases linearly by orders of magnitude with increasing field and decreasing temperature. This is the C-term behavior of paramagnetic systems (Buckingham & Stephens, 1966). Then at very high field and low temperature, the MCD intensity levels off and is saturated (Schatz, Mowery, & Krausz, 1978; Solomon & Hanson, 2006; Stephens, 1976; Thomson & Johnson, 1980). The origin of this saturation magnetization behavior can be understood by first considering the simplest $S = \frac{1}{2}$

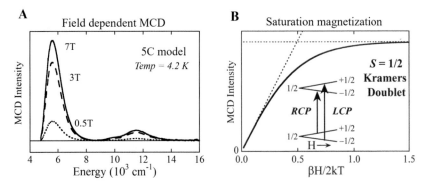

Fig. 6 (A) Field dependence of the MCD signal of a 5C square pyramidal complex as in Fig. 5. (B) Saturation magnetization MCD for an $S = \frac{1}{2}$ Kramers' doublet; signal as a function of field/temperature (H/T). Inset shows Kramers' doublet ground and excited states in a magnetic field and associated MCD allowed transitions. *Adapted with permission from Solomon, E.I., Brunold, T.C., Davis, M.I., Kemsley, J.N., Lee, S.K., Lehnert, N., ... Zhou, J. (2000). Geometric and electronic structure/function correlations in non-heme iron enzymes. Chemical Reviews, 100(1), 235–349. https://doi.org/10.1021/cr9900275. Copyright 2000 American Chemical Society.*

Kramers' doublet system which is included in the inset of Fig. 6B. Both the ground state and the excited state are two-fold spin degenerate ($M_s = \pm \frac{1}{2}$) and this is split in energy by the magnetic Zeeman effect. Two transitions from the ground to the excited state of equal magnitude but opposite sign are allowed by the $\Delta M = \pm 1$ selection rule of MCD. Note that these Zeeman energy splittings at high magnetic fields are a few wavenumbers while the transition band shapes have full-width at half-maxima of $\sim 1000\ \mathrm{cm}^{-1}$. Thus at high temperatures and low magnetic fields, the ground state sublevels are equally populated and these transitions cancel leading to the very low MCD intensity at the left of the saturation magnetization curve in Fig. 6B. As the field increases and the temperature decreases (increasing H/T from left to right), the higher energy sublevel of the ground state loses Boltzmann population; thus, the transition probability from this sublevel decreases, cancellation no longer occurs, and the intensity initially increases linearly with H/T (the C-term MCD mechanism, $\beta H/2kT = 0$ to ~ 0.4 in Fig. 6B). Finally, when the field is further increased and temperature decreased, only the lowest $M_s = -\frac{1}{2}$ sublevel is populated and the MCD intensity is saturated (at $\beta H/2kT \sim 1.0$ in Fig. 6B).

For an $S = \frac{1}{2}$ Kramers' ion, saturation magnetization curves of the MCD intensity of a specific band obtained at different fixed temperatures (with varying fields from 0 to 7T for each temperature) all superimpose when plotted versus H/T as in Fig. 6B, a behavior described by the Brillouin function for an $S = \frac{1}{2}$ system (Eq. 1) (Solomon & Hanson, 2006):

$$I = \frac{1}{2} N g B \, tanh\,(g\beta H/2kT) \tag{1}$$

However, when saturation magnetization curves are obtained for an $S = 2$ non-Kramers' Fe(II) active site, there is a spread or nesting of these curves (each with a fixed temperature and varying field) as shown in Fig. 7A. Insight into the origin of this nesting behavior was obtained by uncoupling the temperature and field variables, thus plotting the MCD intensity with decreasing temperature for fixed magnetic fields that were incrementally increased. This plot is presented in Fig. 7B, which shows that the saturation of MCD intensity at low temperature for each magnetization curve increases with increasing magnetic field eventually leveling off at high field. This increase of intensity at saturation requires that the wavefunction of the lowest component of the ground state changes with

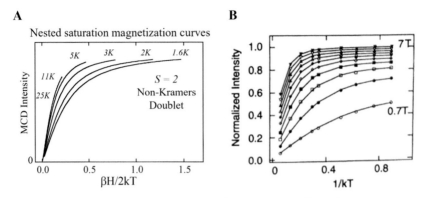

Fig. 7 Saturation magnetization data and fits using Eq. 2 for a ferrous 5C square pyramidal complex as in Fig. 5 and 6 A recorded at 6000 cm^{-1}. (A) Nesting behavior shown in plot of intensity versus $\beta H/2kT$; (B) intensity versus temperature for fixed field increments. Adapted from Solomon, E.I., Pavel, E.G., Loeb, K.E., & Campochiaro, C. (1995). Magnetic circular dichroism spectroscopy as a probe of the geometric and electronic structure of non-heme ferrous enzymes. *Coordination Chemistry Reviews, 144*, 369–460. https://doi.org/10.1016/0010-8545(95)01150-N (Solomon, Pavel, Loeb, & Campochiaro, 1995) with permission from Elsevier.

magnetic field. This behavior led us to the **non-Kramers' doublet model for saturation magnetization** presented below (Whittaker & Solomon, 1988).

When the spin of the ground state is greater than 1/2, its 2S+1 spin degeneracy in M_s (=0, ± 1, ± 2) will split in energy even in the absence of a magnetic field when the metal site symmetry is lower than octahedral or tetrahedral. This zero field splitting (ZFS) of an $S = 2$ with a negative axial (*D*) and then rhombic (*E*) distortion of the LF is shown on the left in Fig. 8. It is important to observe that for an integer spin non-Kramers' ion, the doublets are split in energy by the rhombic LF even in the absence of a magnetic field; the ± 2 doublet is split by δ in the middle of Fig. 8. δ is generally larger than the microwave energy at X-band which is the origin of the lack of an EPR signal for non-Kramers' ions. However, this is also the origin of the nesting in the saturation magnetization curves in Fig. 7. As shown on the right of Fig. 8, in addition to the rhombic splitting of the $M_s = \pm 2$ by δ, the wavefunctions also equally mix eliminating their spin expectations, and therefore their MCD signals which depend on this, at zero and low magnetic fields. However, as the field is increased in addition to the Zeeman splitting of the sub-levels (by $g_\| \beta H \cos\theta$) the wavefunctions also change, becoming pure $M_s = -2$ (and +2) and MCD active at high magnetic fields. The MCD intensity

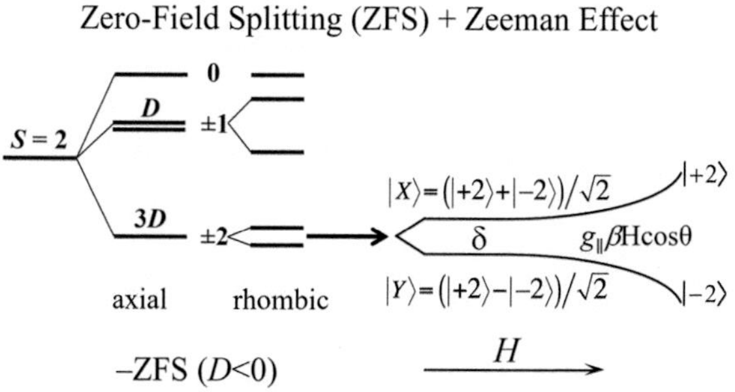

Zero-Field Splitting (ZFS) + Zeeman Effect

Fig. 8 Energy splittings of the $S = 2$ sublevels for an axial −ZFS and a rhombic distortion and then the magnetic field splitting and change in mixing of an $M_s = \pm 2$ non-Kramers' doublet. *Adapted with permission from Campochiaro, C., Pavel, E.G., & Solomon, E.I. (1995). Saturation magnetization magnetic circular dichroism spectroscopy of systems with positive zero-field splittings: Application to FeSiF6·6H2O. Inorganic Chemistry, 34(18), 4669–4675. https://doi.org/10.1021/IC00122A025. Copyright 1995 American Chemical Society.*

expression for this non-Kramers' behavior is given by Eq. 2 (Solomon et al., 1995; Whittaker & Solomon, 1988):

$$
\begin{aligned}
I_{MCD} \\
= A_{SatLim} \int_{0}^{\pi/2} \frac{\cos^2 \theta \sin \theta}{\sqrt{\delta^2 + (g_{\parallel}\beta H \cos \theta)^2}} (g_{\parallel}\beta H) \cdot \tanh \\
\left[\frac{\sqrt{\delta^2 + (g_{\parallel}\beta H \cos \theta)^2}}{2kT} \right] d\theta
\end{aligned}
\tag{2}
$$

The saturation magnetization data in Fig. 7 can be fit using the non-Kramers' doublet Eq. 2 with orientation averaging for a frozen protein solution to obtain experimental values for δ and $g_{\parallel}$ in Fig. 8. Thus, we use an excited state to study the ground state and obtain EPR parameters from an EPR inactive site. Note that Fig. 8 considers the case of a negative ZFS. The positive ZFS case gives equivalent behavior and involves a parallel analysis as presented by Campochiaro, Pavel, and Solomon (1995) and Solomon et al. (1995).

4. LFT of spin-hamiltonian parameters from VTVH MCD

We now relate the δ and $g_{\parallel}$ obtained from the VTVH MCD data in Fig. 7 to the axial (Δ) and rhombic (V) split set of t_2 $d\pi$ orbitals in Fig. 9A.

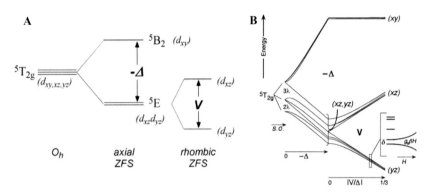

Fig. 9 LF splitting of the t₂ dπ orbitals (A), including SOC (B) and the Zeeman effect (inset). *(A) Adapted from Solomon, E.I., & Hanson, M.A. (2006). Bioinorganic spectroscopy. In E.I. Solomon & A.B. P. Lever (Eds), Inorganic electronic structure and spectroscopy (Vol. 2, pp. 1–129). John Wiley & Sons, Inc. https://www.wiley.com/en-us/Inorganic+Electronic +Structure+and+Spectroscopy%2C+Volume+II-p-9780471971146 with permission from John Wiley & Sons and (B) Solomon, E.I., Pavel, E.G., Loeb, K.E., & Campochiaro, C. (1995). Magnetic circular dichroism spectroscopy as a probe of the geometric and electronic structure of non-heme ferrous enzymes. Coordination Chemistry Reviews, 144, 369–460. https://doi.org/10.1016/0010-8545(95)01150-N with permission from Elsevier.*

Low symmetry distortion from octahedral can only split orbital degeneracy, thus ZFS of a spin degeneracy must derive from spin–orbit coupling (SOC). While for many transition metal sites (e.g. high spin Fe(III)) this involves out-of-state SOC which has a number of contributions limiting its utility, for high spin Fe(II) sites the $^5T_{2g}$ has dominantly in-state SOC and its analysis is direct and insightful. The $^5T_{2g}$ ground state is three-fold orbitally degenerate (extra electron can be in the d_{xy}, d_{yz}, or d_{xz} orbital) and this behaves as having an effective orbital angular momentum of 1. This SOCs to the $S = 2$ giving the spin orbit split (by 2λ and 3λ, where λ is the T_{2g} state SOC parameter that is approximately $-80\,\text{cm}^{-1}$ for Fe(II) (Solomon et al., 1995)) levels at the left in Fig. 9B. Proceeding from left to right in Fig. 9B, SOC is combined with the axial and rhombic splitting of the t₂ dπ orbitals (from Fig. 9A) and finally the Zeeman effect in the expanded inset. Thus, from LFT including in-state SOC we can relate the δ and $g_{\parallel}$ obtained experimentally from the VTVH MCD saturation magnetization curves (Fig. 7) to the axial (Δ) and rhombic (V) splitting of the dπ orbitals (Fig. 9A). These correlations are given in Fig. 10. Thus for the four complexes in Fig. 5, we can experimentally obtain the energy splittings of their five d orbitals (Fig. 11). Thus, we have developed a

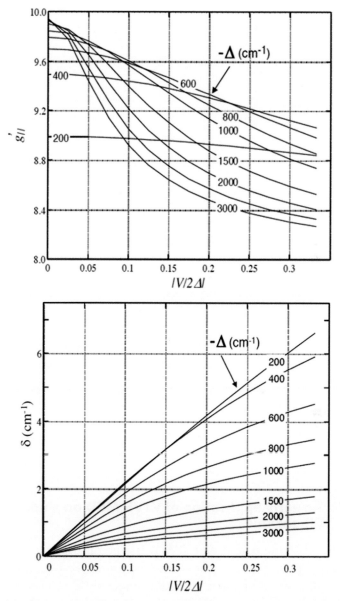

Fig. 10 Correlations of $g_\parallel$ (top) and δ (bottom) of a non-Kramers' doublet with the axial (Δ) and rhombic (V) LF splitting of the t_2 dπ orbitals. Adapted from Solomon, E.I., Pavel, E.G., Loeb, K.E., & Campochiaro, C. (1995). Magnetic circular dichroism spectroscopy as a probe of the geometric and electronic structure of non-heme ferrous enzymes. *Coordination Chemistry Reviews*, 144, 369–460. https://doi.org/10.1016/0010-8545(95)01150-N with permission from Elsevier.

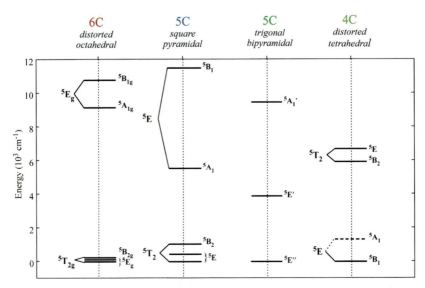

Fig. 11 Experimental ligand-field splittings of the five d orbitals for the complexes in Fig. 5. Adapted with permission from Pavel, E.G., Kitajima, N., & Solomon, E.I. (1998). Magnetic circular dichroism spectroscopic studies of mononuclear non-heme ferrous model complexes. correlation of excited- and ground-state electronic structure with geometry. Journal of the American Chemical Society, 120(16), 3949–3962. https://doi.org/10.1021/JA973735L (Pavel et al., 1998). Copyright 1998 American Chemical Society.

spectroscopic probe of the geometric and electronic structures of non-heme ferrous active sites and can now use NIR VTVH MCD to obtain molecular level insight into catalytic mechanisms.

5. An early application of VTVH MCD on a non-heme Fe (II) enzyme

Phenylalanine hydroxylase (PAH) is a pterin dependent non-heme iron enzyme (Table 1) that is familiar to the reader because of the warning on diet soda cans concerning phenylketonuria (PKU). This is a genetic disease where a mutation in PAH leads to poor catalysis and a buildup of phenylalanine resulting in brain damage (Erlandsen, Patch, Gamez, Straub, & Stevens, 2003). We studied the VTVH MCD of this enzyme in collaboration with J. Caradonna (Kemsley, Mitić, Zaleski, Caradonna, & Solomon, 1999; Kemsley et al., 2003). Fig. 12A panel I (black solid line) gives the LT MCD spectrum of resting Fe(II) PAH; two LF transitions at

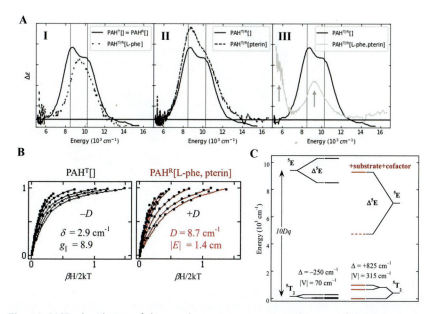

Fig. 12 MCD elucidation of the cosubstrate activation mechanism of the Fe(II) site in PAH. (A) LT MCD, (B) VTVH MCD, and (C) experimentally determined LF splitting of the d orbitals. Copyright 1999 and 1997 American Chemical Society. Adapted with permission from Kemsley, J.N., Mitić, N., Zaleski, K.L., Caradonna, J.P., & Solomon, E.I. (1999). Circular dichroism and magnetic circular dichroism spectroscopy of the catalytically competent ferrous active site of phenylalanine hydroxylase and its interaction with pterin cofactor. Journal of the American Chemical Society, 121(7), 1528–1536. https://doi.org/10.1021/JA9833063 and Loeb, K.E., Westre, T.E., Kappock, T.J., Mitić, N., Glasfeld, E., Caradonna, J.P., ... Solomon, E.I. (1997). Spectroscopic characterization of the catalytically competent ferrous site of the resting, activated, and substrate-bound forms of phenylalanine hydroxylase. Journal of the American Chemical Society, 119(8), 1901–1915. https://doi.org/10.1021/JA962269H (Loeb et al., 1997).

~10,000 cm^{-1}, split by ~2000 cm^{-1} indicating a 6C active site. Substrate binding leads to the black dotted MCD spectrum; this shows only a small perturbation of the LF transitions indicating that the substrate binding in the protein pocket results in only a small change of the 6C Fe(II) site (panel I, dotted). Panel II shows that there is no effect of the pterin on the Fe(II) (panel II, dashed) indicating that this cofactor does not bind to the Fe(II) in the absence of substrate. Importantly in panel III, substrate and cofactor together bind to PAH to produce a large change in the LF MCD spectrum (gray), now with one transition at ~10,000 cm^{-1} and a second lower energy transition at <5500 cm^{-1} indicating that the Fe(II) site in PAH has

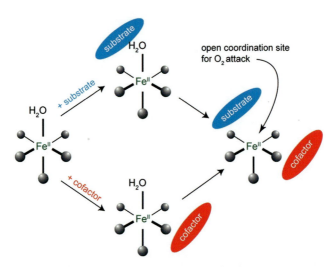

Fig. 13 General mechanistic strategy for cofactor dependent NHFe enzymes. Adapted with permission from Solomon, E.I., Deweese, D.E., & Babicz, J.T. (2021). Mechanisms of O2 activation by mononuclear non-heme iron enzymes. Biochemistry, 60(46), 3497–3506. https://doi.org/10.1021/acs.biochem.1c00370. Copyright 2021 American Chemical Society.

gone 5C. This behavior is also observed in the VTVH MCD data on PAH in Fig. 12B where there is a large change in the saturation magnetization curves with substrate plus cofactor binding to the active site (right). Analyses of the VTVH MCD data in Fig. 12A and B give the experimental splittings of the Fe(II) five 3d orbitals of PAH shown in Fig. 12C. In going from left to right with substrate plus cofactor binding, 10 Dq goes down and there is a large low symmetry splitting of the t_2 and e sets of d orbitals all demonstrating that the PAH active site has become 5C.

We have observed parallel behavior over the five classes of mononuclear non-heme ferrous enzymes in Table 1 (Neidig & Solomon, 2005; Solomon et al., 2000), presented specifically for the cofactor dependent enzymes in Fig. 13. The resting ferrous site, the Fe(II) site with substrate, and the Fe(II) site with cofactor are all 6 C, coordinately saturated and relatively stable in the presence of dioxygen. However, when both the substrate and cofactor are simultaneously bound to the enzyme, the Fe(II) site becomes 5C (via loss of an H_2O ligand (Light, Hangasky, Knapp, & Solomon, 2014)) and able to react with O_2 only in the presence of both co-substrates to generate the Fe(IV)=O intermediate for the efficient coupled reaction with substrate.

6. Perspective

While this *Methods in Enzymology* article has focused on VTVH MCD of mononuclear non-heme ferrous enzymes, we note that VTVH MCD has also provided significant insight in a number of the binuclear non-heme ferrous enzymes that also activate dioxygen (Solomon & Park, 2016). There are two additional features concerning application of this method to binuclear ferrous enzymes. First, it is able to determine the exchange coupling (J) between the ferrous ions that directly defines the bridging ligation. This required the development of J/D diagrams (i.e. exchange/ZFS) that allow analysis of this coupling in the presence of the ZFS of both ferrous sites that can be comparable to J for biferrous sites (Solomon et al., 2000). In addition, each Fe(II) in principle exhibits two LF transitions in the NIR MCD spectrum (total of up to four) that can be associated to each other (i.e. two sets of two) through their VTVH behavior that directly reflects the contribution of each Fe(II) (i.e. its M_s value) to the exchange coupled total ground state wavefunction enabling coordination structural assignment for each Fe(II) center (Neese & Solomon, 1999).

The reaction of the 5C, coordinately unsaturated Fe(II) sites defined for the five classes of mononuclear non-heme ferrous enzymes in Table 1 and for the correctly oriented coordinately unsaturated biferrous sites in the coupled binuclear non-heme iron enzymes (Park et al., 2017) by VTVH MCD generates oxygen intermediates that accomplish the wide range of chemistries noted in Table 1 and by Solomon and Park (2016). To understand the mechanisms of these reactions, it is critical to use stopped-flow and rapid freeze quench methods to trap and then spectroscopically define their associated oxygen intermediates (Babicz et al., 2023; Mitić, Clay, Saleh, Bollinger, & Solomon, 2007). As oxygen intermediates are oxidized and thus more spectroscopically accessible, a number of insightful spectroscopic methods have been and are being applied to elucidate the geometric and electronic structures of these intermediates (Solomon & Hanson, 2006). Here we emphasize that VTVH MCD on oxygen intermediates probe the low lying unoccupied or half-occupied orbitals that serve as the Frontier Molecular Orbitals (FMOs) enabling their overlap with substrate for electrophilic and HAA reactivity. One example is the π^* FMO of the Fe(IV)=O intermediate in the αkg dependent halogenases (Fig. 14) (Srnec et al., 2016) that enables selective halogenation over the thermodynamically favored hydroxylation (Srnec & Solomon, 2017).

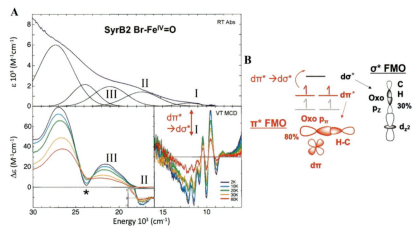

Fig. 14 (A) Absorption (top) and VT MCD (at 7 T) spectra (bottom) of SyrB2 Br–Fe(IV)=O. Transition I is the dπ* →dσ* LF transition, and II and III are oxo to iron charge transfer transitions. The * indicates a minor heme contaminant in the sample. Copyright 2016 American Chemical Society. (B) Schematic showing the relevant FMOs reflecting the orbitals participating in the dπ* →dσ* LF transition. *Adapted with permission from Srnec, M., Wong, S.D., Matthews, M.L., Krebs, C., Bollinger, J.M., & Solomon, E.I. (2016). Electronic structure of the ferryl intermediate in the α-ketoglutarate dependent non-heme iron halogenase SyrB2: Contributions to H atom abstraction reactivity. Journal of the American Chemical Society, 138, 5110–5122. https://doi.org/10.1021/jacs.6b01151.*

Acknowledgments

EIS acknowledges the National Institutes of Health (NIH) Grant R35 GM 145202 for support of this research and his students and collaborators that have greatly contributed to this science.

References

Babicz, J. T., Rogers, M. S., DeWeese, D. E., Sutherlin, K. D., Banerjee, R., Böttger, L. H., ... Solomon, E. I. (2023). Nuclear resonance vibrational spectroscopy definition of peroxy intermediates in catechol dioxygenases: Factors that determine extra- versus intradiol cleavage. *Journal of the American Chemical Society*. https://doi.org/10.1021/JACS.3C02242.

Baldwin, J. E., & Abraham, E. (1988). The biosynthesis of penicillins and cephalosporins. *Natural Product Reports, 5*(2), 129–145. https://doi.org/10.1039/NP9880500129.

Barry, S. M., & Challis, G. L. (2013). Mechanism and catalytic diversity of rieske non-heme iron-dependent oxygenases. *ACS Catalysis, 3*(10), 2362–2370. https://doi.org/10.1021/cs400087p.

Blasiak, L. C., Vaillancourt, F. H., Walsh, C. T., & Drennan, C. L. (2006). Crystal structure of the non-haem iron halogenase SyrB2 in syringomycin biosynthesis. *Nature, 440*(7082), 368–371. https://doi.org/10.1038/nature04544.

Buckingham, A. D., & Stephens, P. J. (1966). Magnetic optical activity. *Annual Review of Physical Chemistry, 17*(1), 399–432. https://doi.org/10.1146/annurev.pc.17.100166.002151.

Campochiaro, C., Pavel, E. G., & Solomon, E. I. (1995). Saturation magnetization magnetic circular dichroism spectroscopy of systems with positive zero-field splittings: Application to FeSiF6·6H2O. *Inorganic Chemistry, 34*(18), 4669–4675. https://doi.org/10.1021/IC00122A025.

Erlandsen, H., Patch, M. G., Gamez, A., Straub, M., & Stevens, R. C. (2003). Structural studies on phenylalanine hydroxylase and implications toward understanding and treating phenylketonuria. *Pediatrics, 112*(Supplement 4), 1557–1565. https://doi.org/10.1542/peds.112.S4.1557.

Fernandez, R. L., Juntunen, N. D., & Brunold, T. C. (2022). Differences in the second coordination sphere tailor the substrate specificity and reactivity of thiol dioxygenases. *Accounts of Chemical Research, 55*(17), 2480–2490. https://doi.org/10.1021/acs.accounts.2c00359.

Fitzpatrick, P. F. (2023). The aromatic amino acid hydroxylases: Structures, catalysis, and regulation of phenylalanine hydroxylase, tyrosine hydroxylase, and tryptophan hydroxylase. *Archives of Biochemistry and Biophysics, 735*, 109518. https://doi.org/10.1016/j.abb.2023.109518.

Gerstman, B. S., & Brill, A. S. (1985). Magnetic circular dichroism of low symmetry cupric sites. *The Journal of Chemical Physics, 82*(3), 1212–1230. https://doi.org/10.1063/1.448494.

Goudarzi, S., Babicz, J. T., Kabil, O., Banerjee, R., & Solomon, E. I. (2018). Spectroscopic and electronic structure study of ETHE1: Elucidating the factors influencing sulfur oxidation and oxygenation in mononuclear nonheme iron enzymes. *Journal of the American Chemical Society, 140*(44), 14887–14902. https://doi.org/10.1021/JACS.8B09022.

Kal, S., & Que, L. (2017). Dioxygen activation by nonheme iron enzymes with the 2-His-1-carboxylate facial triad that generate high-valent oxoiron oxidants. *Journal of Biological Inorganic Chemistry, 22*(2–3), 339–365. https://doi.org/10.1007/s00775-016-1431-2.

Kappock, T. J., & Caradonna, J. P. (1996). Pterin-dependent amino acid hydroxylases. *Chemical Reviews, 96*(7), 2659–2756. https://doi.org/10.1021/cr9402034.

Kemsley, J. N., Mitić, N., Zaleski, K. L., Caradonna, J. P., & Solomon, E. I. (1999). Circular dichroism and magnetic circular dichroism spectroscopy of the catalytically competent ferrous active site of phenylalanine hydroxylase and its interaction with pterin cofactor. *Journal of the American Chemical Society, 121*(7), 1528–1536. https://doi.org/10.1021/JA9833063.

Kemsley, J. N., Wasinger, E. C., Datta, S., Mitić, N., Acharya, T., Hedman, B. P., ... Solomon, E. (2003). Spectroscopic and kinetic studies of PKU−inducing mutants of phenylalanine hydroxylase: Arg158Gln and Glu280Lys. *Journal of the American Chemical Society, 125*(19), 5677–5686. https://doi.org/10.1021/ja029106f.

Krebs, C., Galonić Fujimori, D., Walsh, C. T., & Bollinger, J. M., Jr. (2007). Non-heme Fe (IV)−oxo intermediates. *Accounts of Chemical Research, 40*(7), 484–492. https://doi.org/10.1021/ar700066p.

Light, K. M., Hangasky, J. A., Knapp, M. J., & Solomon, E. I. (2014). First- and second-sphere contributions to Fe(ii) site activation by cosubstrate binding in non-heme Fe enzymes. *Dalton Transactions, 43*(4), 1505–1508. https://doi.org/10.1039/C3DT53201A.

Lipscomb, J. D. (2008). Mechanism of extradiol aromatic ring-cleaving dioxygenases. *Current Opinion in Structural Biology, 18*(6), 644–649. https://doi.org/10.1016/J.SBI.2008.11.001.

Loeb, K. E., Westre, T. E., Kappock, T. J., Mitić, N., Glasfeld, E., Caradonna, J. P., ... Solomon, E. I. (1997). Spectroscopic characterization of the catalytically competent ferrous site of the resting, activated, and substrate-bound forms of phenylalanine hydroxylase. *Journal of the American Chemical Society, 119*(8), 1901–1915. https://doi.org/10.1021/JA962269H.

Mitić, N., Clay, M. D., Saleh, L., Bollinger, J. M., & Solomon, E. I. (2007). Spectroscopic and electronic structure studies of intermediate X in ribonucleotide reductase R2 and two variants: A description of the FeIV-oxo bond in the FeIII—O—FeIV dimer. *Journal of the American Chemical Society, 129*(29), 9049–9065. https://doi.org/10.1021/ja070909i.

Neese, F., & Solomon, E. I. (1999). MCD C-term signs, saturation behavior, and determination of band polarizations in randomly oriented systems with spin S ≥ 1/2. Applications to S = 1/2 and S = 5/2. *Inorganic Chemistry, 38*(8), 1847–1865. https://doi.org/10.1021/IC981264D.

Neidig, M. L., & Solomon, E. I. (2005). Structure–function correlations in oxygen activating non-heme iron enzymes. *Chemical Communications, 47*, 5843–5863. https://doi.org/10.1039/B510233M.

Park, K., Li, N., Kwak, Y., Srnec, M., Bell, C. B., Liu, L. V., ... Solomon, E. I. (2017). Peroxide activation for electrophilic reactivity by the binuclear non-heme iron enzyme AurF. *Journal of the American Chemical Society, 139*(20), 7062–7070. https://doi.org/10.1021/JACS.7B02997.

Pavel, E. G., Kitajima, N., & Solomon, E. I. (1998). Magnetic circular dichroism spectroscopic studies of mononuclear non-heme ferrous model complexes. correlation of excited- and ground-state electronic structure with geometry. *Journal of the American Chemical Society, 120*(16), 3949–3962. https://doi.org/10.1021/JA973735L.

Purpero, V., & Moran, G. R. (2007). The diverse and pervasive chemistries of the α-keto acid dependent enzymes. *Journal of Biological Inorganic Chemistry, 12*(5), 587–601. https://doi.org/10.1007/S00775-007-0231-0.

Schatz, P. N., Mowery, R. L., & Krausz, E. R. (1978). M.C.D./M.C.P.L. saturation theory with application to molecules in D ∞h and its subgroups. *Molecular Physics, 35*(6), 1537–1557. https://doi.org/10.1080/00268977800101151.

Solomon, E. I., & Hanson, M. A. (2006). Bioinorganic spectroscopy. In E. I. Solomon, & A. B. P. Lever (Vol. Eds.), *Inorganic electronic structure and spectroscopy: Vol. 2*, (pp. 1–129). (pp. 1)John Wiley & Sons, Inc. ⟨https://www.wiley.com/en-us/Inorganic+Electronic+Structure+and+Spectroscopy%2C+Volume+II-p-9780471971146⟩.

Solomon, E. I., & Park, K. (2016). Structure/function correlations over binuclear non-heme iron active sites. *Journal of Biological Inorganic Chemistry, 21*(5), 575–588. https://doi.org/10.1007/s00775-016-1372-9.

Solomon, E. I., Brunold, T. C., Davis, M. I., Kemsley, J. N., Lee, S. K., Lehnert, N., ... Zhou, J. (2000). Geometric and electronic structure/function correlations in non-heme iron enzymes. *Chemical Reviews, 100*(1), 235–349. https://doi.org/10.1021/cr9900275.

Solomon, E. I., Deweese, D. E., & Babicz, J. T. (2021). Mechanisms of O_2 activation by mononuclear non-heme iron enzymes. *Biochemistry, 60*(46), 3497–3506. https://doi.org/10.1021/acs.biochem.1c00370.

Solomon, E. I., Pavel, E. G., Loeb, K. E., & Campochiaro, C. (1995). Magnetic circular dichroism spectroscopy as a probe of the geometric and electronic structure of non-heme ferrous enzymes. *Coordination Chemistry Reviews, 144*, 369–460. https://doi.org/10.1016/0010-8545(95)01150-N.

Srnec, M., & Solomon, E. I. (2017). Frontier molecular orbital contributions to chlorination versus hydroxylation selectivity in the non-heme iron halogenase SyrB2. *Journal of the American Chemical Society, 139*(6), 2396–2407. https://doi.org/10.1021/jacs.6b11995.

Srnec, M., Wong, S. D., Matthews, M. L., Krebs, C., Bollinger, J. M., & Solomon, E. I. (2016). Electronic structure of the ferryl intermediate in the α-ketoglutarate dependent non-heme iron halogenase SyrB2: Contributions to H atom abstraction reactivity. *Journal of the American Chemical Society, 138*, 5110–5122. https://doi.org/10.1021/jacs.6b01151.

Stephens, P. J. (1976). Magnetic circular dichroism. *Advances in Chemical Physics*, 197–264. https://doi.org/10.1002/9780470142547.ch4.

Sutherlin, K. D. S., Rivard, B. H., Böttger, L. V., Liu, L. S., Rogers, M., Srnec, M., ... Solomon, E. (2018). NRVS studies of the peroxide shunt intermediate in a rieske dioxygenase and its relation to the native FeII O_2 reaction. *Journal of the American Chemical Society, 140*(16), 5544–5559. https://doi.org/10.1021/jacs.8b01822.

Sutherlin, K. D., Wasada-Tsutsui, Y., Mbughuni, M. M., Rogers, M. S., Park, K., Liu, L. V., ... Solomon, E. I. (2018). Nuclear resonance vibrational spectroscopy definition of O2 intermediates in an extradiol dioxygenase: Correlation to crystallography and reactivity. *Journal of the American Chemical Society, 140*(48), 16495–16513. https://doi.org/10.1021/jacs.8b06517.

Thomson, A. J., & Johnson, M. K. (1980). Magnetization curves of haemoproteins measured by low-temperature magnetic-circular-dichroism spectroscopy. *Biochemical Journal, 191*(2), 411–420. https://doi.org/10.1042/bj1910411.

Vaillancourt, F. H., Bolin, J. T., & Eltis, L. D. (2006). The ins and outs of ring-cleaving dioxygenases. *Critical Reviews in Biochemistry and Molecular Biology, 41*(4), 241–267. https://doi.org/10.1080/10409230600817422.

Whittaker, J. W., & Solomon, E. I. (1988). Spectroscopic studies on ferrous nonheme iron active sites: Magnetic circular dichroism of mononuclear iron sites in superoxide dismutase and lipoxygenase. *Journal of the American Chemical Society, 110*(16), 5329–5339. https://doi.org/10.1021/ja00224a017.

CHAPTER THREE

Equilibrium dialysis with HPLC detection to measure substrate binding affinity of a non-heme iron halogenase

Elizabeth R. Smithwick, Ambika Bhagi-Damodaran*, and Anoop Rama Damodaran*

Department of Chemistry, University of Minnesota – Twin Cities, Minneapolis, MN, United States
*Corresponding authors. e-mail address: ambikab@umn.edu; rdanoop@umn.edu

Contents

1. Introduction		52
2. Materials		53
2.1 Quantitation of ligand and calibration curve		53
2.2 Equilibrium dialysis apparatus setup and initial measurements		54
2.3 Equilibrium dialysis for determination of substrate affinity in BesD		54
3. Methods		54
3.1 Quantitation of ligand and calibration curve		54
3.2 Equilibrium dialysis apparatus setup and initial measurements		55
3.3 Equilibrium dialysis for determination of substrate affinity in BesD		58
4. Conclusions		60
Acknowledgments		62
Author contributions		62
References		62

Abstract

Determination of substrate binding affinity (K_d) is critical to understanding enzyme function. An extensive number of methods have been developed and employed to study ligand/substrate binding, but the best approach depends greatly on the substrate and the enzyme in question. Below we describe how to measure the K_d of BesD, a non-heme iron halogenase, for its native substrate lysine using equilibrium dialysis coupled with High Performance Liquid Chromatography (HPLC) for subsequent detection. This method can be performed in anaerobic glove bag settings. It requires readily available HPLC instrumentation for ligand quantitation and is adaptable to meet the needs of a variety of substrate affinity measurements.

Methods in Enzymology, Volume 703
ISSN 0076-6879, https://doi.org/10.1016/bs.mie.2024.05.004
Copyright © 2024 Elsevier Inc. All rights are reserved, including those for text and data mining, AI training, and similar technologies.

1. Introduction

Understanding and quantifying substrate binding is fundamental to decoding enzyme function. When combined with sequence or structural information, determining the K_d value for a given substrate can accelerate the design of novel biocatalysts and the identification of promising new medically active compounds by allowing researchers to better map structure to function in an enzyme. Currently, there are numerous methods available to measure the binding affinity of a substrate with some popular techniques being isothermal calorimetry (ITC) (Bastos et al., 2023), surface plasmon resonance (SPR) (Olaru, Bala, Jaffrezic-Renault, & Aboul-Enein, 2015), differential scanning fluorimetry (DSF) (Niesen, Berglund, & Vedadi, 2007), and fluorescence polarization (FP) (Rossi & Taylor, 2011). The optimal strategy for substrate affinity determination greatly depends on the enzyme, the substrate being probed, the experimental conditions under which the affinity needs to be measured, and the resources and instrumentation that are readily available.

Non-heme iron halogenases are a family of enzymes capable of directly halogenating an unactivated aliphatic carbon using a mononuclear iron center (Blasiak, Vaillancourt, Walsh, & Drennan, 2006; Voss, Honda Malca, & Buller, 2019; Wilson, Chatterjee, Smithwick, Dalluge, & Bhagi, 2022). Their ability to chlorinate and brominate a variety of substrates including amino acids, nucleotides, and natural products makes them a highly versatile biocatalytic platform that invites further optimization. In addition to halogenation, these enzymes have also been shown to functionalize C–H bonds with azide (Matthews et al., 2014; Neugebauer et al., 2019) and nitrite (Matthews et al., 2014), albeit withlow yields. A variety of methods have been used to determine substrate affinity in these and related classes of enzymes, including tryptophan fluorescence (Martin, Chaplin, Eyles, & Knapp, 2019), SPR (Hu et al., 2015), and UV-Vis spectral shift-based methods (Matthews et al., 2014; Price, Barr, Tirupati, Bollinger, & Krebs, 2003; Ryle, Padmakumar, & Hausinger, 1999). During our investigations of BesD, a non-heme iron halogenase that natively chlorinates lysine at the γ-position (Marchand et al., 2019), we characterized a positive heterotropic cooperativity that exists between the primary substrate lysine and the co-substrate chloride (Smithwick et al., 2023). To quantify how the concentration of lysine influences the binding of chloride, we were able to utilize a UV/Vis spectral shift-based assay which takes advantage of a perturbation that occurs in a metal-to-ligand charge transfer (MLCT) band between the ferrous iron center and the co-substrate 2-oxoglutarate (2OG) upon the direct binding of chloride to iron (Matthews et al., 2014). In other non-heme iron enzymes, the substrate has also been observed to indirectly perturb this MLCT band by inducing the dissociation of an aqua ligand initially bound to the iron center before substrate binding (Ho et al., 2001). Quantifying this shift upon increasing substrate concentrations has also allowed researchers to determine binding affinities for substrates of non-heme iron enzymes (Price et al., 2003; Ryle et al., 1999). However upon the addition of 5 mM lysine to the Fe^{II}-2OG-BesD complex in the absence of chloride, no significant shift in the MLCT was observed. While the absence of any shift could suggest negligible lysine binding at these concentrations, a method that can directly measure lysine binding to the Fe^{II}-2OG-BesD complex under anaerobic settings was needed to confirm such a hypothesis. In turn, we developed an equilibrium dialysis assay to directly measure the apparent K_d of lysine to BesD as a function of total chloride concentrations which is described in this chapter (Smithwick et al., 2023). While the procedure presented here focuses on lysine detection, it is highly adaptable to any amino acid substrate. The assay can be performed anaerobically in a glove bag setting which is a crucial requirement for redox active non-heme iron enzymes, and subsequent HPLC quantification of bound/free lysine fractions can be completed outside the glove bag.

2. Materials

All buffers were filtered through 0.22-micron filters prior to application on the HPLC.

2.1 Quantitation of ligand and calibration curve

1. High-performance liquid chromatography (HPLC) system with photo diode array (PDA) detector
2. C18 column
3. HPLC-grade acetonitrile
4. pH probe
5. 5 mM ammonium acetate solution pH 5.0, pH adjusted using glacial acetic acid (*see* **Note 1**)
6. 30 mM sodium tetraborate pH 10.5 (*see* **Note 2**)
7. 6-Aminoquinolyl-N-hydroxysuccinimidyl Carbamate (AQC)
8. Vortex mixer
9. L-lysine

10. Deactivated clear glass 2 mL vials compatible with autosamplers
11. 300 μL volume polypropylene vials compatible with autosamplers
12. 0.2–10, 10–100, and 100–1000 μL pipettes and compatible tips
13. Software to integrate peak area
14. 0.22-micron filters
15. Volumetric glassware

2.2 Equilibrium dialysis apparatus setup and initial measurements

1. 0.6 mL low adhesion microcentrifuge tubes
2. Scissors
3. Dialysis membrane with 6–8 kDa molecular weight cutoff
4. Parafilm
5. Kimwipes
6. Equipment/materials from Section 2.1.

2.3 Equilibrium dialysis for determination of substrate affinity in BesD

1. Anaerobic glovebag/glovebox
2. UV-Vis spectrometer
3. 0.5 mL Centricons (centrifugal filter devices) with 10-kDa molecular weight cutoff
4. Centrifuge with cooling capabilities or other refrigeration method
5. Purified non-heme iron halogenase
6. Equipment/materials from Sections 2.1 and 2.2.

3. Methods

3.1 Quantitation of ligand and calibration curve

1. Prepare a series of standard solutions of the ligand you intend to measure. In this study, a stock of 30 mM lysine in 50 mM HEPES pH 7.5 was prepared using a 50 mL volumetric flask to adjust to the final volume after pH adjustment. Serial dilutions of 30 mM stock lysine solution were performed in 50 mM HEPES pH 7.5 with pipettes to obtain a set of solutions with concentrations ranging from 2 μM to 300 μM.
2. For analysis of amino acid substrates, prepare the amino acid derivatizing agent solution. Dissolve 3 mgs of AQC in 1 mL of acetonitrile to create a ca. 10 mM AQC solution. Vortex this solution vigorously until all the solid dissolves (see **Note 3**). It is best to prepare this solution on the day

of analysis as AQC is prone to hydrolysis upon exposure to any H_2O contamination.

3. Combine 50 µL of one of your standard lysine solutions prepared above with 40 µL of 30 mM sodium tetraborate pH 10.5 in a deactivated glass vial (*see* **Note 4**). Vortex to combine before adding the derivatizing agent (*see* **Note 5**). Next add 10 µL of freshly prepared AQC solution directly into the sample. Vortex immediately after each addition. Repeat for all concentrations of standard lysine solutions.

4. Transfer derivatized samples to 300 µL polypropylene vials or other vials that work well with the HPLC system of choice. If not immediately analyzing samples, store samples at −20 or −80°C degrees. However, AQC derivatized amino acids have been found to be stable for at least a week (Díaz, Lliberia, Comellas, & Broto-Puig, 1996).

5. Run standard samples on an HPLC to build calibration curve for your ligand of interest. For the analysis of lysine derivatized with AQC, samples were injected onto a Shimadzu Prominence-i LC-2030C 3D Plus system equipped with a PDA detector and a Regis Technologies REXCHROM C18 column (4.60 mm × 250 mm × 5 µm). Solvent A was 5 mM ammonium acetate pH 5.0, and Solvent B was 60% acetonitrile in water. The column temperature was maintained at 35°C. Separation was performed with a gradient method of 0% B to 70% B, 0.0 to 40.0 min; 70% B to 100% B, 40.0 to 42.0 min; 100% B, 42.0 to 50.0 min; 100% B to 0% B, 50.0 to 52.0 min; 0% B, 52.0 to 60.0 min at a flow rate of 0.5 mL/min and an injection volume of 50 µL. All HPLC traces were constructed from extracted absorbance at 254 nm.

6. Build a calibration curve from the ligand specific peak area of the HPLC traces. Peak areas of the three resulting peaks from derivatized lysine products (both mono- and di-derivatized products) were integrated using LabSolutions software (Fig. 1A). The total adjusted peak area of all three AQC-tagged lysine peaks were graphed versus concentration of lysine to create a calibration curve (Fig. 1B, *see* **Note 6**).

3.2 Equilibrium dialysis apparatus setup and initial measurements

1. To prepare an equilibrium dialysis chamber as previously described (Mega & Hase, 1991), cut a 0.6 mL Eppendorf tube approximately 6 mm below the cap to create a tube open on one side (see **Note 7**). Discard the bottom portion of the Eppendorf tube not connected to the cap. Cut the cap off the top of the tube so that the tube portion is

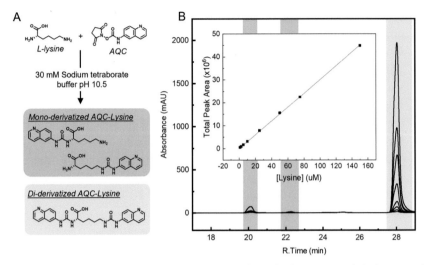

Fig. 1 (A) Reaction between AQC and lysine produces both mono- and di-derivatized lysine products. (B) HPLC traces from AQC-derivatized standard lysine with calibration curve relating relevant peak area to sample concentration shown as an inset. *Reprinted with permission from Smithwick, E. R., Wilson, R. H., Chatterjee, S., Pu, Y., Dalluge, J. J., Damodaran, A. R., & Bhagi-Damodaran, A. (2023). Electrostatically regulated active site assembly governs reactivity in nonheme iron halogenases. ACS Catalysis, 13(20), 13743–13755. Copyright 2023 American Chemical Society.*

completely separated from the cap (Fig. 2A). Cut a second cap from a different Eppendorf tube.

2. Cut dialysis membrane into small ~10–15 mm tall pieces and snip the edges of the tubing so that the membrane will be one layer thick. Wash the membrane pieces thoroughly with water and soak them for at least 30 min in buffer of choice.

3. Taking one of the previously cut caps, place it face up such that it can be used as the bottom chamber for dialysis (Fig. 2B). For initial tests, fill the bottom chamber with 65 μL buffer of choice, in this case, 50 mM HEPES pH 7.5. Retrieve a hydrated dialysis membrane and remove excess liquid on either side with a clean Kimwipe. Do not allow the dialysis membrane to fully dry out. Gently place the dialysis membrane on top of the bottom chamber. Avoid having any solution that is filled in the bottom chamber from being pulled out by capillary action. Once the membrane is well positioned on top, carefully push the previously cut Eppendorf tube down on the bottom chamber to trap the membrane in between them (Fig. 2B). Make sure that the solution in the bottom

Equilibrium dialysis with HPLC detection

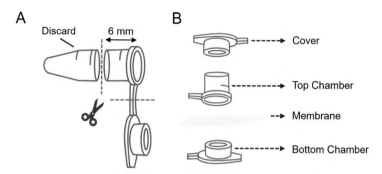

Fig. 2 Description of equilibrium dialysis apparatus setup (A) Cut centrifuge tube approximately 6 mm from the cap and discard the bottom of the tube. Separate cap from the reserved tube portion and keep both pieces for later. (B) Assemble equilibrium dialysis apparatus by adding solution to one of the cap lids facing up which will act as the bottom chamber. Position the dialysis membrane in between the cap lid and the reserved tubing, and then push down to seal the bottom chamber. Add solution into the resulting open tube portion so that it rests on top of the dialysis membrane. Push a second cap down on top of the open tube to seal the top chamber.

chamber is in contact with the dialysis membrane at this point. If the solution does not look flush with the membrane very gently tap the tube apparatus until it makes contact. It is common to see a small air bubble off to one side of the tube cap.

4. The trapped dialysis membrane along with tube pressed down upon it forms the floor of top chamber. Add 65 μL of the ligand solution to the top chamber, and then close it up by pushing the second cap onto the open portion of the tube (*see* **Note 8**) (Fig. 2B). After the top chamber has been capped, wrap parafilm around the entire apparatus to prevent evaporation of samples. Do not allow samples to tip at this point as liquid getting on sides of the tubing can lead to inaccuracy in quantitation.

5. Allow samples to equilibrate at 4 °C. During initial studies, it is best to test a variety of equilibration times to ensure that the highest concentration ligand of interest fully equilibrates between the top and bottom chambers. It was determined that 20 h was sufficient for complete lysine equilibration at the concentrations of lysine used to evaluate K_d of BesD.

6. After equilibration, carefully remove parafilm around the equilibrium dialysis setup and then take off the cap of the top chamber. The cap is sometimes difficult to remove so be careful not to remove the bottom cap at the same time as removing the top cap.

7. Transfer 50 µL from the top chamber to a deactivated glass vial and wipe out excess solution with clean Kimwipe. Once excess liquid from the top chamber is removed, puncture the dialysis membrane with a piece of wire or a needle and transfer 50 µL from bottom chamber to a different deactivated glass vial.

8. Derivatize all lysine samples with AQC and analyze on HPLC as described in the previous section. Integrate the peaks corresponding to derivatized lysine in each of the sample chromatograms. Using the calibration curve constructed in the previous section, convert integrated peak area to concentration. For samples without protein, fully equilibrated samples should have equal concentrations between the top and bottom chambers and sum to the initial concentration of the solution added to the top chamber. Select an equilibration time for future samples in which the highest concentration of substrate you intend to measure has fully equilibrated between the two chambers. A schematic overview of the equilibrium dialysis process is given in Fig. 3A.

3.3 Equilibrium dialysis for determination of substrate affinity in BesD

1. Bring in materials/chemicals required for the equilibrium dialysis experiment into the anaerobic chamber at least 24 h ahead of the experiment to ensure proper equilibration and removal of oxygen. Deoxygenate any necessary solutions using a Schlenk line, cycling between vacuum and argon gas for 15-minute intervals three times. Bring into the anaerobic chamber at least 24 h ahead of the experiment to ensure complete deoxygenation.

2. The day of the experiment, bring in a solution of concentrated stock protein into the anaerobic chamber. Once inside, wash the protein in centricons with deoxygenated buffer by concentrating the protein solution through centrifugation, adding buffer to refill the filter device, and concentrating again. Repeat this process three times to remove oxygen and any other storage buffer molecules. In this study, we examined the substrate affinity of lysine at different concentrations of NaCl in BesD, and our buffer of choice was 50 mM HEPES pH 7.5 for the washing step.

3. After the washing step is complete, invert the filter device in a clean centrifuge tube and centrifuge gently for approximately a minute to reclaim the protein. Assess concentration of protein using the absorbance at 280 nm and the molar absorptivity calculated by ExPASy ProtParam (Gasteiger, 2003).

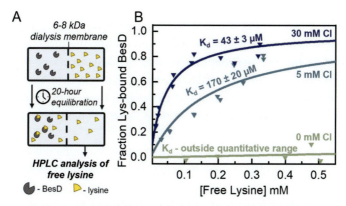

Fig. 3 (A) Diagram of conceptual equilibrium dialysis workflow. (B) Equilibrium dialysis data and derived $K_{d,app}$ for lysine and BesD at different concentrations of chloride. *Reprinted with permission from Smithwick, E. R., Wilson, R. H., Chatterjee, S., Pu, Y., Dalluge, J. J., Damodaran, A. R., & Bhagi-Damodaran, A. (2023). Electrostatically regulated active site assembly governs reactivity in nonheme iron halogenases. ACS Catalysis, 13(20), 13743–13755. Copyright 2023 American Chemical Society.*

4. Prepare the ligand and protein for equilibrium dialysis. In our study, the top and bottom chamber solution consisted of 0.2–1 mM lysine, 4 mM 2OG, 1 mM $(NH_4)_2Fe(SO_4)_2$, 30 mM NaCl, and 35 mM Na_2SO_4 in 50 mM HEPES (pH 7.5) and of 425 ± 25 μM BesD, 4 mM 2OG, 1 mM $(NH_4)_2Fe(SO_4)_2$, 30 mM NaCl, and 35 mM Na_2SO_4 in 50 mM HEPES (pH 7.5), respectively. All solutions were prepared from degassed stock solutions of reagents, except for the stock solution of $(NH_4)_2Fe(SO_4)_2$ which was prepared from dry solid dissolved in degassed MQ water. To assess how the binding affinity of lysine changes with respect to NaCl in BesD, this process was repeated for sample sets containing 0 mM NaCl and 45 mM Na_2SO_4 and sample sets containing 5 mM NaCl and 42.5 mM Na_2SO_4 while keeping all other compound concentrations constant.
5. Prepare the equilibrium dialysis setup as described in the above section with the protein solution in the bottom chamber and the ligand solution in the top chamber.
6. After preparation, place sample in refrigerated area set at 4 °C. Allow samples to equilibrate for 20 h. We recommend stirring liquids in the two chambers using stir bar or an orbital shaker, if feasible.
7. After equilibration, recover samples from the equilibrium dialysis setup as described in Section 3.2. Additionally, take an extra ~5 μL sample of

the top chamber and assess if any protein leaked through to the top chamber using the absorbance at 280 nm. While the protein should not be able to traverse the membrane, this check is necessary to verify the integrity of the equilibrium dialysis setup. If protein is found in the top chamber, identify if this finding is an isolated issue. If the leakage occurs in all samples, consider using a different variety of dialysis membrane (*see* **Note 9**).

8. If no protein leakage has occurred, derivatize the extracted top chamber samples and run HPLC analysis using the method described in previous sections.

9. Convert peak areas of sample to concentration using calibration curve to determine the free lysine concentration in each equilibrium dialysis sample. The Hill equation for single site binding is given as:

$$\textit{Fraction of lysine bound BesD} = \frac{[ES]}{[E]_t} = \frac{[S]}{[S] + K_d} \qquad (1)$$

Where $[S]$, in this study, is the concentration of free lysine in the top chamber post dialysis as determined by HPLC, $[E]_T$ is the concentration of BesD in the bottom chamber, and $[ES]$ is the concentration of BesD-lysine complex. To find $[ES]$, consider that the total amount of ligand in both the top and bottom chambers can be described by the following relationship $[S]_T = [ES] + [S]_{bottom\ chamber} + [S]_{top\ chamber}$, where $[S]_T$ is the pre-equilibrium concentration of the ligand in the top chamber. As the dialysis membrane allows the ligand of interest to exchange freely between two equal volume chambers, the free ligand concentration between the two chambers should be equal $([S]_{bottom\ chamber} = [S]_{top\ chamber} = [S])$, allowing the total ligand concentration relationship to be rewritten as $[S]_T = [ES] + 2[S]$. Therefore, $[ES]$ can be calculated by rearranging the relationship to be $[ES] = [S]_T - 2[S]$. Using values of $\frac{[ES]}{[E]_t}$ and $[S]$, a relationship of points can be constructed and fit to Eq. (1) using Origin or other software capable of performing non-linear regression.

4. Conclusions

Using equilibrium dialysis methods for lysine binding to BesD, we were able to determine the apparent K_d for lysine as $43 \pm 3\,\mu M$ and $170 \pm 20\,\mu M$ at chloride concentrations of 30 mM and 5 mM respectively (Fig. 3B) (Smithwick et al., 2023). In the absence of any chloride, these equilibrium dialysis measurements found negligible binding of lysine to BesD which matched with our observations from UV-Vis spectral shift studies. Overall, the enhanced apparent lysine binding affinity with increasing chloride concentrations measured via equilibrium dialysis approaches in tandem with the enhanced apparent chloride binding affinity with increasing lysine concentrations measured using UV-Vis spectral shift-based approaches demonstrate strong positive heterotropic cooperativity in the binding of lysine and chloride to BesD's catalytic core. Together they provide insight into the unique ways enzymes can regulate the assembly of their catalytic core and ultimately their biochemical activity. Overall, equilibrium dialysis has the potential to be highly adaptable to a variety of enzyme-substrate pairs and can be easily translated to suit a variety of situations, such as an anerobic environment, due to the low barrier of experimental setup.

Notes.

1. The pH of ammonium acetate buffer solution is very important to reliably measure the retention time of analytes, and deviations of even $\pm$ 0.1 pH units can affect retention time of the analytes (Van Wandelen & Cohen, 1997).

2. Sodium tetraborate (also known as Borax) is known to be hazardous to human health. Avoid contact with skin, seek appropriate medical attention if ingested, and follow regional waste disposal guidelines. Wear personal protective equipment like gloves and safety glasses when handling and refer to safety data sheets before using the chemical.

3. This solution is on the border line of AQC solubility but should be able to completely dissolve in acetonitrile at a ratio of 3 mg/mL. Hold the vial in hand to try to warm up the solution to help all the AQC dissolve in solution.

4. We have observed decreased derivatization performance if deactivated glass vials are not used.

5. The pH of the derivatization solution is very important for the AQC derivatization process, so it is important to premix the solutions before adding the AQC. For optimal derivatization, the pH of solution before AQC is added should be between 8.5 and 10.0. When the indicated buffers are combined in the ratios described above, the pH of the solution before derivatization is approximately 9.5.

6. Mono-derivatized AQC-lysine products were approximated to have half the absorbance of the di-derivatized AQC-lysine product and therefore the peak area corresponding to the monoderivative species were multiplied by 2 when summed to create calibration curve.

7. Instead of the setup we describe here, a variety of pre-made equilibrium dialysis chambers are also available for direct use. For example, Thermo Scientific™ Rapid Equilibrium Dialysis (RED) Inserts and Plates.
8. Depending on how rough the edges are, the cap may be difficult to push down but try to push it so it sits flush with the cut side of the tube. Use a new cap if necessary and be gentle because if you apply force in the wrong way, it will flip the samples and you may have to start over.
9. We did not observe any protein leakage to the top chamber, but leaking can be an issue if you are working with a smaller molecular weight protein (MW < 20 kDa). Additionally, protein leakage might be evidence that the membrane is being punctured during the assembly process. Make sure to check the integrity of the membrane when removing samples.

Acknowledgments

ERS acknowledges the support of the National Institute of Health Chemical Biology Training Grant (T32GM132029). This work was supported by NSF CBET and CLP (grant # 2046527).

Author contributions

ERS, ABD, and ARD designed the study. ERS performed the study. ERS, ABD, and ARD wrote the article. All authors have given approval to the final version of the manuscript.

References

Bastos, M., Abian, O., Johnson, C. M., Ferreira-da-Silva, F., Vega, S., Jimenez-Alesanco, A., & Velazquez-Campoy, A. (2023). Isothermal titration calorimetry. *Nature Reviews Methods Primers, 3*(1), 17.

Blasiak, L. C., Vaillancourt, F. H., Walsh, C. T., & Drennan, C. L. (2006). Crystal structure of the non-haem iron halogenase SyrB2 in syringomycin biosynthesis. *Nature, 440*(7082), 368–371.

Díaz, J., Lliberia, J. L., Comellas, L., & Broto-Puig, F. (1996). Amino acid and amino sugar determination by derivatization with 6-aminoquinolyl-N-hydroxysuccinimidyl carbamate followed by high-performance liquid chromatography and fluorescence detection. *Journal of Chromatography A, 719*(1), 171–179.

Gasteiger, E. (2003). ExPASy: The proteomics server for in-depth protein knowledge and analysis. *Nucleic Acids Research, 31*(13), 3784–3788.

Ho, R. Y. N., Mehn, M. P., Hegg, E. L., Liu, A., Ryle, M. J., Hausinger, R. P., & Que, L. (2001). Resonance Raman studies of the iron(II)-α-keto acid chromophore in model and enzyme complexes. *Journal of the American Chemical Society, 123*(21), 5022–5029.

Hu, L., Lu, J., Cheng, J., Rao, Q., Li, Z., Hou, H., ... Xu, Y. (2015). Structural insight into substrate preference for TET-mediated oxidation. *Nature, 527*(7576), 118–122.

Marchand, J. A., Neugebauer, M. E., Ing, M. C., Lin, C. I., Pelton, J. G., & Chang, M. C. Y. (2019). Discovery of a pathway for terminal-alkyne amino acid biosynthesis. *Nature, 567*(7748), 420–424.

Martin, C. B., Chaplin, V. D., Eyles, S., & Knapp, M. J. (2019). Protein flexibility of the α-ketoglutarate-dependent oxygenase factor-inhibiting HIF-1: Implications for substrate binding, catalysis, and regulation. *Biochemistry, 58*(39), 4047–4057.

Matthews, M. L., Chang, W. C., Layne, A. P., Miles, L. A., Krebs, C., & Bollinger, J. M. (2014). Direct nitration and azidation of aliphatic carbons by an iron-dependent halogenase. *Nature Chemical Biology, 10*(3), 209–215.

Mega, T., & Hase, S. (1991). Determination of lectin-sugar binding constants by micro-equilibrium dialysis coupled with high performance liquid chromatography. *Journal of Biochemistry, 109*(4), 600–603.

Neugebauer, M. E., Sumida, K. H., Pelton, J. G., Mcmurry, J. L., Marchand, J. A., & Chang, M. C. Y. (2019). A family of radical halogenases for the engineering of amino-acid-based products. *Nature Chemical Biology, 15*, 1009–1016.

Niesen, F. H., Berglund, H., & Vedadi, M. (2007). The use of differential scanning fluorimetry to detect ligand interactions that promote protein stability. *Nature Protocols, 2*(9), 2212–2221.

Olaru, A., Bala, C., Jaffrezic-Renault, N., & Aboul-Enein, H. Y. (2015). Surface plasmon resonance (SPR) biosensors in pharmaceutical analysis. *Critical Reviews in Analytical Chemistry, 45*(2), 97–105.

Price, J. C., Barr, E. W., Tirupati, B., Bollinger, J. M., & Krebs, C. (2003). The first direct characterization of a high-valent iron intermediate in the reaction of an α-ketoglutarate-dependent dioxygenase: A high-spin Fe(IV) complex in taurine/α-ketoglutarate dioxygenase (TauD) from Escherichia coli. *Biochemistry, 42*(24), 7497–7508.

Rossi, A. M., & Taylor, C. W. (2011). Analysis of protein-ligand interactions by fluorescence polarization. *Nature Protocols, 6*(3), 365–387.

Ryle, M. J., Padmakumar, R., & Hausinger, R. P. (1999). Stopped-flow kinetic analysis of Escherichia coli taurine/α- ketoglutarate dioxygenase: Interactions with α ketoglutarate, taurine, and oxygen. *Biochemistry, 38*(46), 15278–15286.

Smithwick, E. R., Wilson, R. H., Chatterjee, S., Pu, Y., Dalluge, J. J., Damodaran, A. R., & Bhagi-Damodaran, A. (2023). Electrostatically regulated active site assembly governs reactivity in nonheme iron halogenases. *ACS Catalysis, 13*(20), 13743–13755.

Van Wandelen, C., & Cohen, S. A. (1997). Using quaternary high-performance liquid chromatography eluent systems for separating 6-aminoquinolyl-N-hydroxysuccinimidyl carbamate-derivatized amino acid mixtures. *Journal of Chromatography A, 763*(1), 11–22.

Voss, M., Honda Malca, S., & Buller, R. (2019). Exploring the biocatalytic potential of Fe/α-ketoglutarate-dependent halogenases. *Chemistry—A European Journal, 26*(33), 7336–7345.

Wilson, R. H., Chatterjee, S., Smithwick, E. R., Dalluge, J. J., & Bhagi, A. (2022). Role of a secondary coordination sphere residue in halogenation catalysis of non-heme iron enzymes. *ACS Catalysis, 12*(17), 10913–10924.

> CHAPTER FOUR

Preparation of reductases for multicomponent oxygenases

Megan E. Wolf and Lindsay D. Eltis[*]

Microbiology and Immunology, The University of British Columbia, Vancouver, BC, Canada
*Corresponding author. e-mail address: leltis@mail.ubc.ca

Contents

1. Introduction	66
2. General safety	69
3. Reductase production and activity	69
3.1 Overview	69
3.2 Expression vector cloning	69
3.3 Transformation of *E. coli* and RHA1 with expression vectors	70
3.4 Production of PbdB	71
3.5 Activity of lysates	72
3.6 Results	73
4. Codon optimization and protein purification	74
4.1 Overview	74
4.2 Codon optimization	74
4.3 Protein purification	76
4.4 Results	77
5. Protein characterization	77
5.1 Overview	77
5.2 Cofactor analysis – labile sulfide	78
5.3 Cofactor analysis – non-heme iron	78
5.4 Cofactor analysis – FAD	79
5.5 Activity analysis – cytochrome *c* reduction	79
5.6 Results	80
6. Summary and conclusions	81
References	82

Abstract

Oxygenases catalyze crucial reactions throughout all domains of life, cleaving molecular oxygen (O_2) and inserting one or two of its atoms into organic substrates. Many oxygenases, including those in the cytochrome P450 (P450) and Rieske oxygenase enzyme families, function as multicomponent systems, which require one or more redox partners to transfer electrons to the catalytic center. As the identity of the reductase can change the reactivity of the oxygenase, characterization of the latter with its cognate redox partners is critical. However, the isolation of the native redox partner or partners is

Methods in Enzymology, Volume 703
ISSN 0076-6879, https://doi.org/10.1016/bs.mie.2024.05.016
Copyright © 2024 Elsevier Inc. All rights are reserved, including those for text and data mining, AI training, and similar technologies

often challenging. Here, we report the preparation and characterization of PbdB, the native reductase partner of PbdA, a bacterial P450 enzyme that catalyzes the *O*-demethylation of *para*-methoxylated benzoates. Through production in a rhodoccocal host, codon optimization, and anaerobic purification, this procedure overcomes conventional challenges in redox partner production and allows for robust oxygenase characterization with its native redox partner. Key lessons learned here, including the value of production in a related host and rare codon effects are applicable to a broad range of Fe-dependent oxygenases and their components.

1. Introduction

Oxygenases catalyze manifold biological reactions, including mono- and dihydroxylations, epoxidation, dealkylation, dehydrogenation, ring cleavage and closure (Ferraro, Gakhar, & Ramaswamy, 2005; Meunier, de Visser, & Shaik, 2004; Romero, Gómez Castellanos, Gadda, Fraaije, & Mattevi, 2018; Wu, Meng, & Tang, 2016). In all domains of life, these reactions are essential to perform key processes in catabolism, biosynthesis, DNA repair, regulation, signalling and homeostasis (Aravind & Koonin, 2001; Montellano, 2000). In human health, oxygenases are responsible for the majority of drug and other xenobiotic metabolism in the liver, as well as essential steps in the biosynthesis of steroids and other metabolites (Furge & Guengerich, 2006; Nebert & Russell, 2002). In plants, oxygenases perform reactions in the synthesis of secondary metabolites, which protect and defend the plant, and are analogs to many commodity chemicals (Cochrane & Vederas, 2014; Mitchell & Weng, 2019). Microorganisms also utilize a plethora of oxygenases in the catabolism of organic substrates (Cheng, Chen, Parales, & Jiang, 2022; Ten Have & Teunissen, 2001; Van Beilen & Funhoff, 2005). As many of these substrates are otherwise recalcitrant, these enzymes are essential to the global carbon cycle. Additionally, as oxygenation reactions are challenging to perform chemically with the high degree of selectivity enzymes afford, oxygenases are promising biocatalysts (Grogan, 2011; Tian et al., 2023). Due to their important role in biological systems and biotechnological potential, the reconstitution of *in vitro* oxygenase activity and biochemical characterization is imperative to understanding the reactions they catalyze.

Oxygenases catalyze the reductive cleavage of molecular oxygen (O_2) and incorporation of at least one of its two oxygen atoms into the substrates. To overcome the kinetic barrier associated with O_2 cleavage, they often utilize a metal ion or organic cofactor. Two oxygenase families, the

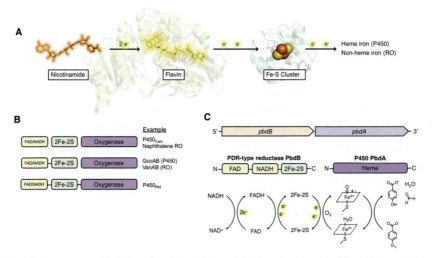

Fig. 1 Summary of P450 reductase activity: (A) The electron transfer chain of the prototypical prokaryotic P450 reductase system from P450$_{cam}$. The representative structures are putidaredoxin reductase (PDB: 1Q1R (Sevrioukova, Li, & Poulos, 2004); yellow ribbon) and putidaredoxin (PDB: 1OQQ (Sevrioukova, Garcia, Li, Bhaskar, & Poulos, 2003); green ribbon). (B) Examples of components and their domain organizations in bacterial P450 and Rieske oxygenase (RO) systems. P450$_{Cam}$ (Kuznetsov, Poulos, & Sevrioukova, 2006) and the naphthalene RO (Jouanneau, Meyer, Jakoncic, Stojanoff, & Gaillard, 2006) are three-component systems. The GcoAB P450 (Mallinson et al., 2018) and VanAB RO (Hibi, Sonoki, & Mori, 2005) are two-component systems. P450$_{Rhf}$ (Klenk et al., 2017) is a one-component system. (C) The operonic structure, domain structure, and reaction cycle of PbdAB, which catalyzes the O-demethylation of para-methoxybenzoates.

cytochrome P450 (P450) and Rieske Oxygenase (RO) families, utilize a heme and non-heme iron ion, respectively. Furthermore, their respective catalytic reactions require electrons provided by a reduced nicotinamide cofactor, NAD(P)H. As such, P450s and ROs are typically part of multicomponent systems, in which one component functions as a reductase to transfer electrons from NAD(P)H to the catalytic center of the oxygenase, either directly or indirectly, via a third, ferredoxin component (Fig. 1A and B) (Kweon et al., 2008; McLean, Luciakova, Belcher, Tee, & Munro, 2015). Moreover, the reductases can have different specificities for oxygenases. Thus, a single diflavin reductase, cytochrome P450 reductase (CPR), supplies electrons to many P450s in eukaryotes whereas in most prokaryotic systems, each P450 is reduced by a cognate reductase. Although oxygenases can function with non-cognate redox partners, recent work with P450s and

ROs has shown that the reaction rate is highly dependent on redox system (Runda, De Kok, & Schmidt, 2023; Tian et al., 2023; Zhang et al., 2018) and that the identity of the redox partners can also affect reaction products and distribution (Li, Du, & Bernhardt, 2020). Therefore, it is optimal to use the cognate redox system to characterize oxygenase activity.

The use of cognate redox partners in the study of oxygenases can be limited by various challenges. Firstly, these partners may be difficult to identify, as the genes encoding them are not always co-localized with the oxygenase in the genome. In such instances, transcriptomic or proteomic studies may be required to identify the cognate reductase system for an oxygenase. Once identified, preparing the redox partners in their active form can be challenging as many contain an iron-sulfur cluster, which can be O_2-labile (Petering, Fee, & Palmer, 1971). Finally, the optimum ratio between all redox partners must be determined to maximize activity while minimizing uncoupling, or adventitious reactivity of the reductase with O_2 (Bopp, Bernet, Kohler, & Hofstetter, 2022; Runda, Bastian, & Fatma Feyza, 2023; Suzuki et al., 2017). Uncoupling produces reactive oxygen species, which can further reduce enzyme activity (Cabiscol, Tamarit, & Ros, 2000).

We recently reported the identification of a P450-reductase system in *Rhodococcus jostii* RHA1 (RHA1 hereafter) which catalyzes the O-deme-thylation of *para*-methoxylated benzoates, and is necessary for growth of the organism on those compounds (Wolf et al.). The *pbdAB* genes, encoding the P450 and the reductase, respectively, are found in a putative operon. Somewhat unusually, *pbdB* occurs upstream of *pbdA*. PbdB has the same domain organization as phthalate dioxygenase reductase (PDR), in which the flavin-containing domain is in the N-terminal region and the [2Fe-2S] cluster-containing domain is in the C-terminal region (Fig. 1C). Although this PDR-type domain organization commonly occurs in two-component RO systems and in self-sufficient P450s, it has rarely been reported in two-component P450 systems. Another unusual feature of *pbdB* is that it is enriched in rare codons, perhaps to regulate the relative level of its translation with respect to *pbdA*.

Here, we report the production, purification, and characterization of the reductase PbdB and reconstitution of the PbdAB system *in vitro*. The activity of PbdB with its cognate oxygenase, but not other electron acceptors, was dependent on production host. The native sequence, which was enriched in rare codons at the 5′ end, was modified for increased protein production. Finally, due to its O_2-induced lability, we purified the enzyme anaerobically. These considerations allowed for the PbdA oxygenase to be characterized with its cognate redox partner. The described methods can be applied to the production, purification, and characterization of other oxygenases and their components.

2. General safety

Standard safety measures for Risk Group 1 organisms and other hazardous materials, including personal protective equipment, are required when performing the following experiments.

3. Reductase production and activity
3.1 Overview
To facilitate their purification and subsequent characterization, many proteins are overproduced in *Escherichia coli* using recombinant DNA technology due to this host's well-studied physiology, easy laboratory cultivation and rapid growth rate (Rosano & Ceccarelli, 2014). However, various actinobacterial oxygenase components, such as monooxygenases from *Mycobacterium* (Dresen et al., 2010; McCarl et al., 2018) as well as desulfurization proteins and ROs from rhodococci (Denome, Olson, & Young, 1993; Iida et al., 2009), are inactive when produced in *E. coli*. Here, we describe the production of PbdB in *E. coli* and RHA1 as expression hosts using the pET28a and pTipQC1 (Nakashima & Tamura, 2004) plasmid systems, respectively. As PbdAB is a *para*-methoxybenzoate O-demethylase, the activity of cell-free extracts was assayed by reconstituting the system *in vitro* with purified PbdA together with the requisite co-substrates.

3.2 Expression vector cloning
3.2.1 Equipment and materials
- Q5 Polymerase and Reaction Buffer (New England Biolabs)
- Gibson Assembly Mix (New England Biolabs)
- T4 Ligase and reaction buffer
- Monarch Plasmid Miniprep Kit (New England Biolabs)
- *E. coli* DH5α competent cells
- RHA1 genomic DNA
- Lysogeny Broth (LB) Media

- Plasmid vector pET28a, digested using *Nde*I and *Hind*III, and gel-purified
- Restriction enzymes *Nde*I, *Hind*III
- Plasmid vector pTipQC1, digested using *Nde*I and *Hind*III, and gel-purified

3.2.2 Procedure

1. Amplify the *pbdB* gene with the following primers and PCR program
   Forward: 5′-ctggtgccgcgcggcagccatatgccacagaacccggtttc-3′
   Reverse: 5′-aacgccagccgcactctgaaagctttcagaggtcgagcacgag-3′
   Underlined sequences denote restriction enzyme sites

Step	Temperature (°C)	Time (s)
Initial denaturation	98	30
29 cycles	98	10
	65	30
	72	30
Final extension	72	120
Hold	10	–

2. Clone the amplicon into digested pET28a backbone using Gibson assembly, regenerating *Nde*I and *Hind*III restriction sites.
3. Transform into chemically competent *E. coli* DH5α competent cells, isolate plasmid using manufacturer's protocol and confirm sequence of isolated plasmid.
4. Using *Nde*I and *Hind*III restriction sites and T4 ligase, subclone *pbdB* into digested pTipQC1 vector.
5. Transform into *E. coli* DH5α cells, isolate plasmid using manufacturer's protocol, and confirm nucleotide sequence of isolated plasmid.

3.3 Transformation of *E. coli* and RHA1 with expression vectors

3.3.1 Equipment and materials

- Bio-Rad MicroPulser Electroporator
- Bio-Rad GenePulser Cuvettes (1 mm gap length)
- *E. coli* BL21 competent cells
- Wild-type RHA1

Preparation of reductases for multicomponent oxygenases

- LB media
- Glycerol
- Plasmids generated in **Section 2.2**

3.3.2 Procedure

1. Transform pET28a_*pbdB* into chemically competent *E. coli* BL21 cells using standard methods. Select for positive transformants with LB agar plates containing 50 µg/mL kanamycin.
2. Generate electrocompetent RHA1
   a. Inoculate 5 mL of non-selective LB with a single colony of RHA1. Grow overnight at 30 °C, shaking at 200 rpm.
   b. Inoculate 500 mL non-selective LB with the overnight culture. Incubate for 16–20 h, until the optical density at 600 nm (OD_{600}) of the culture reaches 0.4–0.7.
   c. Harvest cells by centrifugation at $4000 \times g$, and wash 5–6 times with ice-cold distilled and deionized water (ddH_2O).
   d. Suspend in 5 mL of ice-cold 10% glycerol. Snap freeze in 50 µL aliquots.
3. Combine 50 µL of electrocompetent RHA1 with 100 ng of purified plasmid. Incubate on ice for 30 min
4. Add plasmid and cells to an electroporation cuvette. Electroporate using standard settings for bacteria (1.8 kV).
5. Recover with 950 µL of LB for 5–6 h at 30 °C.
6. Plate on LB agar plates containing 34 µg/mL chloramphenicol to select for positive transformants.

   NOTE: Unless otherwise stated, *E. coli* and RHA1 are incubated at 37 °C and 30 °C, respectively. Liquid cultures are agitated at 200 rpm using a platform shaker.

3.4 Production of PbdB

3.4.1 Equipment and materials

- EmulsiFlex-C5 homogenizer
- Floor centrifuge
- 0.1 mm diameter Zirconia/silica beads
- 0.5 mm diameter glass beads
- FastPrep-24 bead beater
- LB media
- *E. coli* BL21::pET28a_*pbdB* (from **Step 2.3.2**)
- RHA1:pTip_*pbdB* (from **Step 2.3.2**)

- Isopropyl β-d-1-thiogalactopyranoside (IPTG), 0.5 M stock prepared in water
- Thiostreptone, 5 mg/mL stock prepared in dimethyl sulfoxide (DMSO)
- Homogenization buffer, Tris(hydroxymethyl)aminomethane and HCl (Tris-Cl) pH 8, 200 mM NaCl, 10% glycerol, protease inhibitors

3.4.2 Procedure

1. Production of PbdB in *E. coli*
   a. Inoculate 1 mL of LB containing 50 µg/mL kanamycin, incubate overnight.
   b. Inoculate 100 mL of LB containing 50 µg/mL kanamycin, incubate for 3–4 h until OD_{600} reaches 0.4–0.6.
   c. Add IPTG to a concentration of 0.5 mM. Incubate overnight at 20 °C.
   d. Harvest cells by centrifugation, suspend in 10 mL of homogenization buffer.
   e. Lyse cells by three passes of high-pressure homogenization at 15,000 psi.
   f. Remove cell debris from the lysate by centrifugation.
2. Production of PbdB in RHA1
   a. Inoculate 1 mL of LB containing 34 µg/mL chloramphenicol, incubate overnight.
   b. Inoculate 100 mL of LB containing 34 µg/mL chloramphenicol, incubate for 16 h until OD_{600} reaches 0.4–0.6.
   c. Add thiostreptone to a concentration of 5 µg/mL. Incubate for 24–48 h at 20 °C.
   d. Harvest cells by centrifugation, suspend in 10 mL of homogenizer buffer.
   e. Lyse cells by bead-beating homogenization, 3 cycles of 6 m/s for 45 s
   f. Remove cell debris from the lysate by centrifugation.

3.5 Activity of lysates

3.5.1 Equipment and materials

- Lysates (from **Section 2.4.2**)
- Microcentrifuge
- High pressure liquid chromatography (HPLC) separation module equipped with UV–Vis detector module
- Luna 5 µm C18(2) 150 × 4.6 mm column
- 1% formic acid in water
- HPLC grade methanol
- NADH (35 mM stock prepared in water)
- Purified PbdA

- *para*-methoxybenzoate, *p*-MBA (10 mM stock prepared in DMSO)
- *para*-hydroxybenzoate, *p*-HBA (10 mM stock prepared in DMSO)
- Glacial acetic acid

3.5.2 Procedure
1. Add 35 µM NADH and 100 µM *p*-MBA to 1 mL of *E. coli* and RHA1 lysates. Add 1 µM PbdA. Incubate for 10 min at 25 °C.
2. Quench reaction with the addition of 100 µM acetic acid.
3. Centrifuge to remove debris and filter supernatant with syringe driven 0.22 µm filter.
4. Elute samples on a 16.8 mL linear gradient of 1% methanol + 0.1% formic acid in H_2O to 100% methanol. Monitor eluate at 260 nm to detect products.

3.6 Results

Lysates of *E. coli* BL21 over-expressing *pbdB* were unable to confer PbdA O-demethylase activity (Fig. 2). Conversely, in reactions performed using lysates of RHA1 over-expressing *pbdB*, *p*-MBA was completely O-demethylated after 10 min (Fig. 2). No O-demethylation was detected in control reactions of *E. coli* or RHA1 lysates containing the empty vector pET28a or pTipQC1, respectively (data not shown). These results suggest that PbdB is unable to transfer electrons to PbdA when produced in *E. coli* but does so effectively when produced in RHA1.

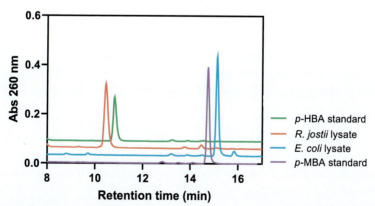

Fig. 2 Activity of lysates expressing *pbdB*. *E. coli* BL21 (blue) and *R. jostii* RHA1 (red) expressing *pbdB via* the pET and pTip expression plasmids, respectively, lysates were combined with 1 µM purified PbdA, 350 µM NADH and 100 µM of *p*-MBA and incubated for 10 min at 25 °C. Metabolites were resolved using HPLC-UV and detected at 260 nm. Substrate and product identities were confirmed by comparison to authentic standards of *p*-MBA (purple) and *p*-HBA (green).

4. Codon optimization and protein purification
4.1 Overview

The 5′-end of the *pbdB* gene is enriched in rare codons, with 10 of the first 28 being rare according to RHA1's codon utilization (Fig. 3A). In *E. coli* and several other species, such rare-codon enrichment has been linked to higher-expressed genes, as the rare codons decrease RNA secondary structure which inhibits translation (Goodman, Church, & Kosuri, 2013). Conversely, other studies have found that N-terminal rare codon enrichment decreases heterologous protein production (Lee et al., 2011; Wu et al., 2009). The effect of rare codon enrichment in bacteria other than *E. coli* is poorly described. As expression is proportional to transcript length (Lim, Lee, & Hussein, 2011), one might expect *pbdB* to be more highly expressed than *pbdA* as the first gene of the operon. It is possible the rare codons in the 5′-end of the gene serve to down-regulate PbdB translation *via* ribosome stalling, thereby ensuring a catalytically optimum ratio of reductase to P450, which is a determinant of P450 activity (Cao & Bernhardt, 1999). Indeed, using the native sequence of *pbdB* in an RHA1 expression vector, production of the reductase was undetectable in cell lysates by sodium dodecyl sulfate–polyacrylamide gel electrophoresis (SDS-PAGE) analysis and yielded only 0.2 mg of purified protein/L culture. Therefore, we optimized the codons at the 5′-terminus (Fig. 3B), and produced PbdB using the codon-optimized sequence.

4.2 Codon optimization
4.2.1 Equipment and materials
- Gibson Assembly Mix (New England Biolabs)
- Monarch Plasmid Miniprep Kit (New England Biolabs)
- *E. coli* DH5α competent cells
- Plasmid vector pTipQC1, digested using *Nde*I and *Hind*III, and gel-purified

4.2.2 Procedure
1. Use the codon optimization tool ATGme (http://atgme.org/) and referencing the codon utilization chart of RHA1, codon optimize the sequence of *pbdB*, manually biasing the adjustment of the codons at the 5′-end of the gene (Fig. 3).
2. Order the codon-optimized sequence for commercial synthesis with Gibson ends for ligation into the pTipQC1 vector: 5′-ggccaccatcac-catcaccat-[synthetic gene]-agatctctcgagccatcac-3′.

Preparation of reductases for multicomponent oxygenases

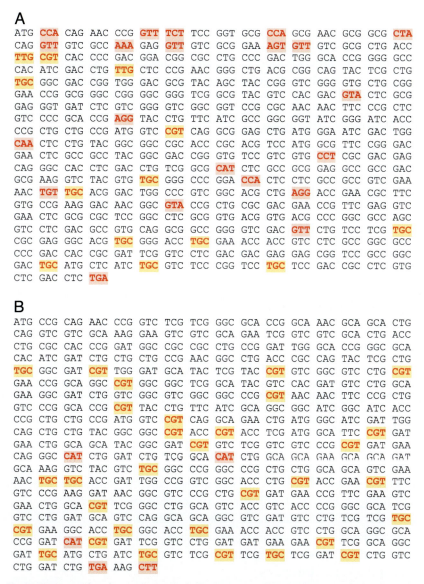

Fig. 3 Native and codon optimized *pbdB* sequences. Rare (<0.5% use) and uncommon codons (0.5–1%) are highlighted in red and yellow, respectively. Figure generated using ATGme.

3. Ligate the synthetic fragment into digested pTipQC1 backbone using standard protocol of the NEB Gibson assembly mix.
4. Transform into chemically competent *E. coli* DH5a, isolate plasmid by standard methods, and confirm nucleotide sequence of isolated plasmid. NOTE: The synthetic fragment was ordered from TWIST Biosciences.

4.3 Protein purification

4.3.1 Equipment and materials
- Labmaster Model 100 glovebox
- Centrifuge
- Centrifugal concentrator
- Stirred cell concentrator
- Nickel-NTA Agarose Resin (Qiagen)
- Lysis buffer (20 mM Tris-Cl pH 8, 300 mM NaCl, protease inhibitor cocktail, 10% glycerol)
- Column equilibration buffer (20 mM Tris-Cl pH 8, 300 mM NaCl, 10 mM imidazole, 10% glycerol)
- Column wash buffer (20 mM Tris-Cl pH 8, 300 mM NaCl, 30 mM imidazole, 10% glycerol)
- Column elution buffer (20 mM Tris-Cl pH 8, 500 mM imidazole, 10% glycerol)
- Protein storage buffer (20 mM Tris-Cl pH 8, 10% glycerol)

4.3.2 Procedure
1. Produce protein in RHA1 according to the protocol of **Section 2.4.2. Step 2**, scaling up 300-fold for a total culture volume of 3 L.
2. *Prepare 2 mL Ni-NTA resin in a gravity column. Rinse with 6–10 mL of ddH$_2$O, then 6–10 mL of column equilibration buffer.
3. Load lysate onto equilibrated column. Wash with 6–10 mL of column equilibration buffer and 6–10 mL column wash buffer. Finally elute in 6–10 mL of elution buffer.
4. Pool the colored fractions.
   a. If aerobic, concentrate and buffer exchange in 10 kDa molecular weight cutoff centrifugal concentration units.
   b. If anaerobic, concentrate and buffer exchange in a stirred cell concentrator with a 10 kDa molecular weight cutoff membrane.
5. Snap freeze aliquots of approximately 10 μL enzyme in liquid N$_2$ for future use.

NOTE: This procedure may be performed aerobically or anaerobically. If anaerobic, perform all steps after the *symbol in a glovebox operated at < 5 ppm O$_2$. Briefly, to help minimize O$_2$ in the glovebox: (a) bubble N$_2$ through all buffers used in the glovebox for at least 20 min prior to introducing them into the glovebox; and (b) exchange the atmosphere in the antechamber at least 3 times whenever introducing materials into the glovebox. We used the same glove box and a similar workflow to that described by Tsai and Tainer for the purification of an iron-sulfur containing protein (Tsai & Tainer, 2018). However, the procedures could be equally well performed using other equipment, such as that described and pictured by Bodea and Balskus (Bodea & Balskus, 2018).

4.4 Results

Substituting all the rare and uncommon codons in the first 150 nt of the *pbdB* gene for more commonly used ones (Fig. 3) increased the yield of purified reductase over 20-fold, from 0.2 culture to 4.8 mg/L. This result suggests that the PbdB production is translationally regulated by the enrichment of rare codons.

5. Protein characterization
5.1 Overview

PDR-type reductases contain two cofactors, a flavin and an iron-sulfur cluster (Correll, Batie, Ballou, & Ludwig, 1992). These cofactors can be detected spectroscopically by characteristic peaks in the UV-Vis spectrum (Eby, Beharry, Coulter, Kurtz, & Neidle, 2001). However, the absorbance of the protein can interfere with the diagnostic regions. Thus, to accurately determine the concentration of each cofactor, precipitation of the protein is necessary. For the bound flavin (in the case of PbdB, flavin adenine dinucleotide (FAD)), after protein precipitation the concentration of flavin can be determined spectroscopically (Sevrioukova & Poulos, 2002). For the iron-sulfur cluster, the content of non-heme iron and labile sulfur can be determined by Ferene-S and methylene blue colorimetric assays, respectively (Hedayati et al., 2018; Rabinowitz, 1978). Additionally, reductase electron transfer activity can be measured using an assay based on the reduction of cytochrome *c*, a small heme-containing protein (Guengerich, Martin, Sohl, & Cheng, 2009). Although this assay is widely used to measure the electron-transfer activity of P450 reductases and the presence of cofactors, it is important to note that the interaction between a reductase and cytochrome c requires different protein-protein interactions to that of a reductase and its cognate P450 (Shen & Kasper, 1995).

5.2 Cofactor analysis – labile sulfide

5.2.1 Equipment and materials

- Cary 5000 Spectrophotometer
- 1 mL plastic cuvettes
- Zinc acetate, 1%, in water
- NaOH, 12%
- N,N-Dimethyl-p-phenylenediamine dihydrochloride, 0.1%, in 5.5 M HCl
- $FeCl_3$, 23 mM, in 1.2 M HCl
- Sodium sulfide
- Purified PbdB (**Section 3.3**)

5.2.2 Procedure

1. Prepare reference sodium sulfide solutions. Load 1.420 g into a 100 mL volumetric flask for a 100 mM solution. Dilute to 5, 10, 25, 40, 50, 70, 90, 100 μM.
2. Place 10 μL of the reductase and each reference sample in 1.3 mL of zinc acetate reagent and add 50 μL of NaOH. Stopper test tubes.
3. Add 0.25 mL of the dimethylphenylenediamine reagent then add 50 μL of the ferric chloride reagent.
4. Incubate at room temperature for 30 min
5. Add 0.85 mL of water to the test tube.
6. Transfer 1 mL to a plastic cuvette and read absorption at 670 nm.
7. Interpolate concentration of labile sulfide in the reductase from a standard curve.

   NOTE: To minimize the loss of detectable sulfide as H_2S, add reagents rapidly and maintain stopper on test tubes.

5.3 Cofactor analysis – non-heme iron

5.3.1 Equipment and materials

- TECAN Spark plate reader
- 96-well plates
- Microcentrifuge
- Concentrated HCl
- 80% Trichloroacetic acid

Preparation of reductases for multicomponent oxygenases

- 45% sodium acetate
- Ferene-S solution (0.75 mM Ferene-S, 10 mM L-ascorbic acid, 45% sodium acetate)
- $FeCl_3$
- Purified PbdB (**Section 3.3**)

5.3.2 Procedure

1. Prepare reference ferric chloride solutions. Load 1.988 g into a 100 mL volumetric flask for a 100 mM solution. Dilute to 5, 10, 25, 40, 50, 70, 90, 100 μM.
2. Aliquot 80 μL of reductase and each reference solution in Eppendorf tubes.
3. Incubate each tube with 10 μL 12 N HCl for 10 mins at 25 °C.
4. Add 10 μL of 80% trichloroacetic acid.
5. Centrifuge to remove precipitant.
6. Add 20 μL of 45% sodium acetate into 96-well plate.
7. Add in 80 μL of sample supernatants, then 100 μL of Ferene-S solution.
8. Measure the absorbance at 595 nm.
9. Interpolate concentration of non-heme iron in the reductase from a standard curve.

5.4 Cofactor analysis – FAD

5.4.1 Equipment and materials

- Cary 5000 Spectrophotometer
- Microcentrifuge
- Saturated ammonium sulfate, pH 1.4 (in 7% H_2SO_4)
- Purified PbdB (**Section 3.3**)

5.4.2 Procedure

1. Add 5 μL saturated ammonium sulfate to 190 μL ddH$_2$O and 10 μL reductase
2. Pellet by centrifugation.
3. Take spectrum, determine concentration using the extinction coefficient of FAD at 454 nm, 11.3 mM^{-1} cm^{-1}.

5.5 Activity analysis – cytochrome *c* reduction

5.5.1 Equipment and materials

- Cary 60 Spectrophotometer
- 1 mL quartz cuvette

- 20 mM MOPS, pH 7.2, $I = 0.1$ M
- cytochrome c from bovine heart (0.5 mM stock solution in water)
- NADH (35 mM stock prepared in water)
- Purified PbdB (**Section 3.3**, 47 μM stock solution)

5.5.2 Procedure

1. Combine 80 μL cytochrome c and 900 μL buffer in a 1 mL cuvette. Add 10 μL NADH stock solution and wait for the signal to stabilize.
2. Initiate reaction by adding 10 μL of reductase solution to a final concentration of 0.47 μM.
3. Monitor absorbance at 550 nm on spectrophotometer using kinetics mode.
4. Calculate rate of cytochrome c reduction using the extinction coefficient at 550 nm, 21 mM^{-1} cm^{-1}.

5.6 Results

PbdB purified from RHA1 had the diagnostic peaks in the spectra of an FAD-containing protein at 454 nm (Fig. 4). The presence and concentration of the FAD cofactor was confirmed by denaturation of the enzyme and quantification by spectroscopy (Table 1). However, peaks corresponding to the iron–sulfur cluster were not detectable in the native spectrum. Non-heme iron and labile sulfide were detected by the Ferene-S and methylene blue assays, respectively. The concentrations of the two elements were in proportions corresponding to the presence of a [2Fe-2S] cluster in the protein (Table 1). Finally, we assayed the reduction of different preparations of reductase by their ability to reduce cytochrome c.

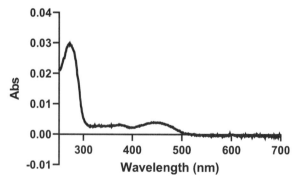

Fig. 4 UV-Vis spectrum of PbdB purified anaerobically from RHA1. The sample contained 0.8 μM PbdB in 20 mM MOPS, pH 7.2, $I = 0.1$ at 25 °C.

Table 1 Cofactor loading of PbdB, purified anaerobically from RHA1.

Analyte	Concentration (μM)	Proportion to (protein)
Protein	47 (5)	–
FAD	43.1	0.9
Non-heme iron	100 (10)	2.1
Labile sulfide	79 (8)	1.7

Table 2 Cytochrome c reduction activity of PbdB preparations.

Host	Atmosphere	Cytochrome c reduction rate (μmol min^{-1} mg protein^{-1})
E. coli	Aerobic	900 (200)
RHA1	Aerobic	Not detected
RHA1	Anaerobic	36 (4)

Despite the inability of PbdB produced in *E. coli* to reduce PbdA and confer *O*-demethylation activity (**Section 2.5.3**), purified PbdB produced in *E. coli* efficiently reduced cytochrome *c* (Table 2). Additionally, it was determined that reductase prepared aerobically from RHA1 did not detectably catalyze cytochrome *c* reduction, in comparison to that produced anaerobically (Table 2), likely indicating that the metal centers in PbdB are O_2-labile.

6. Summary and conclusions

Here, we report the preparation of a reductase, PbdB, which is part of a two-component cytochrome P450 system. Obtaining a preparation of PbdB that confers activity of the P450 required expression in RHA1, a non-model host. Despite rapidly reducing cytochrome *c*, PbdB produced in *E. coli* was inactive as the reductase partner of PbdA. By contrast, when produced in RHA1, the protein effectively reduces PbdA. As the cytochrome *c* assay indicates the presence of an active electron-transfer chain, the dependence of PbdA reductase activity on the identity of the expression host may be due to a factor which disrupts the PbdA-PbdB interaction. Protein production protocols overwhelmingly use *E. coli* as the expression host. Characterization of host-dependent activity and production in non-model hosts expands our ability to characterize enzymes from diverse sources. As illustrated by using rhodococcal strains to produce mycobacterial enzymes (Dresen et al., 2010; McCarl et al., 2018), significant increases in activity and production can be achieved using a host that is more phylogenetically related to the enzymes native host than *E. coli*. Obtaining a sufficient yield of PbdB for PbdA characterization required codon optimization and anaerobic purification. Production of PbdB increased significantly using a gene in which the $5'$-terminal rare codons were optimized, suggesting that codon usage may regulate translation of the reductase. Oxygenases have great potential as biocatalysts, and understanding their regulation is key to developing this potential. Characterization of this reductase and its unique properties advances our ability to characterize multicomponent oxygenases such as cytochrome P450s with their native reductases, expanding our knowledge of their vast biological role and biocatalytic potential. More generally, the principles of protein production in non-model hosts, codon-optimization, and anaerobic purification described here may be used to increase the yield and activity of other oxygenase system components.

References

Aravind, L., & Koonin, E. V. (2001). The DNA-repair protein AlkB, EGL-9, and leprecan define new families of 2-oxoglutarate-and iron-dependent dioxygenases. *Genome Biology, 2*, 1–8.

Bodea, S., & Balskus, E. P. (2018). Chapter two - Purification and characterization of the choline trimethylamine-lyase (CutC)-activating protein CutD. In V. Bandarian (Vol. Ed.), *Methods in enzymology: Vol. 606*, (pp. 73–94). Academic Press. https://doi.org/10.1016/bs.mie.2018.04.012.

Bopp, C. E., Bernet, N. M., Kohler, H. E., & Hofstetter, T. B. (2022). Elucidating the role of O(2) uncoupling in the oxidative biodegradation of organic contaminants by Rieske non-heme iron dioxygenases. *ACS Environmental Au, 2*(5), 428–440. https://doi.org/10.1021/acsenvironau.2c00023.

Cabiscol, E., Tamarit, J., & Ros, J. (2000). Oxidative stress in bacteria and protein damage by reactive oxygen species. *International Microbiology, 3*(1), 3–8.

Cao, P. R., & Bernhardt, R. (1999). Modulation of aldosterone biosynthesis by adrenodoxin mutants with different electron transport efficiencies. *European Journal of Biochemistry, 265*(1), 152–159.

Cheng, M., Chen, D., Parales, R. E., & Jiang, J. (2022). Oxygenases as powerful weapons in the microbial degradation of pesticides. *Annual Review of Microbiology, 76*, 325–348.

Cochrane, R. V. K., & Vederas, J. C. (2014). Highly selective but multifunctional oxygenases in secondary metabolism. *Accounts of Chemical Research, 47*(10), 3148–3161. https://doi.org/10.1021/ar500242c.

Correll, C. C., Batie, C. J., Ballou, D. P., & Ludwig, M. L. (1992). Phthalate dioxygenase reductase: A modular structure for electron transfer from pyridine nucleotides to [2Fe-2S]. *Science, 258*(5088), 1604–1610. https://doi.org/10.1126/science.1280857.

Denome, S. A., Olson, E. S., & Young, K. D. (1993). Identification and cloning of genes involved in specific desulfurization of dibenzothiophene by *Rhodococcus* sp. strain Igts8. *Applied and Environmental Microbiology, 59*(9), 2837–2843. https://doi.org/Doi10.1128/Aem.59.9.

Dresen, C., Lin, L. Y., D'Angelo, I., Tocheva, E. I., Strynadka, N., & Eltis, L. D. (2010). A flavin-dependent monooxygenase from *Mycobacterium tuberculosis* involved in cholesterol catabolism. *Journal of Biological Chemistry, 285*(29), 22264–22275. https://doi.org/10.1074/jbc.M109.099028.

Eby, D. M., Beharry, Z. M., Coulter, E. D., Kurtz, D. M., Jr., & Neidle, E. L. (2001). Characterization and evolution of anthranilate 1,2-dioxygenase from *Acinetobacter* sp. strain ADP1. *Journal of Bacteriology, 183*(1), 109–118. https://doi.org/10.1128/JB.183-1.109-118.2001.

Ferraro, D. J., Gakhar, L., & Ramaswamy, S. (2005). Rieske business: Structure–function of Rieske non-heme oxygenases. *Biochemical and Biophysical Research Communications, 338*(1), 175–190. https://doi.org/10.1016/j.bbrc.2005.08.222.

Furge, L. L., & Guengerich, F. P. (2006). Cytochrome P450 enzymes in drug metabolism and chemical toxicology: An introduction. *Biochemistry and Molecular Biology Education, 34*(2), 66–74.

Goodman, D. B., Church, G. M., & Kosuri, S. (2013). Causes and effects of N-terminal codon bias in bacterial genes. *Science, 342*(6157), 475–479. https://doi.org/10.1126/science.1241934.

Grogan, G. (2011). Cytochromes P450: Exploiting diversity and enabling application as biocatalysts. *Current Opinion in Chemical Biology, 15*(2), 241–248.

Guengerich, F. P., Martin, M. V., Sohl, C. D., & Cheng, Q. (2009). Measurement of cytochrome P450 and NADPH-cytochrome P450 reductase. *Nature Protocols, 4*(9), 1245–1251. https://doi.org/10.1038/nprot.2009.121.

Hedayati, M., Abubaker-Sharif, B., Khattab, M., Razavi, A., Mohammed, I., Nejad, A., ... Ivkov, R. (2018). An optimised spectrophotometric assay for convenient and accurate quantitation of intracellular iron from iron oxide nanoparticles. *International Journal of Hyperthermia, 34*(4), 373–381. https://doi.org/10.1080/02656736.2017.1354403.

Hibi, M., Sonoki, T., & Mori, H. (2005). Functional coupling between vanillate-O-demethylase and formaldehyde detoxification pathway. *FEMS Microbiology Letters, 253*(2), 237–242. https://doi.org/10.1016/j.femsle.2005.09.036.

Iida, T., Moteki, Y., Nakamura, K., Taguchi, K., Otagiri, M., Asanuma, M., ... Kudo, T. (2009). Functional expression of three Rieske non heme iron oxygenases derived from actinomycetes in *Rhodococcus* species for investigation of their degradation capabilities of dibenzofuran and chlorinated dioxins. *Bioscience, Biotechnology, and Biochemistry, 73*(4), 822–827. https://doi.org/10.1271/bbb.80680.

Jouanneau, Y., Meyer, C., Jakoncic, J., Stojanoff, V., & Gaillard, J. (2006). Characterization of a naphthalene dioxygenase endowed with an exceptionally broad substrate specificity toward polycyclic aromatic hydrocarbons. *Biochemistry, 45*(40), 12380–12391. https://doi.org/10.1021/bi0611311.

Klenk, J. M., Nebel, B. A., Porter, J. L., Kulig, J. K., Hussain, S. A., Richter, S. M., ... Flitsch, S. L. (2017). The self-sufficient P450 RhF expressed in a whole cell system selectively catalyses the 5-hydroxylation of diclofenac. *Biotechnology Journal, 12*(3), https://doi.org/10.1002/biot.201600520.

Kuznetsov, V. Y., Poulos, T. L., & Sevrioukova, I. F. (2006). Putidaredoxin-to-cytochrome P450cam electron transfer: Differences between the two reductive steps required for catalysis. *Biochemistry, 45*(39), 11934–11944. https://doi.org/10.1021/bi0611154.

Kweon, O., Kim, S. J., Baek, S., Chae, J. C., Adjei, M. D., Baek, D. H., ... Cerniglia, C. E. (2008). A new classification system for bacterial Rieske non-heme iron aromatic ring-hydroxylating oxygenases. *BMC Biochemistry, 9*, 11. https://doi.org/10.1186/1471-2091-9-11.

Lee, M. S., Hseu, Y. C., Lai, G. H., Chang, W. T., Chen, H. J., Huang, C. H., ... Lin, M. K. (2011). High yield expression in a recombinant *E. coli* of a codon optimized chicken anemia virus capsid protein VP1 useful for vaccine development. *Microbial Cell Factories, 10.* https://doi.org/Artn5610.1186/1475-2859-10-56.

Li, S., Du, L., & Bernhardt, R. (2020). Redox partners: Function modulators of bacterial P450 enzymes. *Trends in Microbiology, 28*(6), 445–454. https://doi.org/10.1016/j.tim. 2020.02.012.

Lim, H. N., Lee, Y., & Hussein, R. (2011). Fundamental relationship between operon organization and gene expression. *Proceedings of the National Academy of Sciences of the United States of America, 108*(26), 10626–10631. https://doi.org/10.1073/pnas.1105692108.

Mallinson, S. J. B., Machovina, M. M., Silveira, R. L., Garcia-Borràs, M., Gallup, N., Johnson, C. W., ... McGeehan, J. E. (2018). A promiscuous cytochrome P450 aromatic O-demethylase for lignin bioconversion. *Nature Communications, 9*(1), 2487. https://doi. org/10.1038/s41467-018-04878-2.

McCarl, V., Somerville, M. V., Ly, M. A., Henry, R., Liew, E. F., Wilson, N. L., ... Coleman, N. V. (2018). Heterologous expression of *Mycobacterium* alkene monooxygenases in gram-positive and gram-negative bacterial hosts. *Applied and Environmental Microbiology, 84*(15), https://doi.org/10.1128/aem.00397-18.

McLean, K. J., Luciakova, D., Belcher, J., Tee, K. L., & Munro, A. W. (2015). Biological diversity of cytochrome P450 redox partner systems. *Advances in Experimental Medicine and Biology, 851*, 299–317. https://doi.org/10.1007/978-3-319-16009-2_11.

Meunier, B., de Visser, S. P., & Shaik, S. (2004). Mechanism of oxidation reactions catalyzed by cytochrome P450 enzymes. *Chemical Reviews, 104*(9), 3947–3980. https://doi. org/10.1021/cr020443g.

Mitchell, A. J., & Weng, J.-K. (2019). Unleashing the synthetic power of plant oxygenases: From mechanism to application. *Plant Physiology, 179*(3), 813–829. https://doi.org/10. 1104/pp.18.01223.

Montellano, P. R. (2000). The mechanism of heme oxygenase. *Current Opinion in Chemical Biology, 4*(2), 221–227. https://doi.org/10.1016/s1367-5931(99)00079-4.

Nakashima, N., & Tamura, T. (2004). Isolation and characterization of a rolling-circle-type plasmid from *Rhodococcus erythropolis* and application of the plasmid to multiple-recombinant-protein expression. *Applied and Environmental Microbiology, 70*(9), 5557–5568.

Nebert, D. W., & Russell, D. W. (2002). Clinical importance of the cytochromes P450. *The Lancet, 360*(9340), 1155–1162.

Petering, D., Fee, J. A., & Palmer, G. (1971). The oxygen sensitivity of spinach ferredoxin and other iron-sulfur proteins. *Journal of Biological Chemistry, 246*(3), 643–653. https:// doi.org/10.1016/S0021-9258(18)62463-9.

Rabinowitz, J. C. (1978). Analysis of acid-labile sulfide and sulfhydryl groups. *Methods in Enzymology, 53*, 275–277. https://doi.org/10.1016/s0076-6879(78)53033-4.

Romero, E., Gómez Castellanos, J. R., Gadda, G., Fraaije, M. W., & Mattevi, A. (2018). Same substrate, many reactions: Oxygen activation in flavoenzymes. *Chemical Reviews, 118*(4), 1742–1769. https://doi.org/10.1021/acs.chemrev.7b00650.

Rosano, G. L., & Ceccarelli, E. A. (2014). Recombinant protein expression in *Escherichia coli*: Advances and challenges. *Frontiers in Microbiology, 5*. https://doi.org/10.3389/fmicb. 2014.00172.

Runda, M. E. K., Bastian, Ö., & Fatma Feyza, S. (2023). An optimized system for the study of rieske oxygenase-catalyzed hydroxylation reactions *in vitro*. *ChemCatChem, 15*(16), e202300371. https://doi.org/10.1002/cctc.202300371.

Runda, M. E., De Kok, N. A. W., & Schmidt, S. (2023). Rieske oxygenases and other ferredoxin-dependent enzymes: Electron transfer principles and catalytic capabilities. e202300078 *ChemBioChem, 24*(15), https://doi.org/10.1002/cbic.202300078.

Sevrioukova, I. F., & Poulos, T. L. (2002). Putidaredoxin reductase, a new function for an old protein. *Journal of Biological Chemistry, 277*(28), 25831–25839. https://doi.org/10.1074/jbc.M201110200.

Sevrioukova, I. F., Garcia, C., Li, H., Bhaskar, B., & Poulos, T. L. (2003). Crystal structure of putidaredoxin, the [2Fe–2S] component of the P450cam monooxygenase system from *Pseudomonas putida*. *Journal of Molecular Biology, 333*(2), 377–392. https://doi.org/10.1016/j.jmb.2003.08.028.

Sevrioukova, I. F., Li, H., & Poulos, T. L. (2004). Crystal structure of putidaredoxin reductase from *Pseudomonas putida*, the final structural component of the cytochrome P450cam monooxygenase. *Journal of Molecular Biology, 336*(4), 889–902. https://doi.org/10.1016/j.jmb.2003.12.067.

Shen, A. L., & Kasper, C. B. (1995). Role of acidic residues in the interaction of NADPH-cytochrome P450 oxidoreductase with cytochrome P450 and cytochrome *c*. *Journal of Biological Chemistry, 270*(46), 27475–27480. https://doi.org/10.1074/jbc.270.46.27475.

Suzuki, H., Inabe, K., Shirakawa, Y., Umezawa, N., Kato, N., & Higuchi, T. (2017). Role of thiolate ligand in spin state and redox switching in the cytochrome P450 catalytic cycle. *Inorganic Chemistry, 56*(8), 4245–4248. https://doi.org/10.1021/acs.inorgchem.6b02499.

Ten Have, R., & Teunissen, P. J. M. (2001). Oxidative mechanisms involved in lignin degradation by white-rot fungi. *Chemical Reviews, 101*(11), 3397–3414. https://doi.org/10.1021/cr000115l.

Tian, J., Liu, J., Knapp, M., Donnan, P. H., Boggs, D. G., & Bridwell-Rabb, J. (2023). Custom tuning of Rieske oxygenase reactivity. *Nature Communications, 14*(1), 5858. https://doi.org/10.1038/s41467-023-41428-x.

Tsai, C.-L., & Tainer, J. A. (2018). Chapter six - Robust production, crystallization, structure determination, and analysis of [Fe–S] proteins: Uncovering control of electron shuttling and gating in the respiratory metabolism of molybdopterin guanine dinucleotide enzymes. In S. S. David (Vol. Ed.), *Methods in enzymology: Vol. 599*, (pp. 157–196). Academic Press. https://doi.org/10.1016/bs.mie.2017.11.006.

Van Beilen, J. B., & Funhoff, E. G. (2005). Expanding the alkane oxygenase toolbox: New enzymes and applications. *Current Opinion in Biotechnology, 16*(3), 308–314. https://doi.org/10.1016/j.copbio.2005.04.005.

Wolf, M.E., Lalande, A.T., Newman, B.L., Bleem, A.C., Palumbo, C.T., Beckham, G.T., & Eltis, L.D. The catabolism of lignin derived p methoxylated aromatic compounds by *Rhodococcus jostii* RHA1. *Applied and Environmental Microbiology, 0*(0), e02155–02123. https://doi.org/10.1128/aem.02155–23.

Wu, L.-F., Meng, S., & Tang, G.-L. (2016). Ferrous iron and α-ketoglutarate-dependent dioxygenases in the biosynthesis of microbial natural products. *Biochimica et Biophysica Acta (BBA) - Proteins and Proteomics, 1864*(5), 453–470. https://doi.org/10.1016/j.bbapap.2016.01.012.

Wu, Z. L., Qiao, J., Zhang, Z. G., Guengerich, F. P., Liu, Y., & Pei, X. Q. (2009). Enhanced bacterial expression of several mammalian cytochrome P450s by codon optimization and chaperone coexpression. *Biotechnology Letters, 31*(10), 1589–1593. https://doi.org/10.1007/s10529-009-0059-5.

Zhang, W., Du, L., Li, F., Zhang, X., Qu, Z., Han, L., ... Li, S. (2018). Mechanistic insights into interactions between bacterial class I P450 enzymes and redox partners. *ACS Catalysis, 8*(11), 9992–10003. https://doi.org/10.1021/acscatal.8b02913.

CHAPTER FIVE

Development of a rapid mass spectrometric method for the analysis of ten-eleven translocation enzymes

Clara Graves and Kabirul Islam*
Department of Chemistry, University of Pittsburgh, Pittsburgh, PA, United States
*Corresponding author. e-mail address: kai27@pitt.edu

Contents

1. Introduction		88
2. Preparation of the materials		90
	2.1 Expression and purification of wild type TET2	90
	2.2 Mutagenesis and expression of V1395A	95
	2.3 Synthesis and characterization of oligonucleotides	99
3. Biochemical assays and results		103
	3.1 Development of a robust *in-vitro* assay	103
	3.2 Measurement of IC_{50} of TET2 inhibitors NOG and 2HG	108
	3.3 Validating the activity of wildtype TET2 and V1395A using BS-seq	111
4. Notes		114
Funding		118
References		118

Abstract

In DNA, methylation at the fifth position of cytosine (5mC) by DNA methyltransferases is essential for eukaryotic gene regulation. Methylation patterns are dynamically controlled by epigenetic machinery. Erasure of 5mC by Fe^{2+} and 2-ketoglutarate (2KG) dependent dioxygenases in the ten-eleven translocation family (TET1–3), plays a key role in nuclear processes. Through the event of active demethylation, TET proteins iteratively oxidize 5mC to 5-hydroxymethyl cytosine (5hmC), 5-formylcytosine (5fC) and 5-carboxycytosine (5caC), each of which has been implicated in numerous diseases when aberrantly generated. A wide range of biochemical assays have been developed to characterize TET activity, many of which require multi-step processing to detect and quantify the 5mC oxidized products. Herein, we describe the development and optimization of a sensitive MALDI mass spectrometry-based technique that directly measures TET activity and eliminates tedious processing steps. Employing optimized assay conditions, we report the steady-state activity of wild type TET2 enzymes to furnish 5hmC, 5fC and 5caC. We next determine IC_{50} values of several

Methods in Enzymology, Volume 703
ISSN 0076-6879, https://doi.org/10.1016/bs.mie.2024.06.001
Copyright © 2024 Elsevier Inc. All rights are reserved, including those for text and data mining, AI training, and similar technologies.

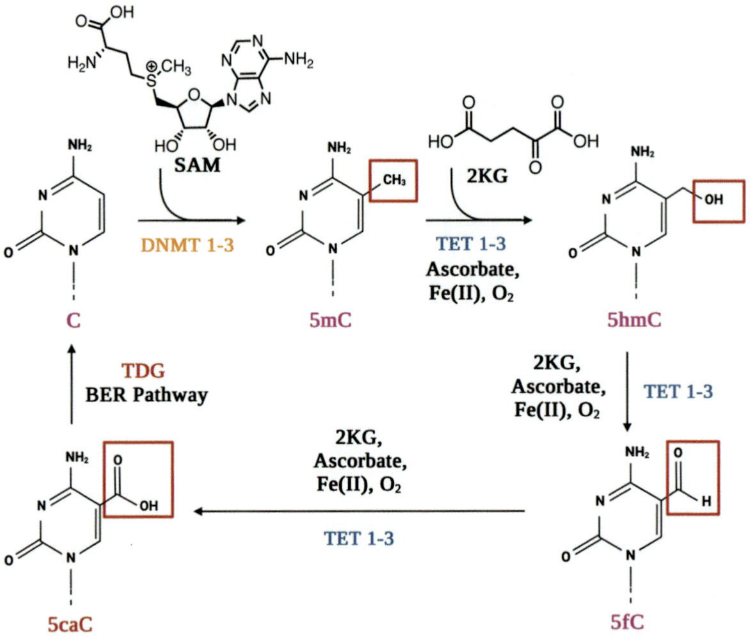

Fig. 1 Generation of 5mC by DNMTs using SAM as methyl donor and its successive oxidation by Fe^{2+}-dependent TET enzymes using 2KG as cofactor.

small-molecule inhibitors of TETs. The utility of this assay is further demonstrated by analyzing the activity of V1395A which is an activating mutant of TET2 that primarily generates 5caC. Lastly, we describe the development of a secondary assay that utilizes bisulfite chemistry to further examine the activity of wildtype TET2 and V1395A in a base-resolution manner. The combined results demonstrate that the activity of TET proteins can be gauged, and their products accurately quantified using our methods.

1. Introduction

All cells in eukaryotic organisms have the same genomic DNA but have different functions and phenotypes due to differential expression (activation or repression) of specific genes (Gibney & Nolan, 2010; Jaenisch & Bird, 2003). Methylation of cytidine at carbon 5 (5mC) by DNA methyltransferases (DNMTs) using S-adenosyl methionine (SAM), constitutes an important mechanism for gene expression (Fig. 1) (Jurkowska, Jurkowski, & Jeltsch, 2011). 5mC is successively oxidized to 5-hydroxymethylcytosine (5hmC), 5-formylcytosine (5fC) and 5-carboxylcytosine (5caC) by the Ten–Eleven Translocation (TET) enzymes, which are members of the Fe (II) and 2- ketoglutaric acid (2KG)-dependent dioxygenases family (Loenarz & Schofield, 2008; Lu, Zhao, & He, 2015). 5caC subsequently undergoes thymine DNA glycosylase (TDG)-mediated base excision followed by repair to unmodified cytidine (Fig. 1), thus providing a biochemical basis for active DNA demethylation which is essential to early mammalian development (He et al., 2011; Ito et al., 2011; Tahiliani et al., 2009). Aside from playing a role as intermediates in the demethylation pathway, growing evidence suggests that each of the TET oxidized products (5hmC, 5fC and 5caC) can independently regulate gene expression (Bachman et al., 2015; Song & He, 2013; Spruijt et al., 2013). TET-mediated oxidation of 5mC is critical to cellular differentiation (Kohli & Zhang, 2013). Consistently, overexpression and catalytically inactive mutations of TET enzymes have also been associated with the development of human diseases such as neurodegenerative disorders and cancers (Huang & Rao, 2014; Joshi, Liu, Breslin, & Zhang, 2022; Ko et al., 2010).

Development of novel methods to detect and quantify the enzymatic activity of TET proteins as well as their oxidized products both *in-vitro* and *in-cellulo* have been driven by efforts to understand the functions of TET proteins in eukaryotic gene regulation (Bhattacharya, Dey, & Mukherji, 2023; Dey, Ayon, Bhattacharya, Gutheil, & Mukherji, 2020; Liu, DeNizio, & Kohli, 2016; Liu, Wang, Liu, Xu, & Zhang, 2022; Liu, Zhang, Hu, & Zhang, 2022; Ma et al., 2024; Marholz, Wang, Zheng, & Wang, 2016; Shen & Zhang, 2012; Sudhamalla, Dey, Breski, & Islam, 2017; Treadway et al., 2024). A majority of current approaches involve multiple enzymatic reactions and purification steps that lead to the loss of the oxidized products and subsequent errors in quantification (Sudhamalla et al., 2017). Fluorescence, quantum dots and NMR-based methods have also been applied to measure TET activity, but these newer assays are limited in scope and practicality due to their complexity often requiring amplification technologies or high DNA substrate and enzyme concentrations in order to obtain a desired sensitivity, and generation of isotope-labeled protein which is expensive and time consuming (Ma et al., 2024; Treadway et al., 2024).

Herein we report the step-by-step development and optimization of a novel assay that facilitates direct and efficient measurement of enzymatic activity of TET proteins (Sudhamalla et al., 2017). The salient features include (i) use of AG® 50W-X8 cation-exchange resin beads to desalt oxidized oligonucleotide products, thus eliminating the need for their subsequent enzymatic digestion and often laborious purification, and (ii) direct analysis of product distribution (5hmC *vs* 5fC *vs* 5caC) by matrix-assisted laser desorption/ionization (MALDI) based mass spectrometry. Employing the optimized assay conditions, we precisely measured steady-state activity of wild type TET2 enzymes to furnish 5hmC, 5fC and 5caC. We next determine IC$_{50}$ values of the selected small-molecule inhibitors of TETs and show that our results are comparable to those obtained by other methods. We further demonstrate the utility of the *in-vitro* assay to analyze the activity of V1395A which is an activating mutant of TET2 known to generate primarily 5caC on short and genomic oligonucleotide substrates (Sappa, Dey, Sudhamalla, & Islam, 2021).

Several chemical approaches coupled with next-generation sequencing techniques have been developed for base-resolution mapping of the 5mC oxidized products in cell- and tissue-specific manners (Booth, Raiber, & Balasubramanian, 2015; Song, Yi, & He, 2012). Herein, we describe the development of a secondary assay to further examine the activity of wildtype TET2 and V1395A in a base-resolution manner on a 76-mer oligonucleotide substrate carrying 5mC at a specific site by treating the oxidized products with bisulfite and coupling with Sanger sequencing (BS-Seq), (Sappa et al., 2021; Sudhamalla et al., 2017). The results corroborate well with the data obtained using MALDI-based *in-vitro* assay, further supporting the ability of our methods to accurately gauge the activity of the TET proteins and likely the related family of 2KG-dependent nucleic acid modifying enzymes.

2. Preparation of the materials
2.1 Expression and purification of wild type TET2
2.1.1 Materials and reagents

Item	Supplier and catalog number
Kanamycin monosulfate	Fisher Cat # K004725G
Escherichia coli strain BL21 star (DE3) competent cells	Fisher Cat # C601003
SOC medium	Corning Cat #46-003-CR
LB agar	Fisher Cat # BP1425-500
LB broth	Fisher Cat # BP9723-2

Isopropyl β-d-1-thiogalactoyranoside (IPTG)	Fisher Cat # BP162010
Tris (hydroxymethyl) aminomethane hydrochloride (Tris-HCl, pH 8.0)	Fisher
Sodium chloride	Fisher Cat #BP358-10
β-Mercaptoethanol	Fisher Cat #BP176-100
Glycerol	Fisher Cat #BP2291
Imidazole	Fisher Cat #AC12202-5000
Lysozyme	Sigma Cat #10837059001
DNase	Fisher Cat #4536282001
Pierce™ protease inhibitor mini tablets, EDTA-free	Thermo Cat #A32955
Ni-NTA agarose resin beads	Thermo Cat #R90101
Bovine serum albumin (BSA) standards	BioRad Cat #5000206
Bradford dye reagent (1 L)	Fisher Cat #AAJ61522K
Cell spreader	Fisher Cat # 50-751-5036
0.2 μm filter	Thermo Cat #CH2225-NH
Syringe needle (18 G × 1 ½″ TW)	BD PrecisionGlide™
10 mL syringe	BD Luer-Lok™ Tip
1.5 mL eppendorf tube	Fisher
Falcon™ round-bottom polystyrene test tubes	Fisher Cat # 14-959-11B
Falcon™ standard tissue culture dishes	Thermo
Semi-micro, 1.5 mL cuvette	Fisher Cat # 14955127
50 mL Falcon™ tube	Corning
SDS-PAGE gel	
Plastic scoopula, 210 mm, blue	SmartSpatula Cat # 17211

2.1.2 Equipment

Item	Supplier
Ultra-10k centrifugal filter device	Amicon
Large centrifuge	Thermo Sorvall Legend XTR
Weighing scale	Mettler Toledo XPE504

CO_2 incubator for bacterial cell culture	Thermo HERATherm
Innova 44® Incubator shaker	New Brunswick Scientific
Vortex mixer	Corning
Water bath	Thermo
BioSpectrometer	Eppendorf
Pulsed sonication device with microtip	Qsonica-Q700
3D shaker rocking plate	Chemglass Life Sciences
Superdex-200 column	GE Healthcare
FPLC instrument	GE Healthcare
0.2 μm filtered water	Thermo Barnstead GenPure
ChemiDoc MP imaging system	BioRad
Ultra-low temperature freezer	New Brunswick

2.1.3 Protocol

1. Bacterial transformation

The catalytic domains of TET enzymes carry a Cys-rich domain and a Double-Stranded Beta Helix (DSβH) motif separated by a large low-complexity insert (Hu et al., 2013, 2015; Tahiliani et al., 2009). The Xu group at Fudan University developed a TET2 construct (1129–1936) in which residues 1481–1843 were replaced with a 15-residue GS-linker (GGGGSGGGGSGGGGS) and found that this construct to be the smallest catalytically proficient enzyme among the larger constructs examined (Sudhamalla et al., 2017; Hu et al., 2013, 2015). A 6xHis tag was introduced at the N-terminus for affinity purification. The TET2 expression construct in pET-28b vector was a kind gift from the Xu group. The plasmid was transformed into *Escherichia coli* strain BL21 star (DE3) competent cells (Fisher) using the following protocol.

a. In a flame sterilized environment, 10 ng of the wild-type TET2 plasmid with pET-28b kanamycin-resistant vector was pipetted into an autoclaved 1.5 mL Eppendorf tube with 9 μL *E. coli* strain BL21 star (DE3) competent cells (Fisher).

b. This mixture was allowed to incubate on ice for 30 min before placed in a 40 °C water bath for 25 s and immediately submerged in ice for 2 min.

c. After the full 2 min had elapsed, 200 μL Super Optimum broth with Catabolite repression (SOC) medium was added to the cells and placed in a shaker at 37 °C for 1 h.

d. The full mixture was spread evenly onto a kanamycin resistant plate and left to incubate overnight in a 37 °C incubator for colony growth.

2. Starter culture inoculation

a. A single colony in close proximity to the edge of the plate away, from colony clusters, was gently picked up with a pipette tip and placed in 10 mL Luria–Bertani (LB) broth and grown for no longer than 16 h at 37 °C in presence of 50 μg/mL kanamycin.

b. It is important to observe the growth of this starter culture. The inoculates should appear significantly foggier than before, but there should be no sign of debris at this stage. Debris is a sign that the cells have begun to die, and the amount of protein obtained at the end of expression will be minimal. If this debris formation is observed, repeat the starter culture inoculation, and incubate for less time.

3. Induction and Expression of Protein

a. The starter culture was diluted 100-fold into 1 L of LB with the addition of 50 μg/mL kanamycin. 1 mL of this solution was put to the side in a cuvette, and the remaining solution was allowed to grow at 37 °C to an optical density at 600 nm (OD_{600}) of 0.8 in an Innova 44® Incubator Shaker. It is important to keep a close eye on the OD_{600} value. If the OD_{600} value increases above 0.8, the amount of protein obtained at the end of expression will be minimal.

b. Once an OD value of 0.8 was reached, protein expression was induced by adding 0.8 mM isopropyl β-d-1-thiogalactoyranoside (IPTG) and incubating for 20 h at 17 °C in an Innova 44® Incubator shaker (New Brunswick Scientific).

4. Affinity Purification

It is important to complete as many of these steps as possible in one day to ensure optimal catalytic activity at the end of this process. Completing each of these steps at 4 °C is crucial for the same reason as TET is a heat sensitive protein.

a. Each 1 L culture was centrifuged at 4 °C for 30 min at 4000 rpm and the supernatant was subsequently discarded.

b. Each liter of harvested cells was resuspended in 15 mL lysis buffer containing 50 mM Tris (hydroxymethyl) aminomethane hydrochloride (Tris-HCl, pH 8.0), 200 mM NaCl, 5 mM β-mercaptoethanol (BME),

10% glycerol, 25 mM imidazole, Lysozyme, DNase, and Roche protease inhibitor cocktail.

c. The cells were lysed in a falcon tube by pulsed sonication (Qsonica-Q700) in a reservoir filled with fresh ice using 10 s 50 mA pulses separated by 10 s pauses.

d. The sonicated cells were then centrifuged at 13,000 rpm for 50 min at 4 °C.

e. While centrifugation was taking place, Ni-NTA agarose resin beads (Thermo) were washed and equilibrated with buffer containing 50 mM Tris-HCl pH 8.0, 200 mM NaCl, 5 mM β-mercaptoethanol, 10% glycerol, and 25 mM imidazole in a 15 mL column. When centrifugation had finished, the equilibration buffer was allowed to elute from this column before pouring in the soluble extracts from the previous step. The column was capped, and the soluble extracts were allowed to incubate with the Ni-NTA resin on a rocking plate (Chemglass Life Sciences) for 40 min at 4 °C.

f. Proteins with non-specific interactions were allowed to drain out of the column leaving proteins containing the 6×-His tag attached to the beads. To displace the His-Ni interaction, beads were incubated for 10 min with 1 mL of an elution buffer containing 50 mM Tris-HCl pH 8.0, 200 mM NaCl, 5 mM β-mercaptoethanol, 10% glycerol, and 400 mM imidazole. Elution fractions were subsequentially captured in a 1.5 mL Eppendorf Tube. This step was repeated two more times, collecting a total of 3 mL from each column.

5. Fast Protein Liquid Chromatography (FPLC)

a. Proteins were further purified by gel filtration chromatography (Superdex-200) using AKTA pure Fast Protein Liquid Chromatography (FPLC) system (GE healthcare) with buffer containing 50 mM Tris-HCl pH 8.0, 150 mM NaCl, and 10% glycerol. Prior to injecting into the FPLC instrument, the protein was filtered by drawing up the liquid through a 10 mL syringe and pushing through a 0.2 μm filter into a 50 mL Falcon tube. The liquid was then drawn up through the syringe once again and injected into the FPLC column for purification. For a protein of this size (~69 kDa), it took approximately 3 h for the protein of interest to elute from the column. To reduce the amount of time the protein was outside of the −80 °C and increase the lifetime of TET, fractions under the UV indication curve were collected as soon as they eluted from the column in 5 mL intervals.

b. FPLC fractions were analyzed *via* sodium dodecyl-sulfate polyacrylamide gel electrophoresis (SDS-PAGE, 4%–12%) to determine which of the 5 mL fractions contained the protein of interest.

c. To concentrate the protein, fractions containing the purified protein of interest were poured into an Amicon Ultra-10k centrifugal filter device (Merck Millipore Ltd.) and centrifuged at 3000 rpm. The volume inside the device was checked every 30 min until 500 µL remained.

d. The protein concentration was initially determined using a Bradford assay kit (BioRad Laboratories). To use this kit, 1 mL of Bradford solution was pipetted into a cuvette. With proteins that were expected to have a high concentration, a 20:2 dilution of water to protein was pipetted into these cuvettes and vortexed thoroughly before sitting at room temperature for 10 min. As this solution is moderately sensitive to light, the Bradford solutions containing the protein were placed inside a covered box during incubation. The darker the blue of the solution was representative of a higher protein concentration. As such, protein concentration was determined *via* UV–VIS spectroscopy and compared to a standard curve. The standard curve was constructed by spiking the Bradford solutions with known concentrations of bovine serum albumin (BSA) and measuring their absorbance.

e. However, the Bradford method measures the total concentration of protein and not just the protein of interest. To determine a more accurate protein concentration of the protein of interest, SDS-PAGE was conducted using three or more BSA standards that had known concentrations to compare with two or more replicates of TET aliquots. TET protein concentrations could then be quantified by comparing and averaging the band signal intensities using the FlexAnalysis program after imaging with the ChemiDoc MP Imaging System.

f. Lastly, 10 µL of the concentrated protein was aliquoted into microcentrifuge tubes and stored at −80 °C for future use to prevent repetitive freeze thaw cycles that would denature the protein, effectively resulting in reduced activity.

2.2 Mutagenesis and expression of V1395A

There is limited knowledge regarding the function of TET oxidation species 5fC and 5caC as they are not as prevalent as 5hmC in the mammalian genome (Sappa et al., 2021). Efforts to modulate the production of these products have been limited. We have recently demonstrated that replacing hydrophobic and polar residues are of particular interest to identify important gatekeeping residues that would maximize oxidation of 5mC, providing a foundation to probe for function of TET oxidation products that are not as abundant in the mammalian genome (Sappa et al., 2021). Through a systematic mutational analysis, we showed that when valine, located at position 1395 in the catalytic pocket of TET2, wass replaced with alanine (V1395A), the resulting mutant displayed a unique ability to produce 5caC as the major enzymatic product as quantitated by the *in-vitro* MALDI-based assay (Sappa et al., 2021).

2.2.1 Materials and reagents

Item	Supplier and catalog number
QuikChange lightning site-directed mutagenesis kit	Agilent Technologies
Nuclease free water	Corning
PCR mutagenesis primers (forward and reverse)	IDT
Deoxynucleotide triphosphate (dNTP) Mix	Fisher Cat # 18427013
E. coli strain XL-10 Gold Ultracompetent cells	Agilent Technologies
Luria-Bertani (LB) broth	Fisher
Kanamycin	Fisher Cat # K004725G
GeneJET plasmid miniprep kit	Thermo Fisher Cat #K0502
PCR tubes	Fisher Cat # 14-222-292
Cell spreader	Fisher Cat # 50-751-5036
1.5 mL eppendorf tube	Fisher
Falcon™ round-bottom polystyrene test tubes	Fisher Cat # 14-959-11B
Falcon™ standard tissue culture dishes	Thermo

2.2.2 Equipment

Item	Supplier
Microspin mini centrifuge	Corning
Thermal cycler	Bio Rad T100
Water bath	Thermo
CO_2 incubator for bacterial cell culture	Thermo HERATherm
Large centrifuge	Thermo Sorvall Legend XTR

2.2.3 Protocol

1. Site-directed mutagenesis

a. This TET2 variant was generated using the QuikChange Lightning site-directed mutagenesis kit (Agilent Technologies).

b. Primers were designed based on the template TET2 plasmid sequence with a focus on the catalytic domain for TET2 activity modulation. General primer design principles regarding primer lengths between 25 and 45 bases were considered and although it was not crucial, designing at least one primer with either a C or G on the 3′ end increased the probability of a successful mutagenesis. These C and G bases make one extra hydrogen bond with its corresponding base on the template strand, creating a tighter bonding interaction. The forward primer had the following sequence: 5′-GGC AGC ACA TTG GCC TGC ACT CTC ACT-3′. The reverse primer was the reverse complement of the forward primer (5′-AGT GAG AGT GCA GGC CAA TGT GCT GCC-3′).

c. The polymerase chain reaction (PCR) mixture included 40 ng of TET2 plasmid DNA template strand, 1.5 μL of 10× reaction buffer, 130 ng of each primer, 0.4 μL of deoxynucleotide triphosphate (dNTP) mix, and 0.5 μL QuikSolution reagent. Nuclease free water was added for a total reaction volume of 14.6 μL. The contents of each tube were mixed gently and briefly centrifuged before adding 0.4 μL QuikChange Lightning (QCL) Enzyme. To ensure mutagenesis was successful, three separate PCR reactions were conducted. In two of the reactions either 130 ng of the forward primer or 130 ng of the reverse primer was added to the reaction mixture. In the third reaction, 130 ng of both primers were added to the reaction mixture. This did not change the cycling parameters, nor did it change the final reaction volume.

d. The PCR was performed using a Thermal Cycler (Bio Rad T100) utilizing conditions provided by the manufacturer. The QCL enzyme was activated at 95 °C for 2 min before denaturation occurring at 95 °C for 20 s. Following denaturation, annealing was carried out at 63.6 °C for 10 s. The annealing temperature depended on the melting temperature of the primers and was the most important part of this step. In the case of TET2 mutants, annealing temperatures that were 3 °C below the primer melting temperature (T_m = 66.6 °C) proved to be the only temperature for a successful

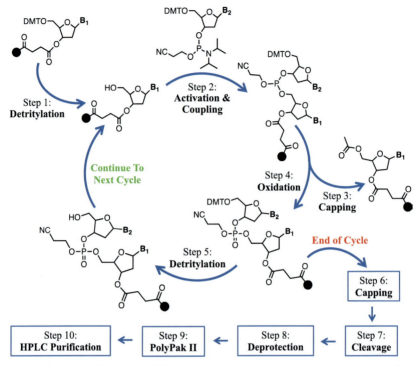

Fig. 2 Schematic showing solid-phase synthesis of oligonucleotides using dimethoxytrityl (DMT)-on protocol.

mutagenesis. Lastly, extension was performed at 68 °C for 30 s/kb of plasmid length. 25 cycles of denaturation, annealing, and extension were performed. The final extension was performed for an additional 5 min at 68 °C.

2. Bacterial Transformation and Sanger Sequencing
   a. The resulting mutant plasmids were transformed into *E. coli* strain XL-10 Gold Ultracompetent cells (Agilent Technologies).
   b. A single colony was picked up and grown overnight at 37 °C in 5 mL of LB broth in presence of 50 μg/mL kanamycin.
   c. The cultures were centrifuged at 3000 rpm for 15 min and DNA was subsequently isolated from bacteria using GeneJET Plasmid Miniprep Kit (Thermo Fisher, Cat #K0502).
   d. 800 ng of plasmid DNA was mixed with 5 μM of either T7, TET middle forward, or T7 reverse primers and sent for Sanger Sequencing through GeneWiz.

e. Resulting sequencing trace files were analyzed using Chromas software and the entire plasmid was compared to the wild-type TET2 construct to verify mutagenesis was successful using Clustal Omega.
3. TET2-V1395A Protein Expression and Purification:
   a. The mutant was transformed into BL21 star (DE3) competent cells and was expressed and purified the same way as the wild-type construct as previously described in Section 2.1.3.

2.3 Synthesis and characterization of oligonucleotides

Using the protocol described below, we synthesized several palindromic DNA oligonucleotides of various lengths to gauge TET2 enzymatic activity starting with a 58-mer DNA with a central 5mC unit: (5′-ACG ATC AGA TCC TAA GGC ATC AGC ACA C(5mC)G GTG TGC TGA TGC CTT AGG ATC TGA TCG T-3′). As the TET oxidative products were primarily confirmed by MALDI-TOF mass spectrometry, the 58-mer DNA substrate was not preferred due to poor signal-to-noise ratio in MALDI-based detection (Sudhamalla et al., 2017). Crystal structures of TET2 revealed that crucial interactions between TET2 and DNA only included a few nucleotides adjacent to the 5mC base which led to the synthesis of the shorter 22-mer, 14-mer, and 8-mer oligonucleotide substrates, using the solid-phase technique (Fig. 2), all which contained a single 5mC group (Hu et al., 2013, 2015). A schematic overview of how oligonucleotides were made using an Expedite 8909 DNA synthesizer is depicted in Fig. 2, where steps one through five were repeated until the desired sequence was obtained and steps six through ten were followed to achieve a pure synthetic oligonucleotide. As subsequent MALDI-TOF analyses utilizing these substrate oligonucleotides demonstrated successful 5mC oxidation by TET2, the 8-mer palindromic DNA was subsequently selected for future studies due to its shortest length (Sudhamalla et al., 2017).

2.3.1 Materials and reagents

Item	Supplier and catalog number
CpG resin	Glen Research
DNA phosphoramidites	Glen Research
Anhydrous acetonitrile	Glen Research
20% acetonitrile/water	Fisher

2% trifluoroacetic acid	Glen Research
2.0 M triethylamine acetate	Glen Research
Oxidant solution – Iodine solution in THF/water/pyridine	Glen Research
0.45 M tetrazole	Glen Research
Capping reagent A – acetic anhydride in pyridine/THF	Glen Research
Capping reagent B – N-methylimidazole in pyridine/THF	Glen Research
Wash A – acetonitrile	Glen Research
Deblock solution – trichloroacetic acid (TCA) in dichloromethane (DCM)	Glen Research
33% ammonium hydroxide/water	Fisher
Auxiliary solution	Glen Research
Activating solution – tetrazole in acetonitrile	Glen Research
Deionized water	
Frit	Glen Research
10 mL syringe	BD Luer-Lok™ Tip
Syringe needle (18 G × 1 ½″ TW)	BD PrecisionGlide™
Argon gas cylinder	

2.3.2 Equipment

Item	Supplier
Expedite 8909 DNA synthesizer	Biolytic
PolyPak II purification cartridge	Glen Research 60-3100-10
Column for DNA synthesis	Glen Research
Ultra-low temperature freezer	New Brunswick
SpeedVac concentrator	Thermo Savant SC210A
HPLC instrument	Agilent 1220 Infinity LC
MALDI-TOF mass spectrometer	Bruker-UltrafleXtreme
UV–Vis biospectrometer	Eppendorf

2.3.3 Protocol

1. Preparation of Expedite 8909 Nucleic Acid Synthesis System (Perspective Biosystems)
   a. The argon gas flow was set to 20 PSI with extra care not to exceed instrument pressure limits.
   b. The Expedite program was opened on the desktop connected to the DNA synthesizer. To prepare the system for synthesis the Run tab at the top of the page was opened before navigating to the 'Synthesis' option.
   c. Either position 1 or 2 was selected which corresponded to the column that would be used.
   d. 'Edit Sequence' was then selected at the bottom of the window to make a new sequence and then the following options were selected in order: 'Sequence', 'New', 'Modify', 'DNA'.
   e. The desired sequence was entered in order from 5′ to 3′. To insert an amidite that was different than the standard A, T, G, and C monomers into the sequence, a different position was selected that indicated the location of this modified phosphoramidite and appeared as a number in the sequence. Once the desired sequence was entered, the sequence was saved, and the sequence editor was closed.
   f. To choose the sequence, the drop-down menu was opened, and 'Database Sequence' was selected.
   g. The first nucleoside is bound to the support and contains a 4,4′-dimethoxytrityl (DMT) protecting group which prevents polymerization during resin functionalization. This protecting group must be removed through a process called detritylation before oligonucleotide synthesis can proceed. However, at the last step of the synthesis, this DMT group can either be left on (DMT-ON) as protection after synthesis or can be removed through a final deprotection (DMT-OFF).
   h. The DMT-ON protocol was selected so that the DMT group could be used as a purification handle due to its hydrophobic properties using a PolyPakII purification kit at the end of the synthesis.
2. Priming Reagents
   a. On the DNA synthesizer, 'Prime' was selected before selecting 'Reagents Only'
   b. The union was tightly placed on column 1 union and 'OK' was selected on the instrument. This step ensures that the reagents are

primed for the column of choice. These syntheses were run under ultra-mild conditions using tetrazole as an activator reagent.

3. Preparation of Phosphoramidites for Synthesis
   a. Synthesis of DNA oligonucleotides with one or more 5mC, 5hmC, 5fC and/or 5caC modifications required using modified phosphoramidites that carried this epigenetic mark. All canonical (A, T, G, and C) and modified phosphoramidites were synthesized using DNA phosphoramidite monomers obtained from Glen Research.
   b. When not in use, bases were stored at $-80\,°C$ in acetonitrile. Bases were allowed to reach room temperature only before they were ready for use.
   c. Each was dissolved with anhydrous acetonitrile diluent using a nitrogen balloon to offset the pressure in the diluent bottle. The volume of acetonitrile added to each base corresponded to the volume recommended by Glen Research.

4. Priming Bases
   a. Each base position was filled with acetonitrile when it was not in use.
   b. On the instrument panel, the following options were selected in order: 'Tools', 'Diagnostics', 'Fluid', 'Column 1', 'More'.
   c. Each base that would be used was selected before choosing the 'Volume' option. The prime was discontinued after 5 clicks from the instrument.
   d. The union was removed and a pre-packed column containing resin from the first base in the sequence was attached in its place.
   e. The column was washed with Wash A by selecting the following options in order: 'Tools', 'Diagnostics', 'Fluid', 'Column 1', 'Wash A', 'Volume'.
   f. The wash was terminated when a steady flow of liquid was observed in the clear tubing above the column.

5. Preparation of Column
   a. One small frit was inserted into one end of a fresh column.
   b. The CpG was chosen that corresponded to the 3′ end of the DNA sequence as the base on the solid support. This CpG was weighed out into the column according to reaction scale that was batch specific. Glen Research had previously supplied these calculations.
   c. Once the appropriate CpG was loaded into the column, a second frit was fitted to the other end of the column.

6. Synthesis
   a. To ensure a successful coupling, modified bases were set to couple for 4.5 min and standard bases were set to couple for 2 min.
   b. Once all conditions were set on the instrument, 'START' was selected on the instrument panel.
7. Post-Synthesis
   a. The crude deoxyoligonucleotide was cleaved from the column resin and deprotected by agitating the resin with ammonium hydroxide (33% v/v) at 25 °C. The ammonium solution containing the oligo is then gently stirred in a closed environment for 24 h. For 5hmC DNA, the deprotection step was heating at 75 °C for 24 h.
   b. Preliminary purification and DMT deprotection were carried out using Poly PakII purification cartridge (Glen Research, 60-3100-10) according to the standard protocol provided by Glen Research.
   c. The purified DNA was concentrated by a SpeedVac concentrator.
8. HPLC Purification
   a. Further purification was conducted utilizing High Pressure Liquid Chromatography (HPLC) by first dissolving the DNA in 950 mL nuclease free water and 50 μL triethylammonium acetate (TEAA).
   b. This mixture was passed through a 0.2 μm filter prior to injecting into the HPLC.
   c. Fractions under each peak were collected and concentrated by SpeedVac for 1–2 h before subjected to lyophilization until only a residue remained.
   d. The residue was then redissolved in nuclease free water and the concentration was measured using UV-Vis (A260/280).
   e. The quality and purity were determined by high resolution MALDI-TOF mass spectrometry (AB SCIEX Voyager DE Pro and Bruker-ultrafleXtreme™ MALDI-TOF/TOF spectrometer).

3. Biochemical assays and results
3.1 Development of a robust *in-vitro* assay

Prior to the development of our assay, exhaustive steps for the detection of TET oxidative products were required as it was necessary to pass the enzymatic products through a nucleotide cleaning kit prior to concentrating the eluent. This served to eliminate salts present in the assay components, improving signal-to-noise ratio in the mass spectrum. However, these steps led

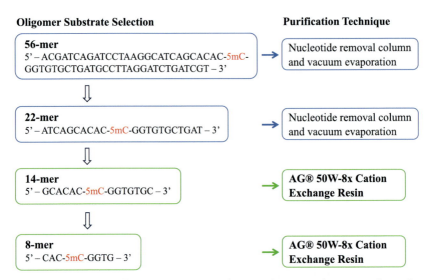

Fig. 3 Optimization of *in-vitro* assay involving substrate selection and product purification.

to the loss of TET oxidative products. With these factors in mind, we sought out a way to streamline post-assay purification procedure and eliminate the need for lengthy, expensive cleanup steps (Fig. 3). As such, we developed a simpler and more straightforward protocol for purifying TET products that greatly reduced post-assay processing (Sudhamalla et al., 2017). We reasoned that upon completion of the TET oxidation reaction, a cation exchange resin could be added directly to the assay mixture to eliminate the need for expensive desalting columns. Resin-treated samples were then spotted on the MALDI plate and subjected to mass spectrometric analysis directly utilizing 3-Hydroxypicolinic Acid (3-HPA) as the matrix. With this simplified process, TET oxidized products were detected with a significantly improved signal-to-noise ratio in the MALDI spectra without requiring multiple enzymatic digestions, liquid chromatographic optimization, and separation of digested products before analysis (Sudhamalla et al., 2017).

3.1.1 Materials and reagents

Item	Supplier and catalog number
Potassium acetate	Fisher
4-(2-Hydroxyethyl)-1-piperazineethanesulfonic acid (HEPES, pH 7.5)	Fisher
Sodium chloride	Fisher Cat #BP358-10

Ammonium iron (II) hexasulfate	Fisher
Ascorbate	Thermo
Dithiothreitol (DTT)	Thermo
Adenosine triphosphate (ATP)	Thermo
2-Ketoglutarate	Fisher
Deionized water	
Agarose	Fisher
QIAquick nucleoside removal kit	QIAGEN
3-Hydroxypicolinic acid	Thermo
AG® 50W-X8 cation exchange resin	BioRad Cat #143-5441
PCR tube	Fisher Cat # 14-222-292
1.5 mL eppendorf tube	Fisher

3.1.2 Equipment

Item	Supplier
Thermocycler	Bio Rad T100
Heat block	Fisher
Medium centrifuge	Eppendorf
ChemiDoc MP imaging system	BioRad
SpeedVac concentrator	Thermo Savant SC210A
MALDI-TOF mass spectrometer	Bruker-UltrafleXtreme
UV-Vis biospectrometer	Eppendorf

3.1.3 Protocol

1. Duplexing Synthesized DNA for *in-vitro* Assays
   a. In order to use synthetic DNA for TET enzymatic assays, single-stranded oligomeric DNA had to be made into double-stranded DNA through a process called duplexing. The single-stranded DNA that we used was palindromic, meaning the sequence was read the same in the forward direction as it did in the reverse direction. It was important to have one methylation mark on this single stranded DNA prior to duplexing as TET is known to oxidize 5mC-containing

double stranded DNA. DNA was duplexed using the following protocol.

b. 100 μM of pre-synthesized, single stranded DNA was added to a PCR tube with duplex buffer which included 100 mM potassium acetate, 30 mM 4-(2-hydroxyethyl)-1-piperazineethanesulfonic acid (HEPES, pH 7.5, Integrated DNA Technology) for a total volume of 100 μL.

c. One option for duplexing the DNA included heating the DNA mixture on a 95 °C heat block for 2 min and gradually cooling down to room temperature. However, this method did not consistently produce duplexed DNA and was discarded. This finding could have been due to the inconsistency in how fast the DNA was cooled to room temperature or that it cooled too fast.

d. Utilizing a Thermal Cycler (Bio Rad T100) for this process proved to be more reliable and consistent. With this method, the DNA was heated to 95 °C for 5 min before decreasing temperature by 1 °C each minute for a total of 40 min. The DNA was then held at 55 °C for 30 min before decreasing the temperature once again by 1 °C each minute for a total of 20 min.

e. To confirm that the DNA was duplexed, the oligonucleotides were separated on a 1% agarose gel and visualized with the ChemiDoc MP Imaging System using ethidium bromide.

2. *In-vitro* Enzymatic Activity Assay

a. 5 μM of double-stranded synthetic DNA substrates containing one methylation mark were incubated with 10 μM TET2 (1099–1936 del-insert) or its mutant (V1395A) in buffer containing 50 mM HEPES (pH 8.0), 100 mM NaCl, 100 μM Fe(NH$_4$)$_2$(SO$_4$)$_2$, 2 mM ascorbate, 1 mM dithiothreitol (DTT), 1 mM adenosine triphosphate (ATP), and 1 mM 2KG for 3 h at 37 °C.

b. After incubation, the reaction was quenched by heating the reaction at 95 °C for 10 min.

3. Product Oligonucleotide Purification

a. *Nucleotide clean up method:* The oligonucleotide products were purified using the QIAquick Nucleotide Removal Kit (QIAGEN) following the manufacturer's instructions prior to denaturation at 100 °C for 10 min. The oligonucleotides were concentrated using SpeedVac concentrator for 10 min and analyzed by MALDI-TOF mass spectrometry (AB SCIEX Voyager DE Pro) by spotting 1 μL of the purified DNA sample with 1 μL of 3-Hydroxypicolinic Acid (3-HPA) matrix on the MALDI plate.

b. *Cation-exchange resin method:* To streamline the assay and eliminate the need for a lengthy, expensive cleanup step, we developed a simpler and more straightforward protocol for purifying TET product oligonucleotides that greatly reduced purification time. To analyze any synthetic DNA oligonucleotide *via* MALDI, reaction mixtures containing this DNA were desalted by adding 8 μL of AG® 50 W-X8 Cation Exchange Resin (BioRad, Cat. #143-5441) directly into the 20 μL reaction mixture and briefly agitating the solution. Samples were subsequentially centrifuged at 10,000 rpm for 2 min. TET oxidized products were then analyzed by MALDI-TOF MS by spotting 1 μL of sample and then mixing with 1 μL of 3-Hydroxypicolinic Acid (3-HPA) matrix on a MALDI plate. The 3-HPA matrix has a low amount of fragmentation for oligonucleotides between 1 and 30 kDa and is suitable for oligonucleotide analysis (Bruker-Daltonics, 2012; Chou & Limbach, 2000).

4. Analysis of TET2 Oxidative Products *via* MALDI

a. MALDI data was collected on a Bruker ultrafleXtreme™ MALDI-TOF/TOF mass spectrometer using the reflectron negative TOF mode (Fig. 4).

b. The background noise was corrected for calculating signal/noise ratios using AB Sciex Data Explorer (4.0). Resulting MALDI spectra could be analyzed quantitatively by normalizing ion intensities to the highest peak and measuring apex peak intensities using GraphPad Prism software.

c. MALDI-based measurement is suitable for providing quantitative information on TET enzymatic activity due to similar ionization potentials of 5mC, 5hmC, 5fC and 5caC containing DNAs as was determined by generating a standard curve (Sappa et al., 2021). For the standard curve, the percentage of 5mC, 5hmC, 5fC and 5caC containing synthetic DNAs was varied from 0% to 100%, while the total DNA concentration remained a consistent 10 μM. Samples were mixed and analyzed by MALDI-TOF. The relative intensities of synthesized DNAs were fitted to a straight line with nonlinear regression using GraphPad Prism. Since standard curves generated from this experiment displayed similar ionization potentials of 5mC, 5hmC, 5fC and 5caC, we determined that the relative abundance of each oxidation product can be directly compared using MALDI.

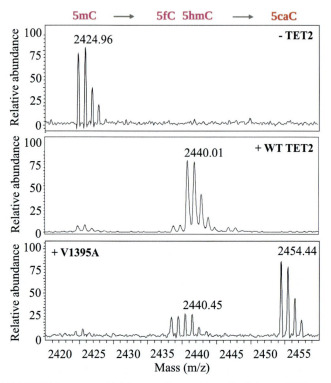

Fig. 4 MALDI-TOF-MS spectra displaying the activity of wild-type TET2 and V1395A mutant.

d. For quantitative analysis of TET enzymatic activity, peaks corresponding to 5mC, 5hmC, 5fC and 5caC were clearly labeled as displayed in Fig. 4. With respect to the wildtype enzyme, V1395A demonstrated that it was indeed acting as a superior dioxygenase, represented by its complete oxidation of 5mC to primarily 5caC. In contrast, wildtype TET2 produced a wide variety of products, but primarily produced 5hmC (Fig. 4).

3.2 Measurement of IC$_{50}$ of TET2 inhibitors NOG and 2HG

We next applied the optimized *in-vitro* assay to measure the potency of two known TET2 inhibitors: 2-hydroxyglutarate (2HG) and *N*-oxalylglycine (NOG) (Fig. 5), (Sudhamalla et al., 2017). These two molecules have been shown to inhibit a range of 2KG-dependent dioxygenases in a cofactor-competitive manner due mainly to its structural similarity with 2KG (Hamada et al., 2009). Cancers including gliomas and acute myeloid

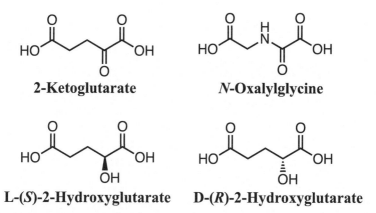

Fig. 5 Cofactor 2KG and competitive inhibitors of TET2: N-oxalylglycine (NOG), l-(S)-2-hydroxylutarate (l-2-HG) and d-(R)-2-hydroxyglutarate (d-2-HG).

leukemia have been associated with not only an increase in 2HG levels, but also with decreased levels of global 5hmC, suggesting that inhibition of TET activity by 2HG is crucial for hematopoietic transformation (Losman & Kaelin, 2013). Utilizing our MALDI-based assay, we show that the inhibitory effect of NOG and 2HG on the human TET2 protein can be reliably determined as it measures the formation of oxidized products. Due to the cofactor-competitive nature of the inhibition, this assay was modified by varying the concentration of 2KG from which we concluded that 80 μM 2KG (5-fold higher than $K_{M,2KG}$) was optimal for TET activity (Sudhamalla et al., 2017). Under such conditions, TET demonstrated maximum activity, but was still competitively inhibited by NOG and 2HG. NOG has been previously reported to inhibit Naegleria Tet1, with an IC_{50} of 49 ± 6 μM in a fluorescence polarization-based competitive binding assay that did not require the cofactor 2KG (Hashimoto et al., 2014; Marholz et al., 2016). However, by optimizing 2KG in our MALDI-based assay, the inhibitory activity of NOG and both enantiomers of 2HG (d-2HG and l-2HG) reflected the ability to modulate the catalytic potential of TET2 in a dose-dependent manner as our assay measures the formation of oxidized products. Under our assay conditions, we determined an IC_{50} for NOG to be 149.5 ± 7.6 μM. Using the dose-response inhibition studies, IC_{50} values were found to be 5.3 ± 0.3 mM and 12.4 ± 2.4 mM for d-2HG and l-2HG, respectively (Sudhamalla et al., 2017). The higher inhibitory effect of NOG towards TET enzymes was linked to the structure of NOG consisting of the 2-keto acid moiety as well as an internal

–NH– group that participates in strong hydrogen bonding with TET enzymes, and which is absent in 2HG (Sudhamalla et al., 2017).

3.2.1 Materials and reagents

Item	Supplier and catalog number
N-oxalylglycine	Sigma Cat #O9390
d-2-Hydroxyglutaric acid	Santa Cruz Cat #sc-227739
l-2-Hydroxyglutaric acid	Santa Cruz Cat #sc-361834
HEPES	Fisher
Sodium chloride	Fisher Cat #BP358-10
Ammonium iron (II) hexasulfate	Fisher
Ascorbate	Thermo
DTT	Thermo
ATP	Thermo
2-Ketoglutarate	Fisher
AG® 50W-X8 cation exchange resin	BioRad Cat #143-5441
Deionized water	
1.5 mL eppendorf tube	Fisher

3.2.2 Equipment

Item	Supplier
Heat block	Fisher
Ultra-low temperature freezer	New Brunswick
MALDI-TOF mass spectrometer	Bruker-UltrafleXtreme

3.2.3 Protocol

a. Assay mixture containing $10\,\mu M$ TET2 was incubated with varying concentrations $(0.05–5\,mM)$ NOG or varying concentrations $(0.5–30\,mM)$ of 2HG in buffer containing $50\,mM$ HEPES (pH 8.0), $100\,mM$ NaCl, $100\,mM$ $Fe(NH_4)_2(SO_4)_2$, $2\,mM$ ascorbate, $1\,mM$ DTT, $1\,mM$ ATP for 5 min on ice.

b. Demethylation was initiated by adding $100\,\mu M$ 2-KG and $10\,\mu M$ 8-mer double stranded 5mC DNA and further incubated at $37\,°C$ for 3 h, (see Section 3.1.3 step 2). Variations to the demethylase assay mentioned in Section 3.1.3 were minimized to ensure that inhibition of this protein was due to NOG and not due to differences in assay conditions. This supports the decision to keep the 8-mer DNA as well as the 3-h incubation time.

c. The product DNA was denatured at 95 °C for 10 min and desalted using cation exchange resin method described above.

d. To accurately determine the inhibitory effects of NOG and 2-HG towards TET2, the concentrated oligonucleotides were analyzed by MALDI-TOF mass spectrometry, and the values were fitted to the 4-parameter non-linear regression algorithm ($Y = Bottom + (Top - Bottom))/(1+10^{((Log\ IC_{50}-x)*Hill\ Slope)})$ of the GraphPad Prism software. X: log of dose or concentration; Y: response, decreasing as X increases; Top and bottom: upper and lower values of a given curve; $logIC_{50}$: same log units as X; Hill Slope: Slope factor or Hill Slope, unitless. With these parameters, the effect of NOG and 2-HG was accurately determined as the oxidation of 5hmC decreased with increasing concentration of NOG or 2-HG in a dose-dependent manner.

3.3 Validating the activity of wildtype TET2 and V1395A using BS-seq

We employed a secondary assay to further examine the activity of wildtype TET2 and V1395A in a base-resolution manner by treating the oxidized products with bisulfite and coupling with Sanger Sequencing (BS-Seq) (Fig. 6) (Lu et al., 2013). Upon treatment with bisulfite under alkaline conditions, 5caC is deaminated and read as T during sequencing. In contrast, 5mC and 5hmC do not react with bisulfite and are subsequentially read as C after Sanger Sequencing (Fig. 6). Employing our optimized *in-vitro* demethylase assay, a duplexed 76-mer DNA carrying a central 5mC unit was subjected to oxidation using wildtype TET2 and the V1395A mutant followed by bisulfite treatment, PCR amplification and Sanger sequencing. While 5mC in samples exposed to either no protein or wildtype TET2 read as C, the equivalent site in the V1395A-treated sample emerged as T demonstrating the ability of the mutant to predominantly generate 5caC (Fig. 6) (Sappa et al., 2021).

3.3.1 Materials and reagents

Item	Supplier and Catalog Number
GeneJET PCR purification kit	Thermo Fisher Cat #K0701
Isopropyl alcohol	Fisher

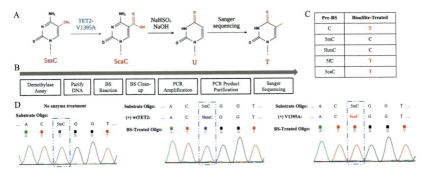

Fig. 6 Analysis of TETs oxidative products using bisulfite (BS) coupled with Sanger sequencing. (A) Scheme showing bisulfitemediated 5caC conversion to uracil which is read as thymine after Sanger sequencing. (B) Overview of steps required to perform bisulfite conversion. (C) Sanger readout of cytidine and its modified congeners upon BS treatment. (D) 76-mer DNA substrate containing a single 5mC modification subjected to bisulfite treatment before and after being oxidized by either wild-type TET2 or V1395A, confirming 5caC as the major product of V1395A utilizing the base-resolution sequencing technique.

76-mer DNA substrate	IDT
EpiTect bisulfite kit	QIAGEN Cat #59104
Hot star taq polymerase	QIAGEN Cat #203207
PCR primers (forward and reverse)	IDT
Agarose	Fisher
Nuclease free water	Corning
GeneJET plasmid miniprep kit	Thermo Fisher Cat #K0502
E. coli strain XL-10 gold ultracompetent cells	Agilent Technologies
1.5 mL eppendorf tube	Fisher
PCR tubes	Fisher Cat # 14-222-292

3.3.2 Equipment

Item	Supplier
Heat block	Fisher
Microspin mini centrifuge	Corning
Centrifuge – medium	Eppendorf
Ultra-low temperature freezer	New Brunswick

SpeedVac concentrator	Thermo Savant SC210A
MALDI-TOF mass spectrometer	Bruker-UltrafleXtreme
Thermocycler	Bio Rad T100

3.3.3 Protocol

1. *In-vitro* Demethylase assay
   a. 5000 ng of double-stranded 76-mer DNA substrate obtained from IDT (5′-CCT CAC CAT CTC AAC CAA TAT TAT ATT A TGT GTA TAC AC 5mC GGT GTT TGT GTT ATA ATA TTG AGG GAG AAG TGG TGA-3′) (Lu et al., 2013), was incubated with 20 μM of V1395A TET2 in demethylase activity buffer for 3 h at 37 °C as previously described in 3.1.3 step 2.
   b. After incubation, the reaction was quenched by heating reaction at 95 °C for 10 min.
2. Initial DNA purification
   a. Following the demethylase assay, 76-mer DNA was purified with the GeneJET PCR Purification Kit (Thermo Fisher, Cat. #K0701) following the supplier's instructions with several modifications.
   b. The reaction mixture was diluted to 100 μL with nuclease free water before adding 100 μL of 100% isopropyl alcohol and 200 μL binding buffer. After addition of the binding buffer, the solution should remain a bright orange color which indicates that the pH of the solution is optimal for DNA binding.
   c. This mixture was transferred into a purification spin column, provided in the kit, and centrifuged at 11,000 rpm for 30 s
   d. After the flowthrough was discarded, 500 μL of Wash Buffer was added to the column and the centrifugation was repeated. This step was repeated twice before the empty centrifuge tube was spun for an additional 30 s to remove any residual wash buffer.
   e. The spin column was placed in a fresh Eppendorf tube and 20 μL of nuclease free water was added. To evaporate residual ethanol left over from the wash solution, the spin column was placed on a 56 °C heat plate for 3 min before centrifuging the column once again.
   f. The 20 μL was pipetted out of the bottom of the Eppendorf tube and incubated on the center of the spin column membrane once again at room temperature. After 5 min, the column was centrifuged and the 20 μL of purified DNA was subjected to bisulfite treatment.
3. Bisulfite Treatment

a. The purified DNA was applied to EpiTect Bisulfite Kit (QIAGEN, Cat. #59104) following the supplier's instructions.

b. The bisulfite treated DNA was used for PCR amplification using Hot Start Taq polymerase (QIAGEN, Cat #203207) (forward primer: 5′-CCC TTT TAT TAT TTT AAT TAA TAT TAT ATT-3′; reverse primer: 5′-CTC CGA CAT TAT CAC TAC CAT CAA CCA CCC ATC CTA CCT GGA CTA CAT TCT TAT TCA GTA TTC ACC ACT TCT CCC TCA AT-3′). The PCR was performed in 2 steps with both primers in the PCR mixture.

c. In this PCR reaction, Hot Start Taq polymerase was activated for 15 min at 95 °C. Following activation, initial denaturation (5 min at 94 °C), denaturation (1 min at 94 °C), annealing (reverse primer, 1 min at 63 °C) and extension (1 min at 72 °C) was carried out. The final extension was carried out for 10 min at 72 °C.

d. Without purification, the mixture proceeded for another PCR round for the forward primer starting with Hot Start Taq polymerase activation (15 min at 95 °C). Following activation, initial denaturation (5 min at 94 °C), denaturation (1 min at 94 °C), annealing (forward primer, 1 min at 44 °C) and extension (1 min at 72 °C) were carried out. The final extension was carried out for 10 min at 72 °C. 30 cycles of PCR amplification were performed for each primer.

e. The resulting DNA was purified using a GeneJET PCR Purification Kit (Thermo Fisher, Cat. #K0701) as previously described in Section 3.4.3, step 2a-f.

f. To ensure the PCR reaction worked, 500 ng of each reaction was run on a 1.5% agarose gel.

4. Sanger Sequencing

a. To verify the oxidation, 800 ng of the DNA was sent for Sanger Sequencing (Genewiz) using the following sequencing primer: 5′-CCC TTT TAT TAT TTT AAT TAA TAT TAT ATT-3′.

4. Notes

We have optimized a majority of the conditions mentioned in the procedures above to offer a robust *in-vitro* assay for examining enzymatic activity of TET proteins and likely other 2KG-dependent dioxygenases. However, below are some of the issues we encountered along with suggestions on how to correct these issues should they arise in future experimentation. We also include safety related information important to these methods. It is important that proper personal protective equipment is worn when conducting any experimentation. Disposable nitrile gloves are suitable for carrying out biochemical methods mentioned in this chapter.

1. Protein expression/purification

   a. In a flame sterilized environment, 10 ng of the wild-type TET2 plasmid with pET-28b kanamycin-resistant vector was pipetted into an autoclaved 1.5 mL Eppendorf tube with 9 μL *E. coli* strain BL21 star (DE3) competent cells (Fisher). To ensure that bacterial transformation is done safely, isopropyl alcohol can be used to sterilize the bench prior to turning on the flame and the flame should not be left unattended. Sterilization is important to prevent transformation of any unwanted DNA as well as to ensure the highest transformation efficiency of the plasmid of interest. The competent cells should be thawed on ice as they are sensitive to changes in temperature. If cells come into contact with the eyes or ingested, rinse impacted area thoroughly with water for 15 min. If cells come into contact with skin, wash with soap and water.

   b. It is important that an agar plate from bacterial transformation is not used for starting a bacterial inoculation if it is older than one week. A risk with using an older plate is that there will be little-to-no growth of the bacterial inoculate after 16 h of incubation at 37 °C.

   c. It is important to observe the growth of the starter culture once a single colony has been inoculated in the LB growth media with antibiotics. After 16 h, the inoculate solution should appear significantly foggier than before, but there should be no sign of debris at this stage. Debris is a sign that the cells have begun to die, and the amount of protein obtained at the end of expression will be minimal. If this debris is observed, repeat the starter culture inoculation, and incubate for less time. This may also occur if more than one colony is inoculated in the same culture tube or if a colony cluster is chosen for inoculation.

   d. Prior to diluting the starter culture into 1 L of LB, the flasks containing LB must be autoclaved. It is important to wear heat-resistant gloves while handling hot flasks. Do not dilute starter cultures or add antibiotics until after these autoclave flasks have cooled down to room temperature. Add 50 μg/mL antibiotic and the starter culture under a flame-sterilized environment.

e. After the starter culture is diluted 100-fold into 1 L of LB with the addition of 50 µg/mL kanamycin and allowed to grow at 37 °C, it is important to keep a close eye on the OD_{600} value. If the OD_{600} value increases above 0.8, the amount of protein obtained at the end of expression will be minimal.

f. Prior to protein purification, the 1 L cultures must be centrifuged at 4000 rpm. To ensure that this process is done safely, the rotor of the high-speed centrifuge must be balanced. This means that the mass of each culture being centrifuged must be the same. Deionized water can be added to increase the mass without decreasing the yield of final protein isolated.

g. TET is a heat sensitive protein meaning that once expressed and purified, it should not remain at room temperature for longer than absolutely necessary. As the number of freeze-thaw cycles increases, the activity of the protein will subsequentially decrease. To avoid this scenario, aliquot 10–20 µL of the purified TET protein into smaller Eppendorf tubes.

2. Site-directed mutagenesis

a. During PCR, the annealing temperature selected for the primers is the most important step and varies depending on the melting temperature of the designed primer as well as the template plasmid construct. In the case of TET2 mutants, annealing temperatures that are 3 °C below the primer melting temperature (T_m = 66.6 °C) are used. If mutagenesis is not successful, repeat the process mentioned above and increase or decrease the temperature by 1 degree. It is important to perform mutagenesis below the melting temperature of the primer to prevent primer degradation.

b. Another technique widely used to overcome difficulties in mutagenesis includes performing mutagenesis with one primer instead of both primers. Although using two primers increases the chance that mutagenesis will occur, using either the forward or reverse primers often yielded a successful mutation when using two primers did not. The overall yield of the plasmid containing the mutation will be significantly decreased with this method, but adding *Dpn*I degrades unreacted plasmid leaving only the mutated plasmid. Utilizing an ultracompetent cell line like XL10-Gold cells increases transformation efficiency and should be used in this case.

3. DNA oligonucleotide synthesis

a. Phosphoramidites are sensitive to water contamination as they are easily hydrolyzed. To prevent hydrolysis, oligonucleotide synthesis with the Expedite 8909 instrument requires a connection to argon gas for an inert environment. When using an argon gas tank, wear personal protective equipment such as gloves and goggles and follow regulations and standards set by OSHA. Use extreme caution not to exceed the instrument pressure limits during synthesis and monitor the pressure throughout use of the instrument. Precautions must also be used to ensure that reagents used are anhydrous as phosphoramidites are easily hydrolyzed.
b. Using the DMT-ON protocol leaves the final DMT on for initial purification purposes using the PolyPakII kit. Using this method serves to isolate the oligonucleotide of interest with ease. Nonetheless, the synthetic oligonucleotide can be directly purified by reverse-phase HPLC if the DMT-OFF setting is chosen by using an ion pairing agent in the mobile phase.
c. To ensure a successful coupling, modified bases are set to couple for 4.5 min and standard bases are set to couple for 2 min. When purchasing modified bases from Glen Research, there are recommended coupling times mentioned both on the website and on the information packet that is received with the phosphoramidites.
d. After synthesis, the crude deoxyoligonucleotide is cleaved from the column resin and deprotected by agitating the resin with ammonium hydroxide (33% v/v) at 25 °C. The ammonium solution containing the oligo is then gently stirred in a closed environment for 24 h. There are extra steps for deprotection that need to be followed when using modified bases. For example, oligonucleotides containing 5hmC must be deprotected by heating at 75 °C for 24 h. Recommendations for deprotections of modified bases are also found on the information packet and on the Glen Research website.
e. If the product oligonucleotide is found by MALDI to still contain the protecting group, repeat the ammonium hydroxide incubation in a closed environment and agitate by stirring.
4. Analysis of TET oxidative products *via* MALDI
a. Preparation of the MALDI plate and proper cleaning is important for visualizing a clear spectrum. To do this, gently wipe 100% isopropyl alcohol horizontally across the MALDI plate using a Kim Wipe. Wiping horizontally across the plate prevents damage while effectively removing dust, old spots, and other unwanted residues. Once

the large particles have been removed, place the MALDI plate in a beaker and fill with 100% isopropyl alcohol until it completely covers the plate and sonicate with pulsed sonication for 10 min to remove smaller particles. The clean plate should then be wiped once more with 100% acetonitrile and then air dried in a closed environment so that dust or other particles do not dry onto the plate. The final wipe with acetonitrile is crucial and improves the resolution of the MALDI peaks.

b. It is encouraged to use fresh 3-HPA matrix when spotting the product oligonucleotide on the MALDI plate to reduce the signal-to-noise ratio and achieve a clear spectrum.

c. In order to achieve a clear spectrum, it is also important to spot the product oligonucleotides without including cation-exchange resin beads and mix thoroughly with the 3-HPA matrix.

Funding

We thank the University of Pittsburgh, the National Science Foundation (CHE-2204114) and the National Institutes of Health (R01GM130752) for financial support.

References

Bachman, M., Uribe-Lewis, S., Yang, X., Burgess, H. E., Iurlaro, M., Reik, W., ... Balasubramanian, S. (2015). 5-Formylcytosine can be a stable DNA modification in mammals. *Nature Chemical Biology, 11*(8), 555–557. https://doi.org/10.1038/nchembio.1848.

Bhattacharya, C., Dey, A. S., & Mukherji, M. (2023). Substrate DNA length regulates the activity of TET 5-methylcytosine dioxygenases. *Cell Biochemistry and Function, 41*(6), 704–712. https://doi.org/10.1002/cbf.3825.

Booth, M. J., Raiber, E. A., & Balasubramanian, S. (2015). Chemical methods for decoding cytosine modifications in DNA. *Chemical Reviews, 115*(6), 2240–2254. https://doi.org/10.1021/cr5002904.

Bruker-Daltonics (2012). *Bruker guide for MALDI sample preparation [Protocol]*, 5–18.

Chou, C.-W., & Limbach, P. A. (2000). Analysis of oligonucleotides by matrix-assisted laser desorption/ionization time-of-flight mass spectrometry. *Current Protocols in Nucleic Acid Chemistry 10.1.1-10.1.25.*

Dey, A. S., Ayon, N. J., Bhattacharya, C., Gutheil, W. G., & Mukherji, M. (2020). Positive/negative ion-switching-based LC-MS/MS method for quantification of cytosine derivatives produced by the TET-family 5-methylcytosine dioxygenases. bpaa019 *Biology Methods and Protocols, 5*(1), https://doi.org/10.1093/biomethods/bpaa019.

Gibney, E. R., & Nolan, C. M. (2010). Epigenetics and gene expression. *Heredity ((Edinb)), 105*(1), 4–13. https://doi.org/10.1038/hdy.2010.54.

Hamada, S., Kim, T. D., Suzuki, T., Itoh, Y., Tsumoto, H., Nakagawa, H., ... Miyata, N. (2009). Synthesis and activity of N-oxalylglycine and its derivatives as Jumonji C-domain-containing histone lysine demethylase inhibitors. *Bioorganic & Medicinal Chemistry Letters, 19*(10), 2852–2855. https://doi.org/10.1016/j.bmcl.2009.03.098.

Hashimoto, H., Pais, J. E., Zhang, X., Saleh, L., Fu, Z. Q., Dai, N., ... Cheng, X. (2014). Structure of a Naegleria Tet-like dioxygenase in complex with 5-methylcytosine DNA. *Nature, 506*(7488), 391–395. https://doi.org/10.1038/nature12905.

He, Y. L. B., Li, Z., Liu, P., Wang, Y., Tang, Q., Ding, J., ... Xu, G. (2011). Tet-mediated formation of 5-carboxylcytosine and its excision by TDG in mammalian DNA. *Science (New York, N. Y.), 333*(6047), 1303–1307. https://doi.org/10.1126/science.1210944.

Huang, Y., & Rao, A. (2014). Connections between TET proteins and aberrant DNA modification in cancer. *Trends in Genetics: TIG, 30*(10), 464–474. https://doi.org/10.1016/j.tig.2014.07.005.

Hu, L., Li, Z., Cheng, J., Rao, Q., Gong, W., Liu, M., ... Xu, Y. (2013). Crystal structure of TET2-DNA complex: Insight into TET-mediated 5mC oxidation. *Cell, 155*(7), 1545–1555. https://doi.org/10.1016/j.cell.2013.11.020.

Hu, L., Lu, J., Cheng, J., Rao, Q., Li, Z., Hou, H., ... Xu, Y. (2015). Structural insight into substrate preference for TET-mediated oxidation. *Nature, 527*(7576), 118–122. https://doi.org/10.1038/nature15713.

Ito, S. S. L., Dai, Q., Wu, S. C., Collins, L. B., Swenberg, J. A., He, C., & Zhang, Y. (2011). Tet proteins can convert 5-methylcytosine to 5-formylcytosine and 5-carboxylcytosine. *Science (New York, N. Y.), 333*(6047), 1300–1303. https://doi.org/10.1126/science.1210597.

Jaenisch, R., & Bird, A. (2003). Epigenetic regulation of gene expression: How the genome integrates intrinsic and environmental signals. *Nature Genetics, 33*(Suppl), 245–254. https://doi.org/10.1038/ng1089.

Joshi, K., Liu, S., Breslin, S. J. P., & Zhang, J. (2022). Mechanisms that regulate the activities of TET proteins. *Cellular and Molecular Life Sciences: CMLS, 79*(7), 363. https://doi.org/10.1007/s00018-022-04396-x.

Jurkowska, R. Z., Jurkowski, T. P., & Jeltsch, A. (2011). Structure and function of mammalian DNA methyltransferases. *Chembiochem: A European Journal of Chemical Biology, 12*(2), 206–222. https://doi.org/10.1002/cbic.201000195.

Kohli, R. M., & Zhang, Y. (2013). TET enzymes, TDG and the dynamics of DNA demethylation. *Nature, 502*(7472), 472–479. https://doi.org/10.1038/nature12750.

Ko, M., Huang, Y., Jankowska, A. M., Pape, U. J., Tahiliani, M., Bandukwala, H. S., ... Rao, A. (2010). Impaired hydroxylation of 5-methylcytosine in myeloid cancers with mutant TET2. *Nature, 468*(7325), 839–843. https://doi.org/10.1038/nature09586.

Liu, M. Y., DeNizio, J. E., & Kohli, R. M. (2016). Quantification of oxidized 5-methylcytosine bases and TET enzyme activity. *Methods in Enzymology, 573*, 365–385. https://doi.org/10.1016/bs.mie.2015.12.006.

Liu, M. H., Wang, C. R., Liu, W. J., Xu, Q., & Zhang, C. Y. (2022). Development of a single quantum dot-mediated FRET biosensor for amplification-free detection of ten-eleven translocation 2. *Talanta, 239*, 123135. https://doi.org/10.1016/j.talanta.2021.123135.

Liu, W. J., Zhang, X., Hu, J., & Zhang, C. Y. (2022). A label-free and self-circulated fluorescent biosensor for sensitive detection of ten-eleven translocation 1 in cancer cells. *Chemical Communications (Camb), 58*(57), 7996–7999. https://doi.org/10.1039/d2cc03019e.

Loenarz, C., & Schofield, C. J. (2008). Expanding chemical biology of 2-oxoglutarate oxygenases. *Nature Chemical Biology, 4*(3), 152–156. https://doi.org/10.1038/nchembio0308-152.

Losman, J. A., & Kaelin, W. G., Jr. (2013). What a difference a hydroxyl makes: Mutant IDH, (R)-2-hydroxyglutarate, and cancer. *Genes & Development, 27*(8), 836–852. https://doi.org/10.1101/gad.217406.113.

Lu, X., Song, C. X., Szulwach, K., Wang, Z., Weidenbacher, P., Jin, P., & He, C. (2013). Chemical modification-assisted bisulfite sequencing (CAB-Seq) for 5-carboxylcytosine detection in DNA. *Journal of the American Chemical Society, 135*(25), 9315–9317. https://doi.org/10.1021/ja4044856.

Lu, X., Zhao, B. S., & He, C. (2015). TET family proteins: Oxidation activity, interacting molecules, and functions in diseases. *Chemical Reviews, 115*(6), 2225–2239. https://doi.org/10.1021/cr500470n.

Marholz, L. J., Wang, W., Zheng, Y., & Wang, X. (2016). A fluorescence polarization biophysical assay for the naegleria DNA hydroxylase Tet1. *ACS Medicinal Chemistry Letters, 7*(2), 167–171. https://doi.org/10.1021/acsmedchemlett.5b00366.

Ma, T., Zhang, Q., Zhang, S., Yue, D., Wang, F., Ren, Y., ... Yu, F. (2024). Research progress of human key DNA and RNA methylation-related enzymes assay. *Talanta, 273*, 125872. https://doi.org/10.1016/j.talanta.2024.125872.

Sappa, S., Dey, D., Sudhamalla, B., & Islam, K. (2021). Catalytic space engineering as a strategy to activate C-H oxidation on 5-methylcytosine in mammalian genome. *Journal of the American Chemical Society, 143*(31), 11891–11896. https://doi.org/10.1021/jacs.1c03815.

Shen, L., & Zhang, Y. (2012). Enzymatic analysis of Tet proteins: Key enzymes in the metabolism of DNA methylation. *Methods in Enzymology, 512*, 93–105. https://doi.org/10.1016/b978-0-12-391940-3.00005-6.

Song, C. X., & He, C. (2013). Potential functional roles of DNA demethylation intermediates. *Trends in Biochemical Sciences, 38*(10), 480–484. https://doi.org/10.1016/j.tibs.2013.07.003.

Song, C. X., Yi, C., & He, C. (2012). Mapping recently identified nucleotide variants in the genome and transcriptome. *Nature Biotechnology, 30*(11), 1107–1116. https://doi.org/10.1038/nbt.2398.

Spruijt, C. G., Gnerlich, F., Smits, A. H., Pfaffeneder, T., Jansen, P. W., Bauer, C., ... Vermeulen, M. (2013). Dynamic readers for 5-(hydroxy)methylcytosine and its oxidized derivatives. *Cell, 152*(5), 1146–1159. https://doi.org/10.1016/j.cell.2013.02.004.

Sudhamalla, B., Dey, D., Breski, M., & Islam, K. (2017). A rapid mass spectrometric method for the measurement of catalytic activity of ten-eleven translocation enzymes. *Analytical Biochemistry, 534*, 28–35. https://doi.org/10.1016/j.ab.2017.06.011.

Tahiliani, M. K. P. K., Shen, Y., Pastor, W. A., Bandukwala, H., Brudno, Y., Agarwal, S., ... Rao, A. (2009). Conversion of 5-methylcytosine to 5-hydroxymethylcytosine in mammalian DNA by MLL partner TET1. *Science (New York, N. Y.), 324*(5929), 930–935. https://doi.org/10.1126/science.1170116.

Treadway, C. J., Boyer, J. A., Yang, S., Yang, H., Liu, M., Li, Z., ... Brown, N. G. (2024). Using NMR to monitor TET-dependent methylcytosine dioxygenase activity and regulation. *ACS Chemical Biology, 19*(1), 15–21. https://doi.org/10.1021/acschembio.3c00619.

CHAPTER SIX

Non-standard amino acid incorporation into thiol dioxygenases

Zachary D. Bennett and Thomas C. Brunold*
Department of Chemistry, University of Wisconsin-Madison, Madison, WI, United States
*Corresponding author. e-mail address: brunold@chem.wisc.edu

Contents

1. Overview	122
2. Eukaryotic thiol dioxygenases	123
2.1 CDO	123
2.2 ADO	126
3. Genetic code expansion	128
3.1 Overview	128
3.2 Suppressor tRNA/aminoacyl-tRNA synthetase pairs and the pEVOL plasmid	129
3.3 Selenocysteine incorporation	131
4. Application to thiol dioxygenases	134
4.1 Fluorotyrosine incorporation into CDO and ADO using pEVOL F2Y	134
4.2 Sec incorporation into ADO	136
5. Conclusions	140
Acknowledgments	140
References	140
Further reading	145

Abstract

Thiol dioxygenases (TDOs) are non-heme Fe(II)-dependent enzymes that catalyze the O_2-dependent oxidation of thiol substrates to their corresponding sulfinic acids. Six classes of TDOs have thus far been identified and two, cysteine dioxygenase (CDO) and cysteamine dioxygenase (ADO), are found in eukaryotes. All TDOs belong to the cupin superfamily of enzymes, which share a common β-barrel fold and two cupin motifs: $G(X)_5HXH(X)_{3-6}E(X)_6G$ and $G(X)_{5-7}PXG(X)_2H(X)_3N$. Crystal structures of TDOs revealed that these enzymes contain a relatively rare, neutral 3-His iron-binding facial triad. Despite this shared metal-binding site, TDOs vary greatly in their secondary coordination spheres. Site-directed mutagenesis has been used extensively to explore the impact of changes in secondary sphere residues on substrate specificity and enzymatic efficiency. This chapter summarizes site-directed mutagenesis studies of eukaryotic TDOs, focusing on the tools and practicality of non-standard amino acid incorporation.

Methods in Enzymology, Volume 703
ISSN 0076-6879, https://doi.org/10.1016/bs.mie.2024.05.022
Copyright © 2024 Elsevier Inc. All rights are reserved, including those for text and data mining, AI training, and similar technologies.

1. Overview

Thiol dioxygenases (TDOs) are a family of non-heme Fe (II)-dependent enzymes that catalyze the conversion of thiol substrates to their corresponding sulfinic acids utilizing molecular oxygen. This reactivity is unusual among mononuclear non-heme iron dioxygenases, which typically perform C-C bond cleavage reactions where the two atoms of molecular oxygen are added to different substrate atoms. TDOs are utilized primarily for the catabolism of thiol-containing molecules. Presently, six classes of TDOs have been identified: cysteine dioxygenase (CDO), mercaptopropionate dioxygenase (MDO), cysteamine dioxygenase (ADO), plant cysteine oxidase (PCO), mercaptosuccinate dioxygenase (MSDO), and the recently discovered ergothionine dioxygenase (ETDO) (Brandt, Galant, Meinert-Berning, & Steinbüchel, 2019; Fernandez, Juntunen, & Brunold, 2022; Nalivaiko, Vasseur, & Seebeck, 2024; Stipanuk, Simmons, Karplus, & Dominy, 2011; Tchesnokov et al., 2015). Of these enzymes, CDO and ADO are the only TDOs found in eukaryotes.

TDOs belong to the cupin superfamily of enzymes. These enzymes display a low overall sequence identity but share a common β-barrel fold and two cupin sequence motifs: $G(X)_5HXH(X)_{3-6}E(X)_6G$ (cupin motif 1) and $G(X)_{5-7}PXG(X)_2H(X)_3N$ (cupin motif 2) (Stipanuk et al., 2011). All TDOs except for MSDO have been structurally characterized (Fig. 1) and feature a rare neutral 3-His iron-binding facial triad, archetypal of the TDO family and only found in four other non-heme iron enzymes outside this family, including SznF, the sulfoxide synthase EgtB, diketone-cleaving dioxygenase, and gentisate 1,2-dioxygenase (Adams, Singh, Keller, & Jia, 2006; Diebold, Neidig, Moran, Straganz, & Solomon, 2010; Goncharenko, Vit, Blankenfeldt, & Seebeck, 2015; Ng, Rohac, Mitchell, Boal, & Balskus, 2019). Despite this shared metal-binding site, TDOs vary greatly in their secondary coordination spheres. This variation has prompted extensive studies involving site-directed mutagenesis to explore the impact of changes in secondary sphere residues on substrate specificity and enzymatic efficiency.

This chapter reviews site-directed mutagenesis studies of eukaryotic TDOs, focusing on the tools and utility of non-standard amino acid incorporation. It also discusses the practicalities of genetic code expansion and summarizes important considerations for utilizing these systems.

2. Eukaryotic thiol dioxygenases
2.1 CDO

CDO was discovered in 1966 as the enzyme that performs the first step of cysteine (Cys) catabolism through the oxidation of Cys to cysteine sulfinic acid utilizing molecular oxygen (Ewetz & Sörbo, 1966). Cys acts as an important canonical amino acid in proteins, the only such amino acid to contain a thiol moiety. This property imparts Cys with unique reactivity and function, such as the ability to form disulfide bonds or to serve as the primary coordinating ligand in iron-sulfur clusters (Rouault, 2015). While Cys is readily found in proteins, there are many essential small molecules that are produced from the transsulfuration pathway and the catabolism of Cys, such as taurine, glutathione, and coenzyme A (Stipanuk & Ueki, 2011). The intracellular Cys concentration must be tightly regulated due to the free thiol toxicity, and high levels of Cys have been implicated in conditions such as Alzheimer's and Parkinson's, as well as vascular diseases (El-Khairy, Ueland, Refsum, Graham, & Vollset, 2001; Heafield et al., 1990).

CDO was the first enzyme in the TDO family to be structurally characterized (McCoy et al., 2006; Simmons et al., 2006). Crystal structures of CDO revealed a unique 3-His iron-binding facial triad, representative of the entire TDO family (Fig. 1). This iron-binding site contrasts with the 2-His-1-carboxylate motif commonly found in other cupin-fold proteins (Stipanuk et al., 2011). The crystal structure of CDO also showed a rare cysteine-tyrosine (Cys-Tyr) crosslink in the secondary coordination sphere of the iron center between C93 and Y157 (*Rattus norvegicus* CDO, *Rn*CDO, numbering). This crosslink is conserved across mammalian CDOs, and its formation, which occurs with repeated substrate turnover, enhances the enzyme's catalytic efficiency at least tenfold (Dominy et al., 2008). The increased catalytic efficiency of crosslinked CDO has been attributed to tighter binding of substrate Cys and suppressing water coordination to the Fe (II) center of CDO (Blaesi, Fox, & Brunold, 2015; Dominy et al., 2008; Fischer et al., 2019). A similar Cys-Tyr crosslink has been identified in a few other enzymes, such as galactose oxidase (GAO), where the crosslinked tyrosine coordinates directly to the active site copper center (Ito et al., 1991). In GAO this crosslink functions differently from that of CDO, with the thioether aiding in the oxidation of the copper-bound tyrosinate to form a tyrosyl radical and enabling catalysis (Cowley et al., 2016).

In CDO, Y157 aids in substrate Cys positioning through hydrogen bonding to the carboxylate moiety of the substrate. The orientation of the

124   Zachary D. Bennett and Thomas C. Brunold

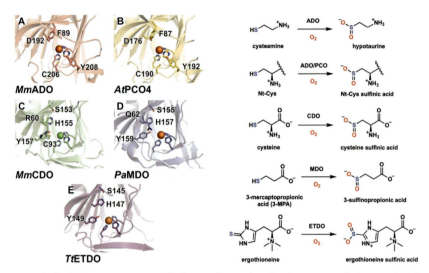

Fig. 1 (Left) Active site regions of all crystallographically characterized classes of TDOs with the metal-binding 3-His triads and important secondary sphere residues shown as sticks. (A) *Mus musculus* cysteamine dioxygenase (PDB: 7LVZ) (Fernandez, Elmendorf, et al., 2021). (B) *Arabidopsis thaliana* plant cysteine oxidase (PDB: 6S7E) (White et al., 2020). (C) *M. musculus* cysteine dioxygenase (PDB: 2ATF) (McCoy et al., 2006). (D) *Pseudomonas aeruginosa* MDO (PDB: 4TLF) (Tchesnokov et al., 2015). (E) *Thermocatellispora tengchongensis* ETDO (PDB: 8QFL) (Nalivaiko et al., 2024). (Right) Oxidation of thiol-containing substrates catalyzed by these TDOs.

hydroxyl group of Y157 is finely tuned via an active-site S153–H155–Y157 hydrogen bonding network (Fig. 1). The importance of Y157 was demonstrated by kinetic studies of the Y157F variant, in which residue 157 was replaced by a phenylalanine that is unable to engage in a hydrogen bond with the Cys carboxylate moiety and to form a crosslink with C93. This variant was found to be inactive or have drastically reduced catalytic efficiency (Driggers et al., 2016; Joseph & Maroney, 2007; Li, Blaesi, Pecore, Crowell, & Pierce, 2013).

Unlike the catalytically relevant high-spin ($S = 2$) Fe(II)-bound form of CDO, oxidized Fe(III)CDO, with a non-integer spin ground state ($S = 5/2$ or $S = 1/2$ depending on the nature of bound substrate [analogues]), can be readily studied with perpendicular-mode electron paramagnetic resonance (EPR) spectroscopy. Cyanide, which serves as a superoxide surrogate, provides a particularly sensitive probe of secondary sphere interactions, since perturbations in the Fe-C-N bond angle affect the extent of π-backbonding and, thus, the EPR g-values. Because as-isolated CDO exists as a mixture of

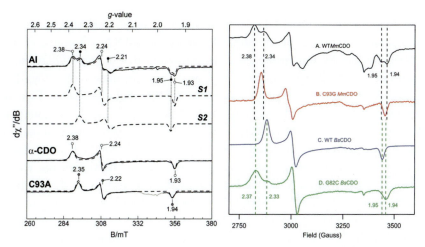

Fig. 2 X-band EPR spectra of low-spin (S = 1/2) Cys/cyanide-bound Fe(III)CDO species. (Left) Spectra of as-isolated (AI) and fully crosslinked (α-CDO) WT CDO and C93A CDO. (Right) Spectra of (A) WT *Mm*CDO, (B) C93G *Mm*CDO, (C) WT *Bs*CDO, and (D) G82C *Bs*CDO demonstrating crosslink formation in the G82C *Bs*CDO variant. Dashed black vertical lines indicate the g₁ and g₃ values for the crosslinked and non-crosslinked fractions of WT *Mm*CDO, and vertical dashed green lines indicate g₁ and g₃ values for crosslinked and non-crosslinked fractions of G82C *Bs*CDO. *The EPR spectral simulations for the crosslinked and non-crosslinked fractions (**S1** and **S2**, respectively) are overlaid on the experimental AI spectrum.* Reproduced with permission from Li, W., Blaesi, E. J., Pecore, M. D., Crowell, J. K., & Pierce, B. S. (2013). Second-sphere interactions between the C93–Y157 cross-link and the substrate-bound Fe site influence the O₂ coupling efficiency in mouse cysteine dioxygenase. Biochemistry, 52(51), 9104–9119. https://doi.org/10.1021/bi4010232; Reproduced with permission from Schultz, R. L., Sabat, G., Fox, B. G., & Brunold, T. C. (2023). A single DNA point mutation leads to the formation of a cysteine-tyrosine crosslink in the cysteine dioxygenase from Bacillus subtilis. Biochemistry, 62(12), 1964–1975. https://doi.org/10.1021/acs.biochem.3c00083.

crosslinked and non-crosslinked isoforms, the EPR spectra of Cys- and cyanide-bound wild-type (WT) Fe(III)CDO exhibit two distinct low-spin (S = 1/2) signals with different g-values (Fig. 2) (Li, Blaesi et al., 2013). Alternatively, in EPR spectra of the Cys/cyanide-Fe(III) adduct of the C93A CDO variant, which is unable to form a crosslink between residue 93 and Y157, a single S = 1/2 signal is observed with g-values identical to those displayed by the non-crosslinked fraction of WT CDO.

Interestingly, while CDOs are found in prokaryotes and eukaryotes, the Cys-Tyr crosslink is only conserved in eukaryotic CDOs (Dominy, Simmons, Karplus, Gehring, & Stipanuk, 2006). In bacterial CDOs, the residue corresponding to C93 of *Rn*CDO is replaced by a conserved glycine (e.g., G82 in *Bacillus subtilis* CDO, *Bs*CDO). Despite the absence of the crosslink in bacterial CDOs, these enzymes show catalytic efficiencies comparable to those of crosslinked eukaryotic CDOs (Dominy et al., 2006). Studies by our group have demonstrated that a single G82C substitution in *Bs*CDO is sufficient to lead to the formation of a Cys-Tyr crosslink analogous to that of eukaryotic CDOs (Schultz, Sabat, Fox, & Brunold, 2023). The catalytic efficiency of the G82C *Bs*CDO variant was found to correlate with the degree of crosslink formation, as observed for eukaryotic CDOs. Collectively, the results obtained for WT and variant CDOs from different species have indicated that to maximize the enzymatic efficiency of CDO, (1) a Tyr residue must be present at the position corresponding to Y157 of *Rn*CDO to help bind and properly orient the substrate Cys and (2) the Cys residue at position 93 must either be crosslinked with Y157 or substituted by a residue that cannot stabilize a water molecule at the metal center of Cys-bound Fe(II)CDO via hydrogen bonding (Miller, Schnorrenberg, Aschenbrener, Fox, & Brunold, 2024).

2.2 ADO

Compared to CDO, ADO remained poorly characterized since its discovery in 1967 (Wood & Cavallini, 1967). In 2007, ADO attracted new interest when a difference between taurine concentrations and levels of *cdo* gene expression in certain tissues was identified (Dominy et al., 2007). Initial work showed that ADO catalyzed the oxidation of cysteamine (2-aminoethanethiol, 2-AET) to hypotaurine, and that this enzyme was the primary producer of hypotaurine and, consequently, taurine in the brain (Dominy et al., 2007). However, more recent studies by the Flashman and Ratcliffe groups demonstrated that ADO also catalyzes the oxidation of N-terminal thiol residues in peptides, analogous to PCO (Masson et al., 2019). PCO initiates protein degradation of the group VII ethylene response factors through the N-degron pathway via N-terminal thiol oxidation of these proteins under low oxygen conditions (hypoxia) (Gunawardana, Heathcote, & Flashman, 2022; White, Kamps, East, Taylor Kearney, & Flashman, 2018). This post-translational modification precedes arginylation and polyubiquitination of the peptide substrates, promoting their proteasomal degradation (Shim et al., 2023). Under hypoxic conditions, group VII ethylene response factors accumulate in the cell, where they function as transcription factors that promote hypoxia-related gene expression. The Flashman and Ratcliffe groups found that ADO can functionally replace native PCO in *Arabidopsis thaliana* (Masson et al., 2019). These data raise the intriguing possibility that ADO may also function as an oxygen sensor in mammals.

ADO's putative N-terminal thiol peptide substrates in eukaryotes are distinct from those of the conserved mammalian oxygen sensors, prolyl hydroxylases. ADO reacts with regulator of G-protein signaling (RGS) proteins 4 and 5, as well as interleukin 32 (Masson et al., 2019). The RGS4 and RGS5 proteins negatively regulate G-protein coupled receptors that are related to cardiovascular development/angiogenesis and hypoxia-induced apoptosis (Vu, Mitchell, Gygi, & Varshavsky, 2020). Alternatively, interleukin 32 is a pro-inflammatory cytokine that is implicated in cancer and pathogenesis of inflammatory disorders, such as inflammatory bowel disease and chronic obstructive pulmonary disease (Shioya et al., 2007; Vu et al., 2020). The nature of these substrates suggests that ADO may function as an oxygen sensor that can modulate cellular signal transduction in response to hypoxia.

Structural data of ADO remained elusive until 2021, when crystal structures of the human and mouse proteins were solved by Liu and coworkers and by our group, respectively (Fernandez, Elmendorf, et al., 2021; Wang, Shin, Li, & Liu, 2021). These structures confirmed the presence of a 3-His facial triad coordinating the Fe center as originally proposed on the basis of sequence alignment and spectroscopic analyses (Dominy et al., 2007; Fernandez, Dillon, Stipanuk, Fox, & Brunold, 2020; Wang et al., 2018) and revealed a large-open active-site conducive of peptide binding to the iron center. Importantly, in the *Mus musculus* ADO (*Mm*ADO) crystal structure, a histidine residue from one ADO monomer protrudes into the active site of another monomer, supporting the hypothesis that this enzyme can oxidize large peptide substrates with N-terminal Cys residues.

Notably, ADO lacks the Cys-Tyr crosslinking residues that are conserved in eukaryotic CDOs. However, research performed by the Liu group provided evidence that a different Cys-Tyr crosslink can be formed in ADO outside of the active site, between C206 and Y208 (*Mm*ADO numbering) (Wang et al., 2018). Although based on mass spectrometry analysis only a small fraction of as-isolated WT ADO appears to be crosslinked, the catalytic activity for 2-AET oxidation (determined via O_2 uptake studies) was found to be reduced by a factor of ~4 in the Y208A ADO variant (Wang et al., 2018).

To assess whether a crosslink has any noticeable effect on the interaction between substrate 2-AET and the Fe(III)ADO active site, we prepared the Y208F *Mm*ADO variant, which is unable to form the Cys-Tyr crosslink (Fernandez et al., 2022). Interestingly, the EPR spectra obtained for the 2-AET/cyanide adducts of WT and Y208F Fe(III)ADO were found to be superimposable, suggesting that either formation of the C206-Y208 crosslink in WT *Mm*ADO has negligible effects on the geometric

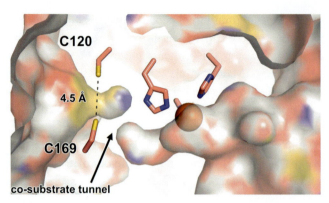

Fig. 3 Active site region of *Mm*ADO showing the main substrate access channel (right) and putative oxygen co-substrate tunnel through the back of the protein (left) that is lined by C120 and C169 in close proximity (PDB: 7LVZ) (Fernandez, Elmendorf, et al., 2021). The Fe atom is shown as an orange sphere and relevant amino acid residues as sticks.

and electronic properties of the ADO active site or only a minor fraction of as-isolated enzyme contains the crosslink (Fernandez, Juntunen, Fox, & Brunold, 2021). Thus, the relevance and exact role of the Cys-Tyr crosslink in ADO remain in question.

While the active sites of ADO and PCO are more open than those of the other TDOs, binding of a large peptide substrate may preclude O_2 access to the Fe(II) center. Inspection of the crystal structure of *Mm*ADO revealed the presence of a secondary, narrower tunnel from the protein surface to the active site (Fig. 3) (Fernandez, Elmendorf, et al., 2021). We hypothesized that this tunnel may act as a co-substrate tunnel for O_2. This secondary tunnel is flanked by two cysteine residues, C120 and C169. With an S⋯S distance of ~4.5 Å, these residues could form a redox active "gate" (Fig. 3). Disulfide formation under conditions of high O_2 levels may serve as a gating mechanism to prevent ADO from depleting organisms of Nt-Cys-containing molecules. Notably, recent work by the Bhagi-Damodaran group has demonstrated how oxygen tunnel bottlenecks have a large effect on catalysis within the oxygen-sensing prolyl hydroxylases (Windsor et al., 2023).

3. Genetic code expansion

3.1 Overview

Protein biosynthesis relies on 22 genetically encoded canonical amino acids. Among these, 20 are in the standard genetic code and an additional two, selenocysteine (Sec) and pyrrolysine, can be incorporated by special translation mechanisms (Gonzalez-Flores, Shetty, Dubey, & Copeland, 2013; Krzycki, 2005). The 20 standard amino acids provide an extensive foundation for the wide diversity of proteins found in living organisms. While useful, this set of amino acids constrains protein engineering to a limited chemical space. For this reason, the incorporation of natural and synthetic non-standard amino acids (nsAAs) into proteins has become the subject of intense research. nsAAs have many desirable applications, ranging from helping create proteins with inaccessible post-translational modifications to designing artificial metalloenzymes and tuning enzymes for biocatalysis (de la Torre & Chin, 2021; Drienovská & Roelfs, 2020; Li, Shi et al., 2013; Liu & Schultz, 2010). While chemical methods have been devised to produce proteins containing nsAAs, these methods can be limited in terms of their site selectivity, yields, and residue accessibility.

To take advantage of the chemical diversity of nsAAs, orthogonal cellular machinery has been engineered to recognize additional signals for their insertion (de la Torre & Chin, 2021). This engineering has been accomplished in many eukaryotic and bacterial cell lines (Liu & Schultz, 2010). In *E. coli,* the recoding of the amber stop codon (UAG) to the sense nsAA codon has become common practice. The amber stop codon is the least abundant stop codon found within *E. coli* and, when recoded as an amino acid insertion sequence, has the least deleterious effects on the host organism (Lajoie et al., 2013).

3.2 Suppressor tRNA/aminoacyl-tRNA synthetase pairs and the pEVOL plasmid

The most common strategy employed to incorporate nsAAs into proteins involves utilizing amber codon suppressor tRNA and an aminoacyl-tRNA synthetase (aaRS) pairs that are orthogonal to the host's protein translation mechanisms (Fig. 4) (de la Torre & Chin, 2021). These suppressor tRNAs need aaRS recognition elements that differ from the host's aaRS recognition elements to prevent misloading of canonical amino acids. Additionally, the orthogonal aaRS must be selective for the amino acid and suppressor tRNA to avoid misloading of the nsAA into native proteins. These methods have facilitated genetic code expansion and thus enabled the production of proteins with new properties and functions.

Many laboratories have identified or evolved orthogonal tRNA/aaRS pairs that are recognized by the host ribosome for nsAA incorporation. To date, more than 200 distinct nsAAs have been incorporated into recombinant

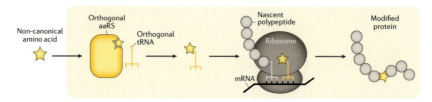

Fig. 4 Overview of genetic code expansion strategy utilizing an orthogonal suppressor tRNA/aminoacyl-tRNA synthetase (aaRS) system. The non-canonical amino acid is either added to the cell culture or biosynthesized by exogenous genes in the organism. The host cell expresses an orthogonal aaRS (yellow box) that aminoacylates the orthogonal tRNA with the non-canonical amino acid. During translation, the orthogonal aminoacyl-tRNA binds to the ribosome in response to an amber codon in the mRNA sequence introduced in the sequence of the protein of interest. Native protein biosynthetic machinery is then used to create a modified protein containing the non-canonical amino acid of interest. *Reproduced with permission from de la Torre, D., & Chin, J. W. (2021). Reprogramming the genetic code. Nature Reviews. Genetics, 22(3), 169–184. https://doi.org/10.1038/s41576-020-00307-7.*

proteins, affording protein engineers an extensive library of nsAA tools at their disposal (Drienovská & Roelfes, 2020). While orthogonal tRNA/aaRS pairs have been identified from a variety of organisms, those from *Methanocaldococcus jannaschii*, *Methanosarcina mazei*, and *Methanosarcina barkeri* have been extensively evolved and derivatized to incorporate a wide variety of nsAAs within *E. coli* (de la Torre & Chin, 2021; Liu & Schultz, 2010). These pairs can be used with the pEVOL plasmid for highly efficient nsAA incorporation.

The pEVOL plasmid was developed by the Schultz laboratory as a flexible backbone optimized for high yields of nsAA incorporated proteins (Young, Ahmad, Yin, & Schultz, 2010). The pEVOL backbone notably contains two copies of the aaRS gene, one constitutively expressed and another under arabinose inducible control. This design affords high levels of nsAA-tRNA production prior to, and throughout, overexpression. Utilizing the pEVOL accessory plasmid may present plasmid compatibility problems with some systems. The pEVOL system confers chloramphenicol resistance and contains the p15A origin of replication. This commonly used medium-copy origin is compatible with typical overexpression plasmids such as pET, but is notably incompatible with the pLys, pLacI, and pRARE accessory plasmids, as they contain the p15A origin and the same antibiotic resistance markers (chloramphenicol). For these latter accessory plasmids, additional steps must be taken into consideration for optimal protein overexpression, such as replacement of the origin and antibiotic resistance cassettes.

The yield of nsAA incorporated protein of interest (POI) will be lower than that of the corresponding WT protein due to a variety of factors, such as host fitness, mistranslation, and efficiency of nsAA incorporation (Chung, Miller, Söll, & Krahn, 2021; de la Torre & Chin, 2021). Host proteins containing the suppressed codon may not be terminated properly during translation, acquiring the encoded nsAA instead. Mistranslated proteins may be detrimental to the host by performing new functions, impose additional burden to host fitness, and lower the pool of available nsAA-tRNA (Lajoie et al., 2013). During protein synthesis, release factors that bind to stop codons within the mRNA transcript may halt elongation by the ribosome, potentially lowering yield and truncating the POI (Mukai et al., 2015). This becomes exacerbated when incorporating multiple nsAAs. Typical expression cell lines such as *E. coli* BL21 may be adequate for single nsAA substitutions depending on incorporation efficiency, but higher numbers of nsAA substitutions become increasingly difficult.

To avoid copurification of truncated proteins, it is recommended to use C-terminal affinity tags (Chung et al., 2021). This design affords the opportunity to purify only fully translated POI, rather than truncated proteins. Additionally, recoded organisms can aid in the production of full-length protein. In recoded *E. coli* strains such as B.95ΔA and C321.ΔA, many (or all) TAG stop codons are replaced with synonymous codons (Lajoie et al., 2013; Mukai et al., 2015). These prevent the mistranslation of essential proteins, aiding in host fitness under amber codon suppression. Additionally, eliminating the need for the TAG codon allows for the knockout of the *prfA* gene, which encodes for the release factor 1 (RF-1). Notably, RF-1 terminates protein translation at the UAG and UGA stop codons. When knocked out in recoded strains, this prevents premature truncation of the POI during translation and may increase the yield of POIs containing multiple nsAA substitutions.

3.3 Selenocysteine incorporation

Sec is a Cys analogue that contains a selenium atom in lieu of sulfur. Discovered in 1951 as the 21st genetically encoded amino acid, this nsAA has since been identified within 25 human proteins that exhibit a wide variety of functions. These selenoproteins play important roles in redox homeostasis or signaling, as well as thyroid hormone metabolism (Gonzalez-Flores et al., 2013; Schmidt & Simonović, 2012). The thiol to selenol substitution, while seemingly a minor change, imparts Sec with many desirable biophysical and chemical properties. Compared to Cys, Sec has a lower pK_a (5.2 vs. 8.3) and redox potential (-381 vs. $-180\,mV$ at pH

7 relative to DTT, Besse, Siedler, Diercks, Kessler, & Moroder, 1997), and displays increased nucleophilicity (Chung & Krahn, 2022; Johansson, Gafvelin, & Arnér, 2005; Metanis & Hilvert, 2014). These properties make Sec extremely desirable for protein engineering, as it can be used to tune redox potentials, form durable diselenide bonds, and may generally enhance reactivity as a potent nucleophile.

While Sec is genetically encoded natively in all three domains of life, the translational machinery used for its incorporation is complex, and not easily coopted for recombinant protein translation. Due to the high reactivity of Sec, Nature has evolved a unique system to avoid cellular damage (Hendrickson, Wood, & Rathnayake, 2021). Unlike other amino acids, Sec is not bioavailable. Cells instead store compounds containing selenium in higher oxidation states, such as selenate, which are reduced and used to synthesize Sec directly on its cognate tRNA as needed (Kang et al., 2020). In this process, the tRNA for Sec (tRNA^{Sec}) is first serylated using the host's native serine RS, producing Ser-tRNA^{Sec}. In bacteria, Ser-tRNA^{Sec} is then directly converted to Sec-tRNA^{Sec} by selenocysteine synthase (SelA), utilizing selenophosphate produced by selenophosphate synthase (SelD) (Fig. 5A). Alternatively, in eukaryotes and archaea, Ser-tRNA^{Sec} is phosphorylated to O-phosphoseryl-tRNA^{Sec} (Sep-tRNA^{Sec}), which is then converted to Sec-tRNA^{Sec} by an additional synthase (Fig. 5B) (Chung & Krahn, 2022; Gonzalez-Flores et al., 2013). Furthermore, Sec incorporation is dependent on a Sec-specific elongation factor (SelB) to mediate Sec delivery to the polypeptide chain. SelB requires an additional noncoding mRNA Sec insertion sequence (SECIS) element encoded directly after the UGA stop codon (bacteria) or in the 3' mRNA untranslated region (eukaryotes and archaea) for Sec incorporation (Berry, Banu, Harney, & Larsen, 1993; Gonzalez-Flores et al., 2013; Schmidt & Simonović, 2012). These prerequisites each contribute to the difficulty of utilizing native Sec incorporation for recombinant systems.

Though complex, protocols have been successfully developed to incorporate Sec into recombinant proteins (Han et al., 2013). These protocols are limited though by their complexity and low POI yields, hindering spectroscopic and X-ray crystallographic studies that typically require relatively large quantities of protein. While solid phase peptide synthesis and native chemical ligation methods have been utilized for Sec incorporation, these are primarily limited by peptide size and yield (Li et al., 2018; Metanis & Hilvert, 2014). To bypass the native Sec incorporation machinery, a variety of genetic code expansion techniques have been developed that rely on orthogonal mechanisms of Sec incorporation for recombinant protein expression.

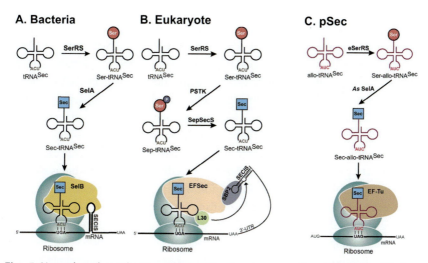

Fig. 5 Natural and synthetic pathways for Sec incorporation into POIs. All three pathways initiate Sec-tRNA^{Sec} production using the host's seryl-tRNA synthetase (SerRS) to produce Ser-tRNA^{Sec}. (A) The bacterial pathway utilizes Sec synthase (SelA) to produce Sec-tRNA^{Sec}. The Sec elongation factor (SelB) then binds to the non-coding Sec insertion sequence (SECIS) in order to initiate Sec insertion into the polypeptide chain. (B) The eukaryotic pathway produces Sec-tRNA^{Sec} in two steps. First, Ser-tRNA^{Sec} is converted to O-phosphoseryl(Sep)-tRNA^{Sec} by Sep-tRNA kinase (PstK). Following this step, selenium is loaded by the Sep-tRNA:Sec-tRNA synthase (SepSecS) to generate Sec-tRNA^{Sec}. The SECIS sequence must be present in the 3′-untranslated region of the mRNA transcript for Sec insertion by EFSec. (C) Engineered Sec biosynthesis and insertion via the pSecUAG-EVOL2 (pSec) plasmid. Allo-tRNA^{Sec} is first loaded by an endogenous SerRS (eSerRS) and then converted to Sec-tRNA^{Sec} by *Aeromonas salmonicida* SelA. Sec-allo-tRNA^{Sec} can bypass SelB-mediated insertion and be incorporated by elongation factor Tu (EF-Tu). *Figure adapted from Chung, C. Z., & Krahn, N. (2022). The selenocysteine toolbox: A guide to studying the 21st amino acid. Archives of Biochemistry and Biophysics, 730(August), 109421. https://doi.org/10.1016/j.abb.2022.109421.*

Two systems have been developed for use in *E. coli* to enable site-specific UAG-dependent Sec incorporation. These systems, devised by the Ellington and Söll laboratories, both utilize the native *E. coli* elongation factor Tu (EF-Tu) to participate in standard protein translation (Mukai, Sevostyanova, Suzuki, Fu, & Söll, 2018; Thyer, Robotham, Brodbelt, & Ellington, 2015). As such, they circumvent the need for SelB-mediated elongation, thus removing the requirement for the SECIS mRNA element. By using orthogonal amber suppressor tRNA, both systems aim to improve the ease and yield of recombinant Sec incorporation.

The Ellington laboratory developed a two-plasmid system optimized for efficient Sec incorporation by creating a biosynthetic Sec-tRNA pathway like that of eukaryotes and archaea (Thyer et al., 2015, 2018). This system utilizes a hybrid evolved tRNA (denoted tRNA^{SecUX}), Sec synthase (SelA), selenophosphate synthase (SelD), and Sep-tRNA^{Sec} kinase (PstK). tRNA^{SecUX} is an evolved tRNA that contains non-standard RNA sequences to efficiently bind EF-Tu through different interactions than canonical tRNAs (Thyer et al., 2015). For high incorporation of Sec, rather than Ser, into POIs, the system encodes for two differentially expressed *selA* genes, one inducible copy and another one that is constitutively expressed, like pEVOL. By producing a Sep-tRNA intermediate, this system helps to avoid Ser misincorporation by EF-Tu. Additionally, Sep-tRNA functions as a more efficient substrate for SelA (Carlson et al., 2004).

The Söll pSecUAG-EVOL2 (pSec) plasmid utilizes elements from *Aeromonas salmonicida* to produce Sec-tRNA^{Sec} (Fig. 5C). This system exploits allo-tRNA^{Sec} to bind to EF-Tu as well as SelA elements (Mukai et al., 2018). Notably this system differs from that of the Ellington laboratory in that it produces Sec-allo-tRNA^{Sec} directly from Ser-tRNA^{Sec}, analogously to bacterial systems. To enable high levels Sec-tRNA^{Sec} production, the gene for the *Treponema denticola* Sec-containing thioredoxin was incorporated into the plasmid. This thioredoxin acts as a potent selenite reduction protein, increasing selenium supply for the production of selenophosphate (Kim, Lee, Hwang, Gladyshev, & Kim, 2015).

The Söll system's advantage is found in its simplicity and availability, as it can be purchased from Addgene. pSec has been used successfully to incorporate Sec into various proteins by the Stubbe and Nocera, Söll, Greene, and Armstrong groups (Cáceres, Bailey, Yokoyama, & Greene, 2022; Chung et al., 2021; Evans et al., 2021; Greene, Stubbe, & Nocera, 2019; Mukai et al., 2018). The pSec plasmid contains all components for Sec incorporation, allowing the gene for the POI to be encoded on a separate, compatible plasmid, which is likely already available. This system is arabinose inducible, confers kanamycin resistance through KanR, and contains the pRSF origin of replication, which makes it compatible with many common plasmid backbones, such as pET.

4. Application to thiol dioxygenases
4.1 Fluorotyrosine incorporation into CDO and ADO using pEVOL F2Y

To study the biogenesis of the Cys-Tyr crosslink in CDO, the Liu group utilized a derivative of the Schultz pEVOL system, namely pEVOL F2Y

(Li et al., 2018). This system, evolved from the *M. jannaschii* TyrRS/tRNA[Tyr] amber suppressor pair, was specifically adapted to incorporate 3,5-difluorotyrosine (F2Y) as an ^{19}F NMR probe for studying the post translational phosphorylation of tyrosine (Li, Shi et al., 2013). Using pEVOL F2Y, the Liu group was able to genetically incorporate various halogen-substituted tyrosine derivatives into CDO at position 157 (Li et al., 2018). Notably, under aerobic conditions, the incorporation of F2Y into CDO resulted in both C93-F2Y157 crosslinked and non-crosslinked isoforms. The ability of CDO to catalyze aromatic C-F bond cleavage is quite remarkable, as the dissociation energy of an aliphatic C-F bond (485 kJ mol^{-1}) is substantially higher than that of the corresponding aliphatic C–H bond (411 kJ mol^{-1}). This work provided the first example of oxidative C-F bond cleavage by a protein–bound iron center, highlighting the potent reactivity and oxidizing potential of TDOs (Li et al., 2018).

In a subsequent study, Liu and coworkers crystallized non–crosslinked F2Y157 CDO in the presence of substrate Cys and nitric oxide (NO•) (Fig. 6) (Li, Koto, Davis, & Liu, 2019). The corresponding X-ray crystal structure revealed a C93 conformer with the S atom closely positioned to the NO• oxygen atom (S···O distance ~3.1 Å). On the basis of this observation and density functional theory (DFT) calculations, it was proposed that crosslink formation may proceed via C93 oxidation by the putative Fe(III)-superoxo intermediate during uncoupled catalysis, producing a highly reactive thiyl radical that then reacts with Y157. This mechanism is atypical as the tyrosyl radical is more commonly observed in biology, yet it aligns with CDO's known reactivity with thiol substrates (Greene et al., 2019; Hoganson & Tommos, 2004).

The Liu group then extended the application of the pEVOL F2Y system to explore the possibility of crosslink formation in ADO (Wang et al., 2018). Despite lacking the Cys-Tyr crosslinking residues that are conserved in eukaryotic CDOs, mass spectrometry studies revealed that WT ADO can form a Cys-Tyr crosslink between C206 and Y208 (*Mm*ADO numbering). Surprisingly, the Y208F2Y ADO variant was found to contain a higher fraction of crosslink than the WT enzyme, despite the fact that in the former crosslink formation requires cleavage of a stronger C-F bond. Subsequent X-ray crystallographic studies of ADO revealed that residues C206 and Y208 are solvent exposed and located near, but not inside, the active site and thus further from the iron center than C93 and Y157 in CDO (Fernandez, Elmendorf, et al., 2021; Wang et al., 2021). Consequently, the biological relevance of the C206-Y208 crosslink in ADO remains uncertain.

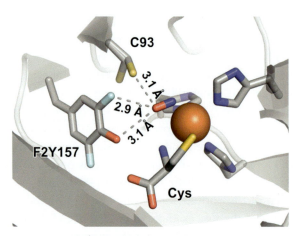

Fig. 6 Active site region of Cys/NO˙-bound Y157F2Y CDO (PDB: 6BPR) (Li et al., 2019). Distances between the NO˙ oxygen and C93 sulfur, F2Y157 fluorine, and F2Y157 oxygen are indicated. The positioning of the iron-bound NO˙ suggests that C93 radical formation by the putative Fe(III)-superoxo intermediate during uncoupled catalysis may represent the first step in Cys-Tyr crosslink formation.

4.2 Sec incorporation into ADO

Due to the unique redox characteristics of Sec, we sought to incorporate this nsAA into ADO to assess the possibility of disulfide formation between C120 and C169 in the putative oxygen tunnel under conditions of high O_2 levels. The lower redox potential of Sec is expected to facilitate the formation of a stable diselenide bond, thus enabling the study of how disulfide/diselenide bond formation affects O_2 access and enzymatic efficiency. Additionally, because Sec is a potent nucleophile and is more easily oxidized than Cys, substitution of C206 by Sec should result in a larger fraction of ADO protein containing the Cys/Sec-Tyr crosslink identified by Liu and coworkers (Wang et al., 2018).

Given its ease of use and availability, we opted to utilize the Söll group's pSec plasmid for site-specific Sec incorporation into ADO. Factors such as plasmid compatibility, cell strain selection, and optimal induction conditions are crucial for the successful application of this system. Contributions from Cáceres and Greene, along with the Söll group, have provided comprehensive methods for utilizing the pSec system (Cáceres et al., 2022; Chung et al., 2021). Below, we aim to offer practical advice for research groups intending to use this system.

Sec-tRNA^{Sec} production is primarily controlled by arabinose induction. The Söll method recommends initiating arabinose induction immediately after inoculation of expression cell cultures (Chung et al., 2021). This allows for adequate buildup of Sec-tRNA^{Sec} levels prior to POI production, avoiding misincorporation of Ser. To achieve high yields of Sec incorporation, the POI production must be controlled separately from that of the tRNA^{Sec} production, which can be achieved using a lactose-inducible T7 RNA polymerase (RNAP), among other approaches. This separate control of Sec-tRNA^{Sec} and POI production aids in alleviating Ser contamination within the POI.

While the DE3/T7 overexpression system is routinely used in many laboratories, additional factors should be taken into consideration to optimize Sec incorporation into POIs. T7 RNAP is extremely active, displaying transcription rates five times higher than those of *E. coli* RNAP; thus, basal expression of T7 RNAP can result in the premature production of target protein (Chamberlin & Ring, 1973; Studier & Moffatt, 1986). In most BL21-derived systems, genomically encoded T7 RNAP production is promoted by the strong, but "leaky", *lacUV5* promoter (Gopal & Kumar, 2013). This leaky expression can be exacerbated when using rich media that may contain lactose contamination from tryptone (a trypsin digest of casein derived from milk). This typically causes expression problems for toxic proteins due to plasmid loss or cell death but can additionally result in lower levels of Sec incorporation due to POI production prior to the buildup of adequate levels of Sec-tRNA^{Sec}. The use of a third compatible plasmid such as pLys can suppress leaky expression, but other methods that allow for tighter transcriptional control may be considered for improved Sec incorporation.

While the pRSF origin of replication found in the pSec plasmid is compatible with many commonly used plasmid backbones such as pET, there are a host of factors to consider when identifying a compatible overexpression plasmid backbone. Notably, the Greene group has developed plug-and-play pCm-based vectors that are useful for the production of POIs (Cáceres et al., 2022). These plasmids are lactose inducible, contain the T7 promoter, confer chloramphenicol resistance, and have variations of genetically encoded solubility tags, affinity tags, and TEV cleavage sites, making them suitable for Sec incorporation into most POIs.

While these plasmids have been reported to be compatible with the pSec system, we note that pCm-based vectors may not be fully compatible with the pSec backbone. Both the pCm and pSec backbones utilize the RSF1030 origin of replication, making them traditionally incompatible.

Plasmids are maintained by the host cell through their origin of replication. The origin of replication a plasmid contains determines which host mechanism regulates the plasmid. This regulation affects how many copies of the plasmid are kept in the host (copy number) and ensures that each plasmid is maintained in the two daughter cells following cell division (Pinto, Pappas, & Winans, 2012). When two plasmids in a cell contain incompatible origins, both contribute to the total copy number and are regulated as one (Scott, 1984). Consequently, this can lead to the loss of either plasmid due to improper partitioning between cells during binary fission, e.g. one daughter cell receives all of plasmid A and the other receives all of plasmid B, or when one plasmid is out-copied by one the other. Although chloramphenicol and kanamycin are effective selection markers that ensure all surviving host cells contain both vectors, the ratio of pSec to pCm plasmids will likely vary from cell to cell. Despite these concerns, experiments reported by Bradbury and coworkers indicate that the pCm/pSec system works. These experiments demonstrated that introduction of high copy number plasmids containing the same origins but different antibiotic selectivity did not result in significant plasmid loss and decreased cell survival rates over the course of a few days (Velappan, Sblattero, Chasteen, Pavlik, & Bradbury, 2007). However, misregulation of plasmid copy number may detrimentally affect Sec incorporation efficiency or yields of the POI.

As an alternative to the pCm-based vectors, our group has successfully employed pQE-derived backbones, which confer β-lactam antibiotic resistance, are lactose inducible, and contain the pSec-compatible pBR322 origin (Fig. 7). These plasmids use the viral T5 promoter, which is not utilized by the T7 RNAP found in *E. coli* DE3 derivatives; instead, mRNA transcripts are produced by the native *E. coli* RNAP (Blommel, Becker, Duvnjak, & Fox, 2007). The T5 promoter system provides several useful benefits for Sec incorporation into POIs. Most importantly, the T5 promoter helps to avoid leaky expression of the POI lacking Sec incorporation and allows for overexpression screening with any *E. coli* cell line. This is beneficial for POIs with overexpression requirements that may include utilizing unique cell lines or when T7 RNAP-deficiency is desired, such as when the recoded C321.ΔA.exp cell line is used. Notably, we have obtained large quantities of Sec-substituted SUMO-tagged ADO with the same pQE backbone using both the C321.ΔA.exp and BL21 AI strains. High levels of Sec incorporation were confirmed by top–down proteomics and inductively coupled plasma mass spectrometry (data not shown).

Non-standard amino acid incorporation into thiol dioxygenases   139

Fig. 7 *E. coli* BL21 AI cell pellet harvested after coexpression the genes for pSec and *Mm*ADO with arabinose and IPTG induction, respectively, in the presence of 150 μM sodium selenite. The red pellet color is indicative of reduced selenium species produced by the *Treponema denticola* thioredoxin.

As demonstrated by Cáceres and Greene, the *E. coli* BL21 AI strain is a promising, commercially available cell line for selenoprotein production (Cáceres et al., 2022). This strain differs from other BL21 strains by the presence of an arabinose inducible T7 RNAP and the inability to metabolize arabinose. This makes arabinose a long-lived expression inducer that promotes high levels of tRNA production (Cáceres et al., 2022). We have achieved high protein yields and levels Sec incorporation using the *E. coli* strains BL21 AI, as well as C321.ΔA.exp, for both the C206U and C(120/169)U *Mm*ADO protein variants (data not shown). As expected, protein yields fell with additional amino acid substitutions, even in recoded strains lacking RF-1. While our constructs contain N-terminal 8 × His tags, we did not observe copurification of truncated protein from either strain, possibly due to aggregation of misfolded truncates into inclusion bodies.

Of note, Greene and coworkers reported a low yield of protein production for a Cys→Sec-substituted class I ribonucleotide reductase in the *E. coli* ME6 cell line after sequential saturation growth, which they attributed to weak antibiotic selectivity with AmpR or instability of the pSec plasmid

(Cáceres et al., 2022; Greene et al., 2019). We have experimented with various overexpression conditions, including sequential saturation (5 mL → 500 mL → 1 L), nonsequential saturation (5 mL → 1 L), and single colony growth, and observed stable antibiotic selection and protein production during 24-h expressions at 18 and 25 °C, with both the pQE and pSec plasmid backbones. However, we encountered difficulties using C321.ΔA.exp cells primarily due to inconsistent growth rates, long doubling times, and the inability to inoculate cultures more than a week after electroporation. We hypothesize that this may be due to loss of the pQE-based vector after β-lactam degradation, or due to reduced fitness of the strain.

5. Conclusions

TDOs stand out among non-heme iron–dependent enzymes due to their unique reactivities and substrate specificities. While TDOs have recently become the focus of significant attention, research relying solely on traditional site-directed mutagenesis efforts have failed to provide conclusive answers to key open questions regarding the structure/function relationships of these enzymes. Initial approaches to incorporate nsAAs into TDOs have been remarkably successful and revealed the tremendous potential that these approaches have to elucidate key mechanistic questions. As such, incorporating non-standard amino acids into TDOs will likely enhance our understanding of the roles and molecular mechanisms of these enzymes.

Acknowledgments

We thank the National Institute of General Medical Sciences of the National Institutes of Health (Grant GM117120) for financial support.

References

Adams, M. A., Singh, V. K., Keller, B. O., & Jia, Z. (2006). Structural and biochemical characterization of gentisate 1,2-dioxygenase from *Escherichia coli* O157:H7. *Molecular Microbiology, 61*(6), 1469–1484. https://doi.org/10.1111/j.1365-2958.2006.05334.x.

Berry, M. J., Banu, L., Harney, J. W., & Larsen, P. R. (1993). Functional characterization of the eukaryotic SECIS elements which direct selenocysteine insertion at UGA codons. *The EMBO Journal, 12*(8), 3315–3322. https://doi.org/10.1002/j.1460-2075.1993.tb06001.x.

Besse, D., Siedler, F., Diercks, T., Kessler, H., & Moroder, L. (1997). The Redox potential of selenocystine in unconstrained cyclic peptides. *Angewandte Chemie (International Edition in English), 36*(8), 883–885. https://doi.org/10.1002/anie.199708831.

Blaesi, E. J., Fox, B. G., & Brunold, T. C. (2015). Spectroscopic and computational Investigation of the H155A variant of cysteine dioxygenase: Geometric and electronic consequences of a third-sphere amino acid substitution. *Biochemistry, 54*(18), 2874–2884. https://doi.org/10.1021/acs.biochem.5b00171.

Blommel, P. G., Becker, K. J., Duvnjak, P., & Fox, B. G. (2007). Enhanced bacterial protein expression during auto-induction obtained by alteration of lac repressor dosage and medium composition. *Biotechnology Progress, 23*(3), 585–598. https://doi.org/10.1021/bp070011x.

Brandt, U., Galant, G., Meinert-Berning, C., & Steinbüchel, A. (2019). Functional analysis of active amino acid residues of the mercaptosuccinate dioxygenase of *Variovorax paradoxus* B4. *Enzyme and Microbial Technology, 120*(August 2018)), 61–68. https://doi.org/10.1016/j.enzmictec.2018.09.007.

Cáceres, J. C., Bailey, C. A., Yokoyama, K., & Greene, B. L. (2022). Selenocysteine substitutions in thiyl radical enzymes. *Methods in enzymology, 662*, Elsevier Inc. https://doi.org/10.1016/bs.mie.2021.10.014.

Carlson, B. A., Xu, X. M., Kryukov, G. V., Rao, M., Berry, M. J., Gladyshev, V. N., & Hatfield, D. L. (2004). Identification and characterization of phosphoseryl-tRNA[Ser]Sec kinase. *Proceedings of the National Academy of Sciences of the United States of America, 101*(35), 12848–12853. https://doi.org/10.1073/pnas.0402636101.

Chamberlin, M., & Ring, J. (1973). Characterization of T7-specific ribonucleic acid polymerase. *Journal of Biological Chemistry, 248*(6), 2235–2244. https://doi.org/10.1016/S0021-9258(19)44211-7.

Chung, C. Z., & Krahn, N. (2022). The selenocysteine toolbox: A guide to studying the 21st amino acid. *Archives of Biochemistry and Biophysics, 730*(August)), 109421. https://doi.org/10.1016/j.abb.2022.109421.

Chung, C. Z., Miller, C., Söll, D., & Krahn, N. (2021). Introducing selenocysteine into recombinant proteins in *Escherichia coli. Current Protocols, 1*(2), e54. https://doi.org/10.1002/cpz1.54.

Cowley, R. E., Cirera, J., Qayyum, M. F., Rokhsana, D., Hedman, B., Hodgson, K. O., ... Solomon, E. I. (2016). Structure of the reduced copper active site in preprocessed galactose oxidase: Ligand tuning for one-electron O_2 activation in cofactor biogenesis. *Journal of the American Chemical Society, 138*(40), 13219–13229. https://doi.org/10.1021/jacs.6b05792.

de la Torre, D., & Chin, J. W. (2021). Reprogramming the genetic code. *Nature Reviews. Genetics, 22*(3), 169–184. https://doi.org/10.1038/s41576-020-00307-7.

Diebold, A. R., Neidig, M. L., Moran, G. R., Straganz, G. D., & Solomon, E. I. (2010). The three-his triad in Dke1: Comparisons to the classical facial triad. *Biochemistry, 49*(32), 6945–6952. https://doi.org/10.1021/bi100892w.

Dominy, J. E., Hwang, J., Guo, S., Hirschberger, L. L., Zhang, S., & Stipanuk, M. H. (2008). Synthesis of amino acid cofactor in cysteine dioxygenase is regulated by substrate and represents a novel post-translational regulation of activity. *Journal of Biological Chemistry, 283*(18), 12188–12201. https://doi.org/10.1074/jbc.M800044200.

Dominy, J. E., Simmons, C. R., Hirschberger, L. L., Hwang, J., Coloso, R. M., & Stipanuk, M. H. (2007). Discovery and characterization of a second mammalian thiol dioxygenase, cysteamine dioxygenase. *Journal of Biological Chemistry, 282*(35), 25189–25198. https://doi.org/10.1074/jbc.M703089200.

Dominy, J. E., Simmons, C. R., Karplus, P. A., Gehring, A. M., & Stipanuk, M. H. (2006). Identification and characterization of bacterial cysteine dioxygenases: A new route of cysteine degradation for eubacteria. *Journal of Bacteriology, 188*(15), 5561–5569. https://doi.org/10.1128/JB.00291-06.

Drienovská, I., & Roelfes, G. (2020). Expanding the enzyme universe with genetically encoded unnatural amino acids. *Nature Catalysis, 3*(3), 193–202. https://doi.org/10.1038/s41929-019-0410-8.

Driggers, C. M., Kean, K. M., Hirschberger, L. L., Cooley, R. B., Stipanuk, M. H., & Karplus, P. A. (2016). Structure-based insights into the role of the Cys–Tyr crosslink and inhibitor recognition by mammalian cysteine dioxygenase. *Journal of Molecular Biology, 428*(20), 3999–4012. https://doi.org/10.1016/j.jmb.2016.07.012.

El-Khairy, L., Ueland, P. M., Refsum, H., Graham, I. M., & Vollset, S. E. (2001). Plasma total cysteine as a risk factor for vascular disease: The European concerted action project. *Circulation, 103*(21), 2544–2549. https://doi.org/10.1161/01.cir.103.21.2544.

Evans, R. M., Krahn, N., Murphy, B. J., Lee, H., Armstrong, F. A., & Söll, D. (2021). Selective cysteine-to-selenocysteine changes in a [NiFe]-hydrogenase confirm a special position for catalysis and oxygen tolerance. *Proceedings of the National Academy of Sciences of the United States of America, 118*(13), 1–9. https://doi.org/10.1073/pnas.2100921118.

Ewetz, L., & Sörbo, B. (1966). Characteristics of the cysteinesulfinate forming enzyme system in rat liver. *Biochimica et Biophysica Acta, 112*(296), 74–80. https://doi.org/10.1016/0926-6593(66)90176-7.

Fernandez, R. L., Dillon, S. L., Stipanuk, M. H., Fox, B. G., & Brunold, T. C. (2020). Spectroscopic investigation of cysteamine dioxygenase. *Biochemistry, 59*(26), 2450–2458. https://doi.org/10.1021/acs.biochem.0c00267.

Fernandez, R. L., Elmendorf, L. D., Smith, R. W., Bingman, C. A., Fox, B. G., & Brunold, T. C. (2021). The crystal structure of cysteamine dioxygenase reveals the origin of the large substrate scope of this vital mammalian enzyme. *Biochemistry, 60*(48), 3728–3737. https://doi.org/10.1021/acs.biochem.1c00463.

Fernandez, R. L., Juntunen, N. D., & Brunold, T. C. (2022). Differences in the second coordination sphere tailor the substrate specificity and reactivity of thiol dioxygenases. *Accounts of Chemical Research, 55*(17), 2480–2490. https://doi.org/10.1021/acs.accounts.2c00359.

Fernandez, R. L., Juntunen, N. D., Fox, B. G., & Brunold, T. C. (2021). Spectroscopic investigation of iron(III) cysteamine dioxygenase in the presence of substrate (analogs): Implications for the nature of substrate-bound reaction intermediates. *JBIC Journal of Biological Inorganic Chemistry, 26*(8), 947–955. https://doi.org/10.1007/s00775-021-01904-5.

Fischer, A. A., Miller, J. R., Jodts, R. J., Ekanayake, D. M., Lindeman, S. V., Brunold, T. C., & Fiedler, A. T. (2019). Spectroscopic and computational comparisons of thiolate-ligated ferric nonheme complexes to cysteine dioxygenase: Second-sphere effects on substrate (analogue) positioning. *Inorganic Chemistry, 58*(24), 16487–16499. https://doi.org/10.1021/acs.inorgchem.9b02432.

Goncharenko, K. V., Vit, A., Blankenfeldt, W., & Seebeck, F. P. (2015). Structure of the sulfoxide synthase EgtB from the ergothioneine biosynthetic pathway. *Angewandte Chemie International Edition, 54*(9), 2821–2824. https://doi.org/10.1002/anie.201410045.

Gonzalez-Flores, J. N., Shetty, S. P., Dubey, A., & Copeland, P. R. (2013). The molecular biology of selenocysteine. *Biomolecular Concepts, 4*(4), 349–365. https://doi.org/10.1515/bmc-2013-0007.

Gopal, G. J., & Kumar, A. (2013). Strategies for the production of recombinant protein in *Escherichia coli*. *The Protein Journal, 32*(6), 419–425. https://doi.org/10.1007/s10930-013-9502-5.

Greene, B. L., Stubbe, J., & Nocera, D. G. (2019). Selenocysteine substitution in a class I ribonucleotide reductase. *Biochemistry, 58*(50), 5074–5084. https://doi.org/10.1021/acs.biochem.9b00973.

Gunawardana, D. M., Heathcote, K. C., & Flashman, E. (2022). Emerging roles for thiol dioxygenases as oxygen sensors. *The FEBS Journal, 289*(18), 5426–5439. https://doi.org/10.1111/febs.16147.

Han, X., Fan, Z., Yu, Y., Liu, S., Hao, Y., Huo, R., & Wei, J. (2013). Expression and characterization of recombinant human phospholipid hydroperoxide glutathione peroxidase. *IUBMB Life, 65*(11), 951–956. https://doi.org/10.1002/iub.1220.

Heafield, M. T., Fearn, S., Steventon, G. B., Waring, R. H., Williams, A. C., & Sturman, S. G. (1990). Plasma cysteine and sulphate levels in patients with motor neurone, Parkinson's and Alzheimer's disease. *Neuroscience Letters, 110*(1–2), 216–220. https://doi.org/10.1016/0304-3940(90)90814-P.

Hendrickson, T. L., Wood, W. N., & Rathnayake, U. M. (2021). Did amino acid side chain reactivity dictate the composition and timing of aminoacyl-trna synthetase evolution? *Genes, 12*(3), https://doi.org/10.3390/genes12030409.

Hoganson, C. W., & Tommos, C. (2004). The function and characteristics of tyrosyl radical cofactors. *Biochimica et Biophysica Acta - Bioenergetics, 1655*(1–3), 116–122. https://doi.org/10.1016/j.bbabio.2003.10.017.

Ito, N., Phillips, S. E. V., Stevens, C., Ogel, Z. B., McPherson, M. J., Keen, J. N., ... Knowles, P. F. (1991). Novel thioether bond revealed by a 1.7 Å crystal structure of galactose oxidase. *Nature, 350*(6313), 87–90. https://doi.org/10.1038/350087a0.

Johansson, L., Gafvelin, G., & Arnér, E. S. J. (2005). Selenocysteine in proteins - Properties and biotechnological use. *Biochimica et Biophysica Acta - General Subjects, 1726*(1), 1–13. https://doi.org/10.1016/j.bbagen.2005.05.010.

Joseph, C. A., & Maroney, M. J. (2007). Cysteine dioxygenase: Structure and mechanism. *Chemical Communications, (32)*, 3338. https://doi.org/10.1039/b702158e.

Kang, D., Lee, J., Wu, C., Guo, X., Lee, B. J., Chun, J. S., & Kim, J. H. (2020). The role of selenium metabolism and selenoproteins in cartilage homeostasis and arthropathies. *Experimental and Molecular Medicine, 52*(8), 1198–1208. https://doi.org/10.1038/s12276-020-0408-y.

Kim, M.-J., Lee, B. C., Hwang, K. Y., Gladyshev, V. N., & Kim, H.-Y. (2015). Selenium utilization in thioredoxin and catalytic advantage provided by selenocysteine. *Biochemical and Biophysical Research Communications, 461*(4), 648–652. https://doi.org/10.1016/j.bbrc.2015.04.082.

Krzycki, J. A. (2005). The direct genetic encoding of pyrrolysine. *Current Opinion in Microbiology, 8*(6), 706–712. https://doi.org/10.1016/j.mib.2005.10.009.

Lajoie, M. J., Rovner, A. J., Goodman, D. B., Aerni, H. R., Haimovich, A. D., Kuznetsov, G., ... Isaacs, F. J. (2013). Genomically recoded organisms expand biological functions. *Science (New York, N. Y.), 342*(6156), 357–360. https://doi.org/10.1126/science.1241459.

Li, W., Blaesi, E. J., Pecore, M. D., Crowell, J. K., & Pierce, B. S. (2013). Second-sphere interactions between the C93–Y157 cross-link and the substrate-bound Fe site influence the O_2 coupling efficiency in mouse cysteine dioxygenase. *Biochemistry, 52*(51), 9104–9119. https://doi.org/10.1021/bi4010232.

Li, J., Griffith, W. P., Davis, I., Shin, I., Wang, J., Li, F., ... Liu, A. (2018). Cleavage of a carbon–fluorine bond by an engineered cysteine dioxygenase. *Nature Chemical Biology, 14*(9), 853–860. https://doi.org/10.1038/s41589-018-0085-5.

Li, J., Koto, T., Davis, I., & Liu, A. (2019). Probing the Cys-Tyr cofactor biogenesis in cysteine dioxygenase by the genetic incorporation of fluorotyrosine. *Biochemistry, 58*(17), 2218–2227. https://doi.org/10.1021/acs.biochem.9b00006.

Li, F., Shi, P., Li, J., Yang, F., Wang, T., Zhang, W., ... Wang, J. (2013). A genetically encoded ^{19}F NMR probe for tyrosine phosphorylation. *Angewandte Chemie - International Edition, 52*(14), 3958–3962. https://doi.org/10.1002/anie.201300463.

Liu, C. C., & Schultz, P. G. (2010). Adding new chemistries to the genetic code. *Annual Review of Biochemistry, 79*(1), 413–444. https://doi.org/10.1146/annurev.biochem.052308.105824.

Masson, N., Keeley, T. P., Giuntoli, B., White, M. D., Lavilla Puerta, M., Perata, P., ... Ratcliffe, P. J. (2019). Conserved N-terminal cysteine dioxygenases transduce responses to hypoxia in animals and plants. *Science (New York, N. Y.), 364*(6448), 65–69. https://doi.org/10.1126/science.aaw0112.

McCoy, J. G., Bailey, L. J., Bitto, E., Bingman, C. A., Aceti, D. J., Fox, B. G., & Phillips, G. N. (2006). Structure and mechanism of mouse cysteine dioxygenase. *Proceedings of the National Academy of Sciences of the United States of America, 103*(9), 3084–3089. https://doi.org/10.1073/pnas.0509262103.

Metanis, N., & Hilvert, D. (2014). Natural and synthetic selenoproteins. *Current Opinion in Chemical Biology, 22*, 27–34. https://doi.org/10.1016/j.cbpa.2014.09.010.

Miller, J. R., Schnorrenberg, E. C., Aschenbrener, C., Fox, B. G., & Brunold, T. C. (2024). Kinetic and spectroscopic investigation of the Y157F and C93G/Y157F variants of cysteine dioxygenase: Dissecting the roles of the second-sphere residues C93 and Y157. Biochemistry, in press. https://doi.org/10.1021/acs.biochem.4c00177.

Mukai, T., Hoshi, H., Ohtake, K., Takahashi, M., Yamaguchi, A., Hayashi, A., ... Sakamoto, K. (2015). Highly reproductive *Escherichia coli* cells with no specific assignment to the UAG codon. *Scientific Reports, 5*(1), 9. https://doi.org/10.1038/srep09699.

Mukai, T., Sevostyanova, A., Suzuki, T., Fu, X., & Söll, D. (2018). A facile method for producing selenocysteine-containing proteins. *Angewandte Chemie International Edition, 57*(24), 7215–7219. https://doi.org/10.1002/anie.201713215.

Nalivaiko, E. Y., Vasseur, C. M., & Seebeck, F. P. (2024). Enzyme-catalyzed oxidative degradation of ergothioneine. *Angewandte Chemie - International Edition, 63*(8), https://doi.org/10.1002/anie.202318445.

Ng, T. L., Rohac, R., Mitchell, A. J., Boal, A. K., & Balskus, E. P. (2019). An N-nitrosating metalloenzyme constructs the pharmacophore of streptozotocin. *Nature, 566*(7742), 94–99. https://doi.org/10.1038/s41586-019-0894-z.

Pinto, U. M., Pappas, K. M., & Winans, S. C. (2012). The ABCs of plasmid replication and segregation. *Nature Reviews. Microbiology, 10*(11), 755–765. https://doi.org/10.1038/nrmicro2882.

Rouault, T. A. (2015). Mammalian iron–sulphur proteins: Novel insights into biogenesis and function. *Nature Reviews. Molecular Cell Biology, 16*(1), 45–55. https://doi.org/10.1038/nrm3909.

Schmidt, R. L., & Simonović, M. (2012). Synthesis and decoding of selenocysteine and human health. *Croatian Medical Journal, 53*(6), 535–550. https://doi.org/10.3325/cmj.2012.53.535.

Schultz, R. L., Sabat, G., Fox, B. G., & Brunold, T. C. (2023). A single DNA point mutation leads to the formation of a cysteine-tyrosine crosslink in the cysteine dioxygenase from *Bacillus subtilis. Biochemistry, 62*(12), 1964–1975. https://doi.org/10.1021/acs.biochem.3c00083.

Scott, J. R. (1984). Regulation of plasmid replication. *Microbiological Reviews, 48*(1), 1–23. https://doi.org/10.1128/MMBR.48.1.1-23.1984.

Shim, S. M., Choi, H. R., Kwon, S. C., Kim, H. Y., Sung, K. W., Jung, E. J., ... Kwon, Y. T. (2023). The Cys-N-degron pathway modulates pexophagy through the N-terminal oxidation and arginylation of ACAD10. *Autophagy, 19*(6), 1642–1661. https://doi.org/10.1080/15548627.2022.2126617.

Shioya, M., Nishida, A., Yagi, Y., Ogawa, A., Tsujikawa, T., Kim-Mitsuyama, S., ... Andoh, A. (2007). Epithelial overexpression of interleukin-32α in inflammatory bowel disease. *Clinical and Experimental Immunology, 149*(3), 480–486. https://doi.org/10.1111/j.1365-2249.2007.03439.x.

Simmons, C. R., Liu, Q., Huang, Q., Hao, Q., Begley, T. P., Karplus, P. A., & Stipanuk, M. H. (2006). Crystal structure of mammalian cysteine dioxygenase: A novel mononuclear iron center for cysteine thiol oxidation. *Journal of Biological Chemistry, 281*(27), 18723–18733. https://doi.org/10.1074/jbc.M601555200.

Stipanuk, M. H., Simmons, C. R., Karplus, P. A., & Dominy, J. E. (2011). Thiol dioxygenases: Unique families of cupin proteins. *In Amino Acids, 41*(1), 91–102. https://doi.org/10.1007/s00726-010-0518-2.

Stipanuk, M. H., & Ueki, I. (2011). Dealing with methionine/homocysteine sulfur: Cysteine metabolism to taurine and inorganic sulfur. *Journal of Inherited Metabolic Disease, 34*(1), 17–32. https://doi.org/10.1007/s10545-009-9006-9.

Studier, F. W., & Moffatt, B. A. (1986). Use of bacteriophage T7 RNA polymerase to direct selective high-level expression of cloned genes. *Journal of Molecular Biology, 189*(1), 113–130. https://doi.org/10.1016/0022-2836(86)90385-2.

Tchesnokov, E. P., Fellner, M., Siakkou, E., Kleffmann, T., Martin, L. W., Aloi, S., ... Jameson, G. N. L. (2015). The cysteine dioxygenase homologue from *Pseudomonas aeruginosa* is a 3-mercaptopropionate dioxygenase. *Journal of Biological Chemistry, 290*(40), 24424–24437. https://doi.org/10.1074/jbc.m114.635672.

Thyer, R., Robotham, S. A., Brodbelt, J. S., & Ellington, A. D. (2015). Evolving tRNA[Sec] for efficient canonical incorporation of selenocysteine. *Journal of the American Chemical Society, 137*(1), 46–49. https://doi.org/10.1021/ja510695g.

Thyer, R., Shroff, R., Klein, D. R., D'Oelsnitz, S., Cotham, V. C., Byrom, M., ... Ellington, A. D. (2018). Custom selenoprotein production enabled by laboratory evolution of recoded bacterial strains. *Nature Biotechnology, 36*(7), 624–631. https://doi.org/10.1038/nbt.4154.

Velappan, N., Sblattero, D., Chasteen, L., Pavlik, P., & Bradbury, A. R. M. (2007). Plasmid incompatibility: More compatible than previously thought? *Protein Engineering, Design and Selection, 20*(7), 309–313. https://doi.org/10.1093/protein/gzm005.

Vu, T. T. M., Mitchell, D. C., Gygi, S. P., & Varshavsky, A. (2020). The Arg/N-degron pathway targets transcription factors and regulates specific genes. *Proceedings of the National Academy of Sciences of the United States of America, 117*(49), 31094–31104. https://doi.org/10.1073/pnas.2020124117.

Wang, Y., Griffith, W. P., Li, J., Koto, T., Wherritt, D. J., Fritz, E., & Liu, A. (2018). Cofactor biogenesis in cysteamine dioxygenase: C−F bond cleavage with genetically incorporated unnatural tyrosine. *Angewandte Chemie - International Edition, 57*(27), 8149–8153. https://doi.org/10.1002/anie.201803907.

Wang, Y., Shin, I., Li, J., & Liu, A. (2021). Crystal structure of human cysteamine dioxygenase provides a structural rationale for its function as an oxygen sensor. *Journal of Biological Chemistry, 279*(4), 101176. https://doi.org/10.1016/j.jbc.2021.101176.

White, M. D., Dalle Carbonare, L., Lavilla Puerta, M., Iacopino, S., Edwards, M., Dunne, K., ... Flashman, E. (2020). Structures of *Arabidopsis thaliana* oxygen-sensing plant cysteine oxidases 4 and 5 enable targeted manipulation of their activity. *Proceedings of the National Academy of Sciences, 117*(37), 23140–23147. https://doi.org/10.1073/pnas.2000206117.

White, M. D., Kamps, J. J. A. G., East, S., Taylor Kearney, L. J., & Flashman, E. (2018). The plant cysteine oxidases from *Arabidopsis thaliana* are kinetically tailored to act as oxygen sensors. *Journal of Biological Chemistry, 293*(30), 11786–11795. https://doi.org/10.1074/jbc.RA118.003496.

Windsor, P., Ouyang, H., Costa, J. A. G., da, Damodaran, A. R., Chen, Y., & Bhagi-Damodaran, A. (2023). Gas tunnel engineering of prolyl hydroxylase reprograms hypoxia signaling in cells. *BioRxiv*2023.08.07.552357. https://doi.org/10.1101/2023.08.07.552357.

Wood, J. L., & Cavallini, D. (1967). Enzymic oxidation of cysteamine to hypotaurine in the absence of a cofactor. *Archives of Biochemistry and Biophysics, 119*(C), 368–372. https://doi.org/10.1016/0003-9861(67)90467-5.

Young, T. S., Ahmad, I., Yin, J. A., & Schultz, P. G. (2010). An enhanced system for unnatural amino acid mutagenesis in *E. coli*. *Journal of Molecular Biology, 395*(2), 361–374. https://doi.org/10.1016/j.jmb.2009.10.030.

Further reading

Ishida, S., Ngo, P.H.T., Gundlach, A., & Ellington, A. (2024). Engineering ribosomal machinery for noncanonical amino acid incorporation. Chemical Reviews, in press. https://doi.org/10.1021/acs.chemrev.3c00912.

CHAPTER SEVEN

Unveiling the mechanism of cysteamine dioxygenase: A combined HPLC-MS assay and metal-substitution approach

Ran Duan, Jiasong Li, and Aimin Liu*
Department of Chemistry, University of Texas at San Antonio, San Antonio, TX, United States
*Corresponding author. e-mail address: Feradical@utsa.edu

Contents

1. Introduction	148
2. Protein expression, purification, and crystallization	151
2.1 Equipment	152
2.2 Reagents	153
2.3 Procedure	153
2.4 Note	155
3. Spectral characterization of Co-ADO	155
3.1 Equipment	156
3.2 Reagents	157
3.3 Optical spectral characterization	157
3.4 EPR spectral characterization	157
4. Cobalt reconstitution in ADO	157
4.1 Equipment	158
4.2 Reagents	158
4.3 Preparation of "apo-ADO" through 1,10-phenanthroline assay	158
4.4 Evaluation of the "apo-ADO" using ferrozine assay	159
4.5 Reconstitution of ADO enzyme by adding divalent metal ions	159
4.6 Note	159
5. HPLC-MS analysis of the hypotaurine formation by ADO	160
5.1 Equipment	161
5.2 Reagents	162
5.3 Procedure	162
5.4 Note	163
6. Summary and conclusions	164
Acknowledgments	164
References	164

Abstract

Mammalian cysteamine dioxygenase (ADO), a mononuclear non-heme Fe(II) enzyme with three histidine ligands, plays a key role in cysteamine catabolism and regulation of the N-degron signaling pathway. Despite its importance, the catalytic mechanism of ADO remains elusive. Here, we describe an HPLC-MS assay for characterizing thiol dioxygenase catalytic activities and a metal-substitution approach for mechanistic investigation using human ADO as a model. Two proposed mechanisms for ADO differ in oxygen activation: one involving a high-valent ferryl-oxo intermediate. We hypothesized that substituting iron with a metal that has a disfavored tendency to form high-valent states would discriminate between mechanisms. This chapter details the expression, purification, preparation, and characterization of cobalt-substituted ADO. The new HPLC-MS assay precisely measures enzymatic activity, revealing retained reactivity in the cobalt-substituted enzyme. The results obtained favor the concurrent dioxygen transfer mechanism in ADO. This combined approach provides a powerful tool for studying other non-heme iron thiol oxidizing enzymes.

1. Introduction

Thiol dioxygenases (TDOs) are a critical family of enzymes that play a vital role in regulating cellular thiol metabolism. These non-heme iron-dependent oxygenases, belonging to the Cupin superfamily, catalyze the oxidation of thiol-containing molecules into sulfinic acids (Aloi, Davies, Karplus, Wilbanks, & Jameson, 2019; Stipanuk, Simmons, Andrew Karplus, & Dominy, 2011). Cysteamine (2-aminoethanethiol, 2-AET) dioxygenase (ADO) and cysteine dioxygenase (CDO) are the only two enzymes that regulate thiol metabolism in mammalian cells by oxidizing thiol-bearing small molecules (Dominy, Simmons, Karplus, Gehring, & Stipanuk, 2006; Dominy et al., 2007). Cysteamine dioxygenase (ADO) stands out for its diverse substrate range. Unlike most TDOs, ADO oxidizes small organic thiols like cysteamine and targets N-terminal cysteine residues in signaling proteins or peptides involved in oxygen sensing (Gunawardana, Heathcote, & Flashman, 2022; Masson et al., 2019). This unique ability highlights the remarkable adaptability of ADO and underscores the need to further understand its reaction mechanism and dynamic structure.

Despite performing similar functions, ADO belongs to the PFam family PF07847 (PCO_ADO), which is distinct from PFam family PF05995 (CDO_I) that includes CDO and 3-mercaptopropionate dioxygenase (MDO). Like plant cysteine oxidases (PCO) in the PCO_ADO family, ADO also oxidizes N-terminal cysteine-containing signaling peptides that

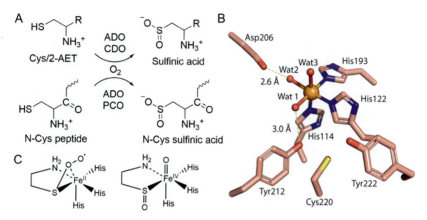

Fig. 1 The reaction, structure, and catalytic mechanism of ADO. (A) Thiol dioxygenase catalyzed reactions. R = H (ADO), −COO⁻ (CDO); ADO and PCO catalyze certain N-terminal Cys-containing peptides and proteins. (B) Active site architecture of human ADO (Wang, Shin, Li, & Liu, 2021); (C) The key reaction intermediates in the two proposed mechanisms. Left: Concurrent dioxygen transfer intermediate. Right: High-valent ferryl-oxo intermediate.

are involved in oxygen-sensing (Gunawardana et al., 2022; Masson et al., 2019) (Fig. 1A). Therefore, ADO distinguishes itself by its broad spectrum of substrates, inserting oxygen atoms from the dioxygen molecule (O₂) into small organic thiol substrates, cysteamine, and large protein substrates. The dynamic structure of ADO, which accommodates two kinds of substrates with distinct sizes and structures, and the reaction mechanism are attractive topics in the emerging field of thiol dioxygenases.

Even though ADO has been purified from horse kidneys for more than 60 years (Cavallini, Scandurra, & De Marco, 1963; Cavallini, de Marco, Scandurra, Dupré, & Graziani, 1966), it was not genetically identified and heterologously expressed until 2007 (Dominy et al., 2007). Due to a relatively slow autooxidation, the mononuclear iron center in as-isolated ADO proteins has been shown to be present in mixed ferrous and ferric forms by EPR spectroscopy (Fernandez, Dillon, Stipanuk, Fox, & Brunold, 2020; Rotilio, Federici, Calabrese, Costa, & Cavallini, 1970; Wang et al., 2020). A Mössbauer spectroscopy study observed that 23% of iron is in the Fe(III) form in as-isolated mouse ADO (Wang et al., 2020). The heterogeneous metal oxidation state inevitably introduces an inhomogeneous coordination environment and corresponding variations in protein structure that are likely responsible for obstructing the crystallographic characterization of this protein for a long period of time. Even though the Fe

(II)-bearing mouse ADO has been crystallized (Fernandez et al., 2021), it has been challenging to obtain a high-quality crystal structure for Fe(II)-bearing human ADO. One method to overcome this technical challenge associated with the metal oxidation state is to swap the Fe center with a Ni ion, as we have done in a previous structural study (Wang, Li, & Liu, 2021). Along with the mutations of surface cysteine residues on ADO to suppress the expected nonhomogeneous disulfide bond formation, the crystal structure of human ADO has been successfully determined at 1.78 Å resolution (Wang, Li, & Liu, 2021) (Fig. 1B). This success inspired us to generate ADO proteins with different metal centers and take advantage of their unique chemical and physical properties to decipher the catalytic mechanism of ADO.

The catalytic mechanism of TDOs remains debatable despite substantial structural and spectroscopic research. Currently, two mechanisms have been proposed based on the extensive study of CDO, the founding member of this group of enzymes. The first proceeds via a non-ferryl-oxo-dependent concurrent dioxygen transfer to the thiol group of the substrate, while the second undergoes a more traditional oxygen activation mechanism involving a high-valent ferryl-oxo species and stepwise O atom transfers (Fig. 1C). The first mechanism was proposed based on a persulfenate intermediate (Driggers et al., 2013; Simmons et al., 2008). The second mechanism was solely proposed based on computational studies (Kumar, Thiel, & de Visser, 2011), while the attempts to trap the ferryl intermediate were not successful (Tchesnokov et al., 2016). Since a high-valent ferryl-oxo species is critical in the second mechanism, substituting iron for other metals that barely present a high-valent state in the enzyme is attractive for discerning between these two mechanisms.

Swapping Fe(II) for other divalent metals has been an important strategy to investigate the catalytic mechanisms of dioxygenases. In an iron-dependent dioxygenase, homoprotocatechuate 2,3-dioxygenase (HPCD), the cobalt-substituted enzyme maintains ring cleaving activity (Fielding, Lipscomb, & Que, 2012), suggesting that the oxidation state of cobalt during the catalysis also remains the same, like the natural iron enzymes (Lipscomb, 2014; Traore & Liu, 2022). In another Cupin-type enzyme, acireductone dioxygenase (ARD), Fe-ARD and Ni-ARD both present catalytic activity with different ARD isomers (Dai, Pochapsky, & Abeles, 2001). These studies imply that non-ferryl-oxo-dependent dioxygenases can maintain reactivity after proper metal swapping from Fe(II) to Co(II) or Ni(II).

In this chapter, we describe a new HPLC-mass spectrometry (HPLC-MS)-based ADO assay protocol in addition to the commonly used oxygen electrode-based ADO assay and the methods to prepare and study cobalt(II)-substituted ADO, Co(II)-ADO. The experimental methods include protein expression, purification, crystallization, spectral characterization, cobalt reconstitution, and a new high-performance liquid chromatography–mass spectrometry (HPLC-MS) analysis of the reaction product. Cell culture for Co(II)-ADO was conducted in minimal media supplemented with $CoCl_2$ during expression. As-purified Co(II)-ADO was characterized by optical and electron paramagnetic resonance (EPR) spectroscopy to confirm the presence of cobalt. We developed a method to prepare a close to metal-free "apo-ADO" form of the protein that uses chelating reagent treatment to eliminate the interference of trace amounts of metals. The cobalt-reconstituted ADO was prepared by incubating "apo-ADO" with $CoCl_2$. A ferrozine assay determined iron content in cobalt-reconstituted ADO, and the reactivity was precisely evaluated by an HPLC-MS method using isocratic elution. This cobalt-substitution method helps us gain insights into the natural iron-dependent thiol dioxygenases (Li, Duan, & Liu, 2024). A Cobalt(IV)-oxo species has not been observed in biology. All synthetic Co(IV)-oxo complexes require the support of a macrocyclic ligand set (Wang et al., 2017). Thus, if a cobalt-substituted non-heme enzyme is catalytically active, it would favor the pathway not involving a high-valent metal-oxo. The same strategy can be applied to the structural and mechanistic study of many other non-heme iron-dependent enzymes.

2. Protein expression, purification, and crystallization

This section describes the expression and purification of human ADO (hADO) with an iron or cobalt metal center. The wild-type (WT) hADO and hADO C18S/C239S were cloned to pET-28a-TEV expression vector plasmid with a cleavable His_6-tag containing 30 additional amino acids at the N-terminus in our previous work (Wang, Li, & Liu, 2021). *Escherichia coli* BL21 (DE3) competent cells were transformed with the plasmid. Cell culture was conducted in M9 medium with the supplement of Fe(II) or Co(II) to minimize the contamination of other metal ions. Fe-ADO was purified using an immobilized metal affinity chromatography (IMAC) column charged with nickel, while Co-ADO was purified using a cobalt IMAC column to prevent contamination with nickel ions. The N-terminal His-tagged ADO was mixed

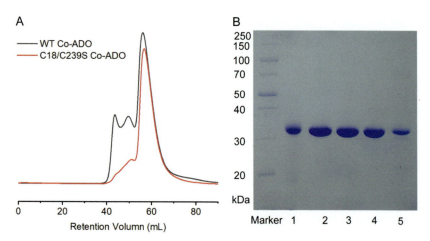

Fig. 2 Comparison of wild-type Co-ADO and C18S/C239S variant by gel filtration chromatography (A) and SDS-PAGE (B). (1) C18S/C239S Fe-ADO; (2) WT Fe-ADO; (3) C18S/C239S Co-ADO; (4) WT Co-ADO; and (5) "Apo-ADO".

with tobacco etch virus (TEV) protease and dialyzed against the dialysis buffer containing 10 mM tris(hydroxymethyl)aminomethane-HCl (Tris-HCl) (pH 8.0). The His_6-tag-cleaved and un-cleaved proteins were separated using the IMAC column. The non-tagged hADO proteins were further purified by gel filtration chromatography using Superdex 75. The purity of the proteins was evaluated using sodium dodecyl sulfate–polyacrylamide gel electrophoresis (SDS-PAGE). The hADO C18S/C239S mutant can be expressed, purified, and crystallized using the same method. ADO forms inter- and intra-disulfide bonds among subunits with two surface located cysteines, Cys18 and Cys239, causing heterogeneity (Wang et al., 2021). These two cysteine residues are not conserved in other thiol dioxygenases, and their mutation to serine does not affect the enzymatic activity of ADO. Compared to wild-type (WT) hADO, the C18S/C239S variant is more homogeneous and predominant in a monomeric form (Fig. 2A). After gel filtration chromatography, the purity of ADO proteins is suitable for future experiments (Fig. 2B).

2.1 Equipment
- Shaking incubator (Excella E25, New Brunswick Scientific).
- Cell disruptor (LM20 Microfluidizer Processor, Microfluidics International Corporation).
- Centrifuge (Avanti JXN-26, Beckman Coulter).

- Fast protein liquid chromatography (FPLC) (ÄKTA pure chromatography system, Cytiva).
- Ni-IMAC column (HisTrap HP, Cytiva).
- Gel filtration chromatography column (HiLoad 16/600 Superdex 75 prep grade, Cytiva).

2.2 Reagents

- Luria Broth (LB) Broth (Miller) medium: 10.0 g/L tryptone, 10.0 g/L NaCl, and 5.0 g/L yeast extract, sterilized by autoclave.
- M9 salt solution: 12.8 g/L Na_2HPO_4, 3.0 g/L KH_2PO_4, 0.5 g/L NaCl, and 1.0 g/L NH_4Cl, dissolved in double-distilled water (ddH_2O), sterilized by autoclave.
- 1 M $MgSO_4$ stock solution, sterilized by 0.22 μm filtration and stored at 4 °C.
- 1 M $CaCl_2$ stock solution, sterilized by 0.22 μm filtration and stored at 4 °C.
- M9 supplement buffer (50×): 200 g/L glucose, 20 g/L casamino acid, and 100 mg/L thiamine, sterilized by 0.22 μm filtration and stored at 4 °C
- Kanamycin stock solution: 50 mg/mL kanamycin dissolved in ddH_2O, sterilized by 0.22 μm filtration and stored at −20 °C.
- $CoCl_2$ stock solution: 0.1 M $CoCl_2$, dissolved in ddH_2O.
- $Fe(NH_4)_2(SO_4)_2$ stock solution: 0.1 M $Fe(NH_4)_2(SO_4)_2$ dissolved in ddH_2O, freshly prepared.
- Isopropyl β-d-1-thiogalactopyranoside (IPTG) stock solution: 1 M IPTG dissolved in ddH_2O, sterilized by 0.22 μm filtration and stored at −20 °C.
- Phenylmethylsulfonyl fluoride (PMSF).
- Lysing buffer: 50 mM Tris-HCl, 200 mM NaCl, pH 8.0, stored at 4 °C.
- Elution buffer: 50 mM Tris-HCl, 200 mM NaCl, 500 mM imidazole, pH 8.0, stored at 4 °C.
- Storage buffer: 50 mM Tris-HCl, 50 mM NaCl, pH 7.6, stored at 4 °C.
- Dialysis buffer: 10 mM Tris-HCl, pH 8.0, stored at 4 °C.
- TEV protease stock solution (2 mg/mL).
- Crystallization buffer: 100 mM Bis-Tris (pH 5.5), 200 mM $(NH_4)_2SO_4$, and 20% (w/v) PEG3350.

2.3 Procedure

1. Inoculate a single colony of competent cells to 5 mL LB medium with 50 mg/L kanamycin. Incubate this preculture at 37 °C, 220 rpm for 8 h.

2. Prepare M9 medium by mixing 200 mL of M9 salt solution, 4 mL of M9 supplement buffer, 0.2 mL of 1 M $MgSO_4$ solution, 0.02 mL of 1 M $CaCl_2$ solution, and 0.2 mL of kanamycin stock solution in a 1 L flask. Transfer 1 mL of preculture to the M9 medium and incubate at 37 °C, 220 rpm for 12 h.

3. Prepare M9 medium by mixing 1 L of M9 salt solution, 20 mL of M9 supplement buffer, 1 mL of 1 M $MgSO_4$ solution, 0.1 mL of 1 M $CaCl_2$ solution, and 1 mL of kanamycin stock solution in a 2 L baffled flask. Transfer 20 mL of preculture to the M9 medium and incubate at 37 °C, 220 rpm.

4. When the optical density at 600 nm (OD_{600}) reaches 0.4, add 0.2 mL of $Fe(NH_4)_2(SO_4)_2$ or $CoCl_2$ stock solution (20 μM final concentration) solution to the 1 L culture.

5. When the OD_{600} reaches 0.8, add 0.5 mL of 1 M IPTG solution (0.5 mM final concentration) to induce gene expression. Incubate at 37 °C, 220 rpm for another 4 h.

6. Harvest cells by centrifugation for 10 min at 8000g, 4 °C. Freeze cell pellet at −80 °C for future use.

7. Resuspend the cell pellet in the lysing buffer with 2 mg PMSF supplement. Disrupt cells at 30,000 psi in the ice bath. Centrifuge for 1 h at 34,000g, 4 °C, to remove cell debris.

8. Assign lysing buffer as buffer A, and elution buffer as buffer B. Load the supernatant to the Ni or Co-IMAC column at 1 mL/min speed in the FPLC.

9. When loading is done, wash the column with two column volumes (CVs) of 100% buffer A. Then, wash the column with two CVs of 1% buffer B. Elute with 100% buffer B and collect ADO protein.

10. Concentrate down ADO protein and dilute in the lysing buffer to remove excess imidazole. Mix ADO protein and TEV protease in a 10:1 w/w ratio to cut the His_6 tag. Dialyze in 2 L dialysis buffer for 12 h at 4 °C.

11. Remove precipitate by centrifugation at 30,000g for 10 min. Load cut protein onto Ni or Co-IMAC columns at 1 mL/min speed. Collect the flow-through solution.

12. Concentrate down ADO protein to 2 mL aliquot. Equilibrate Superdex 75 column with the storage buffer. Purify ADO protein through Superdex 75 column using 1.5 mL/min speed at 4 °C. Collect purified ADO protein and freeze at −80 °C for future use.

13. Load 2 µg of protein in the well of a SDS-PAGE gel. Evaluate the purity of ADO protein by SDS-PAGE.

14. For protein crystallization, hADO C18S/C239S protein was concentrated to approximately 30 mg/mL and mixed at 1:1 (v/v) with a crystallization buffer using the hanging drop, vapor-diffusion method at 289 K.

2.4 Note

1. The cell membrane can be disrupted by sonication without influencing quality or yield.

2. The Co-IMAC column can be prepared by treating a regular Ni-IMAC column with the following solutions in sequence: 2 CV of ddH$_2$O, 4 CV of 0.1 M ethylenediaminetetraacetic acid (EDTA) solution, 2 CV of ddH$_2$O, 4 CV of 0.1 M CoCl$_2$ solution, and 2 CV of ddH$_2$O.

3. ADO purified through a Co-IMAC or Ni-IMAC column may contain cobalt and nickel ions because the iron binding is less tight than that of heme enzymes. However, this metal substitution has a negligible influence on preparing metal-substituted ADO proteins with the procedure of stripping metal ions prior to metal reconstitution.

3. Spectral characterization of Co-ADO

The Co-ADO from the M9 medium exhibits distinct spectroscopic properties. Co-ADO shows a light yellow color, whereas the "apo-ADO" is colorless. The Co(II) is EPR active with its d^7 electron configuration. In contrast, the Fe(II) in WT ADO is EPR silent with its d^6 electron configuration. In this section, we conducted a spectroscopic characterization of the Co(II)-ADO. The optical spectrum of Co(II)-ADO exhibits two strong shoulders at approximately 310 and 420 nm, and a much weaker shoulder in the 450–700 nm region. The spectrum is consistent with spectral features of Co(II)-ACMSD (Li, Walker, Iwaki, Hasegawa, & Liu, 2005), which is assigned as a pentacoordinate Co(II) center ligated via three His, one Asp, and one water. The optical spectral features are much more pronounced when shown as difference spectra over "apo-ADO" (Fig. 3A). The two stronger shoulders are due to the ligand-to-metal charge transfer (LMCT), which gives Co-ADO a light yellow color. The weaker shoulder in the 450–700 nm region is caused by the d-d transitions of the Co(II) ion. The relationship between the coordination number and the extinction

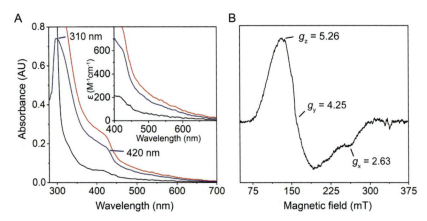

Fig. 3 The optical and EPR spectra of Co-ADO. (A) The optical spectra of 300 μM apo-ADO (black), Co-ADO (red), and their difference spectrum (blue). (B) The EPR spectrum of 100 μM Co-ADO. The EPR data was collected at 30 K with a microwave power of 3.17 mW.

coefficients of the d–d transitions in the visible region (450–750 nm) of high-spin Co(II) model complexes have been extensively discussed (Bertini & Luchinat, 1984; Horrocks, Ishley, Holmquist, & Thompson, 1980; Sellin, Eriksson, Aronsson, & Mannervik, 1983). General guidelines of this literature indicate that extinction coefficients for peaks between 400 and 900 nm are usually below 50 for six-coordination geometry, between 50 and 300 for five-coordination geometry, and greater than 300 for four-coordination geometry. The molar extinction coefficient at 550 nm is approximately 100 $M^{-1} cm^{-1}$, indicating a five-coordinated or highly distorted six-coordinate high-spin Co(II) center. These results revealed a distortion of the cobalt ligand environment away from an idealized six-coordinate geometry. As depicted in Fig. 3B, an as-isolated Co(II)-ADO exhibited a high-spin (HS) center, characteristic of a typical ground spin state of $S = 3/2$ Co(II) species with effective g values (g_x, g_y, g_z) of 2.63, 4.25, and 5.26. HS EPR spectra have been noted in Co(II)-ACMSD (Li et al., 2005) and Co(II)-QueD (Merkens, Kappl, Jakob, Schmid, & Fetzner, 2008), the effective g values of which differ from Co-ADO owing to the variation in ligated residues and coordination geometry.

3.1 Equipment

- X-band EPR spectrometer (E560 EPR/ENDOR spectrometer with a cryogen-free 4 K temperature system, Bruker Corporation).
- UV–vis spectrometer (Evolution Pro UV–vis Spectrophotometer, Thermo Fisher Scientific).

3.2 Reagents
- Storage buffer: 50 mM Tris-HCl, 50 mM NaCl, pH 7.6, stored at 4 °C.
- Co-ADO: As purified Co-ADO from M9 medium, dissolved in the storage buffer in aerobic conditions.
- Metal-strapped ADO: "apo-ADO" from Section 4.3, dissolved in the storage buffer in aerobic conditions after three rounds of buffer changing.

3.3 Optical spectral characterization
1. Turn on the UV–vis spectrometer and wait 30 min until the lamps are warmed up.
2. Set scan range from 250 to 800 nm. Scan the storage buffer as blank for all samples.
3. Scan the UV–vis spectrum of 300 µM Co-ADO or "apo ADO".
4. Determine the molar extinction coefficient.

3.4 EPR spectral characterization
1. Add 200 µL of 100 µM Co-ADO to an EPR quartz tube. Slowly freeze down the sample in liquid nitrogen. Store in a liquid nitrogen tank. Prepare a sample with storage buffer using the same method as the blank.
2. Set the EPR spectrometer at 100 kHz modulation frequency, 0.6 mT modulation amplitude, and wait for the temperature to decrease to 4.5 K.
3. Scan the blank sample's EPR spectrum using 3.17 mW microwave power at 30 K. Each spectrum averages four scans.
4. Scan the EPR spectrum of Co-ADO using 3.17 mW microwave power at 30 K. Each spectrum averages four scans.

4. Cobalt reconstitution in ADO

Even though Co-ADO is cultured in the M9 medium with minor metal ions, trace amounts of iron can still be detected in the Co-ADO from the EPR spectrum. To evaluate the role of cobalt in the ADO reactivity and spectroscopic properties, the proper method is to prepare near metal-free "apo-ADO" and compare the results before and after cobalt incorporation. However, removing metal ions from a non–heme iron-dependent protein is challenging. The three ligated histidine residues make the metal ion resistant to removal by regular EDTA treatment. Here we describe a method to carry out cobalt reconstitution on WT ADO using

1,10-phenanthroline as a chelating reagent (Ren, Lee, Wang, & Liu, 2022). The iron content in "apo-ADO" and Co-ADO was characterized using the ferrozine assay (Tchesnokov, Wilbanks, & Jameson, 2012). The reconstitution process was carried out in 4-(2-hydroxyethyl)-1-piper-azineethanesulfonic acid (HEPES) buffer to reduce the interaction between buffer (i.e., Tris-HCl buffer) and metal ions.

4.1 Equipment

- Centrifuge (Avanti JXN-26, Beckman Coulter).
- FPLC (ÄKTA pure™ chromatography system, Cytiva).
- Desalting column (HiTrap Desalting, Cytiva).
- Hotplate and stirrer (IKA® C-MAG HS hotplate stirrers, IKA Inc.).
- UV–vis spectrometer (Evolution™ Pro UV–vis spectrophotometers, Thermo Scientific).
- Schlenk line.
- Vacuum pump.

4.2 Reagents

- HEPES buffer: 50 mM HEPES buffer, pH 8, stored at 4 °C.
- 1,10-phenanthroline stock solution: 0.1 M 1,10-phenanthroline in 0.1 M HCl.
- EDTA stock solution: 0.1 M EDTA solution, pH 8.
- $Na_2S_2O_4$ stock solution: 1 M $Na_2S_2O_4$, dissolved in ddH_2O, freshly prepared.
- $CoCl_2$ stock solution: 0.1 M $CoCl_2$, dissolved in ddH_2O.
- Ferrozine assay mixture: 5.56 mM ferrozine, 2.9% ascorbic acid, 430 mM ammonium acetate, pH 9.
- 80% H_2SO_4.
- Iron standard: 10 µL $Fe(NH_4)_2(SO_4)_2$ in HEPES buffer; 5 µL 80% H_2SO_4.

4.3 Preparation of "apo-ADO" through 1,10-phenanthroline assay

1. Degas 1.8 mL of 300 µM WT ADO protein with a Schlenk line for 30 min in the ice bath.
2. Add 4 µL of 1 M $Na_2S_2O_4$, 100 µL of EDTA stock solution, and 100 µL of 1,10-phenanthroline stock solution to the ADO protein. The final concentrations of reagents are 2 mM $Na_2S_2O_4$, 5 mM EDTA, and 5 mM 1,10-phenanthroline. Leave the ADO protein solution at 4 °C for 1 h until it reacts thoroughly. The solution should turn to an orange-red color.

3. Fill the desalting column with HEPES buffer. Load ADO protein to the desalting column at 5 mL/min speed and collect the first peak. Concentrate down protein solution into a 1.8 mL aliquot.
4. Repeat Steps 2 and 3 two more times.

4.4 Evaluation of the "apo-ADO" using ferrozine assay

1. Prepare the "apo-ADO" stock solution by concentrating the "apo-ADO" protein from Section 4.3 to millimolar concentration. Calculate the final concentration of the "apo-ADO" sample (1/15 of the apo-ADO stock solution).
2. Mix 10 μL of "apo-ADO" stock solution with 5 μL 80% H_2SO_4 in a centrifuge tube. Heat to 95 °C for 30 min using the hot plate.
3. After cooling down, add 135 μL of ferrozine assay mixture to the sample and mix well. Remove precipitate by centrifugation at 20,000g for 5 min
4. Add 10 μL of HEPES buffer and 5 μL of 80% H_2SO_4 to 135 μL of ferrozine assay mixture to make a blank sample. Measure the UV–vis absorbance at 562 and 750 nm. Use the absorbance at 750 nm as the baseline for all samples.
5. Add 15 μL of iron standard with different $Fe(NH_4)_2(SO_4)_2$ concentrations to 135 μL of ferrozine assay mixture. Measure the UV–vis absorbance at 562 nm (A_{562}). Repeat three times for each concentration of standard. Generate a standard curve of A_{562} vs. iron concentration.
6. Measure the UV–vis absorbance of the "apo-ADO" mixture at 562 nm. Fit the absorbance to the standard curve and determine the iron concentration in the "apo-ADO" mixture. Calculate the iron concentration and percentage in protein following the example in Fig. 4.

4.5 Reconstitution of ADO enzyme by adding divalent metal ions

1. Dilute "apo-ADO" to 50 μM in the desalting buffer.
2. Metal reconstitution method (using cobalt ion as an example): Add 0.1 M $CoCl_2$ stock solution to "apo-ADO" in drops while stirring at 4 °C, until the concentration of $CoCl_2$ reaches 50 μM.
3. Incubate the mixture at 4 °C for 30 min. Centrifuge at 20,000g for 2 min to remove precipitate. Use immediately for the HPLC-MS assay.

4.6 Note

- "Apo-ADO" protein is not stable. It is recommended to use it as soon as possible.
- The 1,10-phenanthroline stock solution should be made freshly within a month.

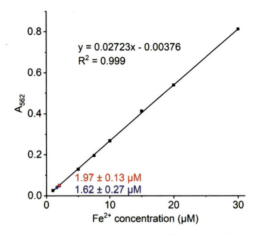

Fig. 4 Iron concentration determination of apo-ADO and Co-ADO produced from M9 media. The black trace shows the standard curve fitted by various concentrations of Fe(NH$_4$)$_2$(SO$_4$)$_2$. The iron concentration of 1.62 μM in 125 μM apo-ADO indicates that the iron occupancy is 1.30% (blue); The iron concentration of 1.97 μM in 114 μM Co-ADO indicates a 1.73% iron occupancy (red).

- To reduce the contamination of metal ions in the ferrozine assay, it is recommended that all glassware be acid-washed. The desalting buffer can be further purified using a Chelex 100 column.
- Each experiment should be performed in triplicate.
- The metal reconstitution method can be applied to other metal ions like Ni(II) or Fe(II). The Fe(II) reconstitution should be carried out in anerobic conditions.

5. HPLC-MS analysis of the hypotaurine formation by ADO

To determine the reactivity of "apo-ADO" and study the influence of cobalt on ADO, a precise method is required to identify the reaction product, hypotaurine, from its analogs in the activity assay. Even though the oxygen consumption of ADO reaction can be recorded using an oxygen electrode (Wang et al., 2018), the self-oxidization of cysteamine into the dimer form, cystamine, cannot be excluded from this method. A previously reported HPLC method can separate and detect hypotaurine efficiently (Coloso, Hirschberger, Dominy, Lee, & Stipanuk, 2006; Dominy et al., 2007). However, this coupled assay requires pre-treatment

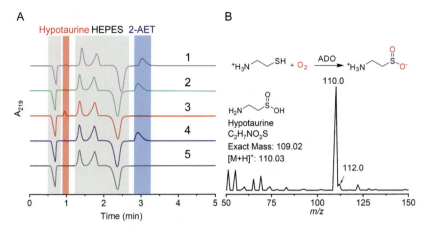

Fig. 5 (A) Demonstration of Ni(II)-ADO's dioxygenase catalytic activity and HPLC profile for ADO activity assay. (1) 50 μM reconstituted Ni(II)-ADO + 10 mM cysteamine; (2) 50 μM "apo-ADO" + 10 mM cysteamine; (3) HPLC hypotaurine sample; (4) HPLC cysteamine sample; and (5) HPLC blank sample. The details of the reaction setup are described in the text. The negative peak in the HPLC elution profiles emerged due to the difference between the reaction HEPES buffer in the samples and the HPLC solvent. (B) Reaction equation of ADO with cysteamine as a substrate and MS spectrum of the hypotaurine peak in Sample (3 of panel A).

with o-phthaldialdehyde and relies on an expensive fluorescence detector for detection. Inspired by the HPLC method used in the CDO activity assays (Li et al., 2018; McCoy et al., 2006), we developed a method using isocratic elution to separate hypotaurine from cysteamine directly. The hypotaurine product was verified using the retention time (Fig. 5A) and mass spectrum (Fig. 5B) of a commercially purchased standard. The reactivity of "apo-ADO" and Co-ADO can be determined through peak area integration.

5.1 Equipment
- HPLC system (UltiMate 3000 with diode array detector, Thermo Fisher Scientific).
- Analytical C18 column (Inertsil ODS-3 HPLC Column, 3 μm, 100 × 4.6 mm, GL Sciences Inc.).
- Ultra-centrifugal filter (Amicon Ultra Centrifugal Filter, 10 kDa MWCO, 0.5 mL, MilliporeSigma).
- Thermomixer (Thermomixer R, Eppendorf).

5.2 Reagents

- HPLC running buffer: 0.6% methanol, 0.3% Heptafluorobutyric acid (HFBA), and 99.1% dd H_2O.
- 6 M hydrochloride acid (HCl)
- HEPES buffer: 50 mM HEPES buffer, pH 8.
- Cysteamine stock solution: 1 M cysteamine (Purchased from Acros Organics), freshly prepared with degassed water (ddH_2O bubbled with N_2 gas).
- Hypotaurine stock solution: 50 mM hypotaurine dissolved in ddH_2O, freshly prepared.
- Hypotaurine standards: 200 μL of hypotaurine solutions at various concentrations, freshly prepared with ddH_2O and hypotaurine stock solution, filtered with 0.22 μm filter.
- HPLC blank sample: 198 μL of HEPES buffer; 2 μL 6 M HCl.
- HPLC cysteamine sample (10 mM): 196 μL HEPES buffer; 2 μL 6 M HCl; and 2 μL of 2-AET (1 M).
- HPLC hypotaurine sample (0.25 mM): 197 μL HEPES buffer; 2 μL 6 M HCl; and 1 μL 50 mM hypotaurine.
- "Apo-ADO" stock solution: 100 μM "apo-ADO", dissolved in HEPES buffer.
- Co-ADO stock solution: 100 μM Co-ADO generated from cobalt reconstitution (Section 4.5), dissolved in HEPES buffer.

5.3 Procedure

1. Wash the C18 column using the HPLC running buffer until the baseline turns flat.
2. Set up HPLC assay condition. Buffer: HPLC running buffer; Flow rate: 1.5 mL/min; Run time: 10 min; Wavelength: 219 nm; Injection volume: 50 μL.
3. Inject 50 μL of a hypotaurine standard and run HPLC assay. Repeat three times for each concentration of standard. Fit the peak area to the concentration as a standard curve.
4. Mix 100 μL of "apo-ADO" stock solution, and 96 μL of HEPES buffer in a microcentrifuge tube (tube A). Preheat to 37 °C on the thermomixer.
5. Add 2 μL of 1 M cysteamine to a 1.5 mL microcentrifuge tube (tube B), preheat to 37 °C on the thermomixer.
6. Add 2 μL of 6 M HCl to a 1.5 mL EP tube (tube C).
7. Transfer 196 μL of "apo-ADO" from tubes A to B to start the reaction. Set the thermomixer at 37 °C, 500 rpm. React for 2 min.

8. Transfer 198 μL of mixture from tubes B to C to quench the reaction.

9. Filter the reaction mixture with ultra-centrifugal filters at 10,000g for 10 min. Collect the flow-through as the "apo-ADO" sample. Final concentration: 50 μM "apo-ADO", 10 mM cysteamine.

10. Repeat steps 3–8 using Co-ADO stock solution. Collect the flow-through as the Co-ADO sample. Final concentration: 50 μM reconstituted Co-ADO, 10 mM cysteamine.

11. Inject 50 μL of HPLC blank sample and run HPLC assay.

12. Inject 50 μL of HPLC hypotaurine sample and run the HPLC assay. Determine the retention time of hypotaurine.

13. Inject 50 μL of HPLC cysteamine sample and run the HPLC assay. Calculate the peak area for hypotaurine. Set this peak area as the baseline for all samples.

14. Inject 50 μL of "apo-ADO" sample and run HPLC assay. Calculate the peak area for hypotaurine.

15. Inject 50 μL of Co-ADO sample and run HPLC assay. Calculate the peak area for hypotaurine.

16. Calculate the k_{obs} from the standard curve. Using the peak area values calculated from Steps 14 to 15, subtract the baseline from Step 13. Then fit the resultant values to the standard curve from Step 3 and convert them to the hypotaurine concentrations. The k_{obs} can be calculated from the formula below, where enzyme concentration is 50 μM and the reaction time is 2 min.

$$k_{obs} = \frac{\text{Hypoaturine concentartion}}{\text{Enzyme concentration} \times \text{Reaction time}}$$

5.4 Note

1. Make sure there are no leaks in the ultra-centrifugal filters. The unfiltered protein may clog the C18 column.

2. Cysteamine is hygroscopic and very easy to oxidize. It is recommended that cysteamine stock solution be stored in an O_2-free anaerobic chamber.

3. The 6 M hydrochloride acid is corrosive. Personal protective equipment (PPE) is required when handling it.

4. Each experiment should be performed in triplicate.

5. Regarding the iron-reconstituted ADO, the concentration of the ADO enzyme and Fe(II) needs to be reduced to 5 μM. Otherwise, most of the dissolved oxygen in the reaction mixture will be consumed, and the reactivity will be underestimated.

6. Summary and conclusions

This chapter presents a novel approach for investigating the catalytic mechanism of cysteamine dioxygenase (ADO) using metal substitution. We successfully generated cobalt(II)- and Ni-(II)-substituted ADO (Co-ADO and Ni-ADO) and characterized their properties. Optical and EPR spectroscopy confirmed cobalt incorporation into the enzyme, and the ferrozine assay verified the removal of the metal in apo-ADO after treatment. Importantly, the HPLC-MS method for activity assay demonstrates Co-ADO and Ni-ADO are catalytically competent and the dioxygenase activity of which is proportional to the amount of Co(II) or Ni(II) present. These combined results strongly support a non-ferryl-oxo-dependent, concurrent dioxygen transfer mechanism during ADO-mediated oxygenation. This work sheds light on thiolate-bound metalloenzyme oxygenation reactions and establishes a powerful method for studying other three-histidine non-heme iron enzymes oxidizing thiol-bearing substrates.

Acknowledgments

This work was supported by NSF award CHE-2204225. The use of HPLC was also partially supported by an administrative supplement of the NIH grant R01GM108988. A.L. acknowledges the support of Lutcher Brown's endowment and the Welch Foundation grant Welch AX-2110-20220331.

References

Aloi, S., Davies, C. G., Karplus, P. A., Wilbanks, S. M., & Jameson, G. N. L. (2019). Substrate specificity in thiol dioxygenases. *Biochemistry, 58*(19), 2398–2407.

Bertini, I., & Luchinat, C. (1984). High spin cobalt(II) as a probe for the investigation of metalloproteins. *Advances in Inorganic Biochemistry, 6*, 71–111.

Cavallini, D., de Marco, C., Scandurra, R., Dupré, S., & Graziani, M. T. (1966). The enzymatic oxidation of cysteamine to hypotaurine: Purification and properties of the enzyme. *Journal of Biological Chemistry, 241*(13), 3189–3196.

Cavallini, D., Scandurra, R., & De Marco, C. (1963). The enzymatic oxidation of cysteamine to hypotaurine in the presence of sulfide. *Journal of Biological Chemistry, 238*(9), 2999–3005.

Coloso, R. M., Hirschberger, L. L., Dominy, J. E., Lee, J.-I., & Stipanuk, M. H. (2006). Cysteamine dioxygenase: Evidence for the physiological conversion of cysteamine to hypotaurine in rat and mouse tissues. *Taurine, 6* (Boston, MA).

Dai, Y., Pochapsky, T. C., & Abeles, R. H. (2001). Mechanistic studies of two dioxygenases in the methionine salvage pathway of *Klebsiella pneumoniae*. *Biochemistry, 40*(21), 6379–6387.

Dominy, J. E., Simmons, C. R., Hirschberger, L. L., Hwang, J., Coloso, R. M., & Stipanuk, M. H. (2007). Discovery and characterization of a second mammalian thiol dioxygenase, cysteamine dioxygenase. *Journal of Biological Chemistry, 282*(35), 25189–25198.

Dominy, J. E., Simmons, C. R., Karplus, P. A., Gehring, A. M., & Stipanuk, M. H. (2006). Identification and characterization of bacterial cysteine dioxygenases: A new route of cysteine degradation for Eubacteria. *Journal of Bacteriology, 188*(15), 5561–5569.

Driggers, C. M., Cooley, R. B., Sankaran, B., Hirschberger, L. L., Stipanuk, M. H., & Karplus, P. A. (2013). Cysteine dioxygenase structures from pH4 to 9: Consistent cyspersulfenate formation at intermediate pH and a cys-bound enzyme at higher pH. *Journal of Molecular Biology, 425*(17), 3121–3136.

Fernandez, R. L., Dillon, S. L., Stipanuk, M. H., Fox, B. G., & Brunold, T. C. (2020). Spectroscopic investigation of cysteamine dioxygenase. *Biochemistry, 59*(26), 2450–2458.

Fernandez, R. L., Elmendorf, L. D., Smith, R. W., Bingman, C. A., Fox, B. G., & Brunold, T. C. (2021). The crystal structure of cysteamine dioxygenase reveals the origin of the large substrate scope of this vital mammalian enzyme. *Biochemistry, 60*(48), 3728–3737.

Fielding, A. J., Lipscomb, J. D., & Que, L. (2012). Characterization of an O_2 adduct of an active cobalt-substituted extradiol-cleaving catechol dioxygenase. *Journal of the American Chemical Society, 134*(2), 796–799.

Gunawardana, D. M., Heathcote, K. C., & Flashman, E. (2022). Emerging roles for thiol dioxygenases as oxygen sensors. *The FEBS Journal, 289*(18), 5426–5439.

Horrocks, W. D., Ishley, J. N., Holmquist, B., & Thompson, J. S. (1980). Structural and electronic mimics of the active site of cobalt(II)-substituted zinc metalloenzymes. *Journal of Inorganic Biochemistry, 12*(2), 131–141.

Kumar, D., Thiel, W., & de Visser, S. P. (2011). Theoretical study on the mechanism of the oxygen activation process in cysteine dioxygenase enzymes. *Journal of the American Chemical Society, 133*(11), 3869–3882.

Li, J., Griffith, W. P., Davis, I., Shin, I., Wang, J., Li, F., ... Liu, A. (2018). Cleavage of a carbon-fluorine bond by an engineered cysteine dioxygenase. *Nature Chemical Biology, 14*(9), 853–860.

Li, J., Duan, R., & Liu, A. (2024). Cobalt(II)-substituted cysteamine dioxygenase oxygenation proceeds through a cobalt(III)-superoxo complex. *Journal of the American Chemical Society* in press. https://doi.org/10.1021/jacs.4c01871.

Li, T., Walker, A. L., Iwaki, H., Hasegawa, Y., & Liu, A. (2005). Kinetic and spectroscopic characterization of ACMSD from *Pseudomonas fluorescens* reveals a pentacoordinate mononuclear metallocofactor. *Journal of the American Chemical Society, 127*(35), 12282–12290.

Lipscomb, J. D. (2014). Life in a sea of oxygen. *Journal of Biological Chemistry, 289*(22), 15141–15153.

Masson, N., Keeley, T. P., Giuntoli, B., White, M. D., Puerta, M. L., Perata, P., ... Ratcliffe, P. J. (2019). Conserved N-terminal cysteine dioxygenases transduce responses to hypoxia in animals and plants. *Science (New York, N. Y.), 365*(6448)), 65–69.

McCoy, J. G., Bailey, L. J., Bitto, E., Bingman, C. A., Aceti, D. J., Fox, B. G., & Phillips, G. N. (2006). Structure and mechanism of mouse cysteine dioxygenase. *Proceedings of the National Academy of Sciences, 103*(9), 3084–3089.

Merkens, H., Kappl, R., Jakob, R. P., Schmid, F. X., & Fetzner, S. (2008). Quercetinase QueD of *Streptomyces* sp. FLA, a monocupin dioxygenase with a preference for nickel and cobalt. *Biochemistry, 47*(46), 12185–12196.

Ren, D., Lee, Y.-H., Wang, S.-A., & Liu, H.-W. (2022). Characterization of the oxazinomycin biosynthetic pathway revealing the key role of a nonheme iron-dependent mono-oxygenase. *Journal of the American Chemical Society, 144*(24), 10968–10977.

Rotilio, G., Federici, G., Calabrese, L., Costa, M., & Cavallini, D. (1970). An electron paramagnetic resonance study of the nonheme Iron of cysteamine oxygenase. *Journal of Biological Chemistry, 245*(22), 6235–6236.

Sellin, S., Eriksson, L. E., Aronsson, A. C., & Mannervik, B. (1983). Octahedral metal coordination in the active site of glyoxalase I as evidenced by the properties of Co(II)-glyoxalase I. *Journal of Biological Chemistry, 258*(4), 2091–2093.

Simmons, C. R., Krishnamoorthy, K., Granett, S. L., Schuller, D. J., Dominy, J. E., Jr., Begley, T. P., ... Karplus, P. A. (2008). A putative Fe^{2+}-bound persulfenate intermediate in cysteine dioxygenase. *Biochemistry, 47*(44), 11390–11392.

Stipanuk, M. H., Simmons, C. R., Andrew Karplus, P., & Dominy, J. E. (2011). Thiol dioxygenases: Unique families of cupin proteins. *Amino Acids, 41*(1), 91–102.

Tchesnokov, E. P., Faponle, A. S., Davies, C. G., Quesne, M. G., Turner, R., Fellner, M., ... Jameson, G. N. L. (2016). An iron–oxygen intermediate formed during the catalytic cycle of cysteine dioxygenase. *Chemical Communications, 52*(57), 8814–8817. https://doi.org/10.1039/C6CC03904A.

Tchesnokov, E. P., Wilbanks, S. M., & Jameson, G. N. L. (2012). A strongly bound high-spin iron(II) coordinates cysteine and homocysteine in cysteine dioxygenase. *Biochemistry, 51*(1), 257–264.

Traore, E. S., & Liu, A. (2022). Charge maintenance during catalysis in nonheme iron oxygenases. *ACS Catalysis, 12*(10), 6191–6208.

Wang, B., Lee, Y.-M., Tcho, W.-Y., Tussupbayev, S., Kim, S.-T., Kim, Y., ... Nam, W. (2017). Synthesis and reactivity of a mononuclear non-haem cobalt(IV)-oxo complex. *Nature Communications, 8*, 14839.

Wang, Y., Davis, I., Chan, Y., Naik, S. G., Griffith, W. P., & Liu, A. (2020). Characterization of the nonheme iron center of cysteamine dioxygenase and its interaction with substrates. *Journal of Biological Chemistry, 295*(33), 11789–11802.

Wang, Y., Griffith, W. P., Li, J., Koto, T., Wherritt, D. J., Fritz, E., & Liu, A. (2018). Cofactor biogenesis in cysteamine dioxygenase: C−F bond cleavage with genetically incorporated unnatural tyrosine. *Angewandte Chemie International Edition, 57*(27), 8149–8153.

Wang, Y., Shin, I., Li, J., & Liu, A. (2021). *PDB entry 7REI: The crystal structure of nickel-bound human ADO C18S C239S variant.* https://doi.org/10.2210/pdb7REI/pdb.

Wang, Y., Shin, I., Li, J., & Liu, A. (2021). Crystal structure of human cysteamine dioxygenase provides a structural rationale for its function as an oxygen sensor. *Journal of Biological Chemistry, 297*(4), 101176.

CHAPTER EIGHT

In vitro analysis of the three-component Rieske oxygenase cumene dioxygenase from *Pseudomonas fluorescens* IP01

Niels A.W. de Kok, Hui Miao, and Sandy Schmidt***

Department of Chemical and Pharmaceutical Biology, Groningen Research Institute of Pharmacy, University of Groningen, Groningen, The Netherlands
*Corresponding author. e-mail address: s.schmidt@rug.nl

Contents

1.	Introduction	168
2.	*In vitro* analysis of Rieske oxygenases	171
3.	Expression of Cumene dioxygenase	173
	3.1 Materials	173
	3.2 Buffers and reagents	174
	3.3 Equipment	174
4.	Step-by-step method details	174
5.	General considerations	175
6.	Cell lysis and protein purification	177
	6.1 Materials and equipment	177
	6.2 Equipment	177
7.	Step-by-step method details	177
	7.1 Cell lysis by sonication	177
	7.2 Purification by immobilized metal ion chromatography (IMAC)	178
	7.3 Desalting by size exclusion chromatography (SEC)	179
	7.4 Concentration by centrifugal ultrafiltration	180
8.	General considerations	180
9.	Enzymatic activity assay	182
	9.1 Materials and equipment	182
	9.2 Equipment	183
10.	Step-by-step method details	183
	10.1 Enzymatic reaction	183
	10.2 Sample extraction	183
	10.3 Non-chiral GC-MS and chiral GC-FID analysis	184
11.	General considerations	185
12.	Summary and conclusions	187
	Acknowledgments	187
	References	187

Methods in Enzymology, Volume 703
ISSN 0076-6879, https://doi.org/10.1016/bs.mie.2024.05.013
Copyright © 2024 Elsevier Inc. All rights are reserved, including those for text and data mining, AI training, and similar technologies.

Abstract

Rieske non-heme iron-dependent oxygenases (ROs) are a versatile group of enzymes traditionally associated with the degradation of aromatic xenobiotics. In addition, ROs have been found to play key roles in natural product biosynthesis, displaying a wide catalytic diversity with typically high regio- and stereo- selectivity. However, the detailed characterization of ROs presents formidable challenges due to their complex structural and functional properties, including their multi-component composition, cofactor dependence, and susceptibility to reactive oxygen species. In addition, the substrate availability of natural product biosynthetic intermediates, the limited solubility of aromatic hydrocarbons, and the radical-mediated reaction mechanism can further complicate functional assays. Despite these challenges, ROs hold immense potential as biocatalysts for pharmaceutical applications and bioremediation. Using cumene dioxygenase (CDO) from *Pseudomonas fluorescens* IP01 as a model enzyme, this chapter details techniques for characterizing ROs that oxyfunctionalize aromatic hydrocarbons. Moreover, potential pitfalls, anticipated complications, and proposed solutions for the characterization of novel ROs are described, providing a framework for future RO research and strategies for studying this enzyme class. In particular, we describe the methods used to obtain CDO, from construct design to expression conditions, followed by a purification procedure, and ultimately activity determination through various activity assays.

1. Introduction

Rieske non–heme iron–dependent oxygenases (ROs) have attracted considerable interest as biocatalysts due to their ability to harness the reactivity of iron (Fe) to catalyze chemically diverse and challenging reactions in a regio- and stereo-specific manner. In particular, the formation of asymmetric vicinal *cis*-diols in a single dioxygenation step is of great interest, as this reactivity cannot be conveniently achieved by chemical means. ROs are multi-component enzymes that were originally found to be involved in the degradation of aromatic xenobiotics, forming *cis*-dihydrodiols (Jerina, Daly, Jeffrey, & Gibson, 1971; Wackett, Kwart, & Gibson, 1988; Yeh, Gibson, & Liu, 1977). Moreover, these enzymes are known to exist as either two- or three-component systems (Ferraro, Gakhar, & Ramaswamy, 2005; Kweon et al., 2008; Nam et al., 2001; Runda, de Kok, & Schmidt, 2023; Runda, Kremser, Özgen, & Schmidt, 2023). Both two- and three- component systems contain a reductase component (RO-Red) and an oxygenase component (RO-Oxy). In two-component systems, the RO-Red delivers electrons directly to the RO-Oxy. The three-component system additionally consists of a Rieske-type ferredoxin (RO-Fd) which facilitates electron transfer, acting as an intermediate electron

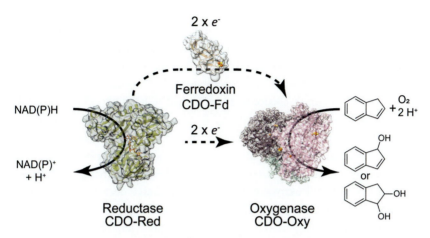

Fig. 1 A schematic representation of the electron transfer chain in two- and three-component ROs with indene as the model substrate. The CDO-Oxy α-subunits are shown in different shades of purple and the CDO-Oxy β-subunits in mint-green. The FAD-cofactor of CDO-Red is shown as orange sticks. The Fe-S clusters in the Rieske centers (CDO-Fd and CDO-Oxy) and Fe ions at mononuclear iron centers (CDO-Oxy) are represented as orange and yellow spheres for iron and sulfur atoms respectively. CDO-Red (CumA4) is a SWISS-MODEL model based on TDO-Red (PDB: 3EF6 (Friemann et al., 2009; Waterhouse et al., 2018)); CDO-Fd (CumA3) is a SWISS-MODEL model based on TDO-Fd (PDB: 3DQY (Friemann et al., 2009; Waterhouse et al., 2018)); CDO-Oxy (CumA1+CumA2) is a model based on the crystal structure (PDB: 1WQL (Dong et al., 2005)).

mediator between the RO-Red and RO-Oxy. The RO-Oxy is dependent on electron transfer from its redox partners for catalysis (Barry & Challis, 2013; Runda, de Kok et al., 2023; Runda, Kremser et al., 2023). Since both RO-Red and RO-Oxy catalyze a half-reaction, and interprotein electron transfer is essential for this process, the RO system as a whole is considered as a single functional unit. Thereby, the flavin-dependent RO-Red liberates single electrons from nicotinamide adenine dinucleotide (NAD(P)H) cofactors. Two single electrons are transferred successively, one-by-one, either indirectly as in three-component systems via a ferredoxin (RO-Fd), or directly as in two-component systems to the Rieske cluster of an oxygenase component (RO-Oxy) (Fig. 1). From there, single electrons are transferred via a bridging aspartate or glutamate residue to the non-heme iron center (Parales, Parales, & Gibson, 1999). At the non-heme iron center, the electrons are used to activate molecular oxygen in order to catalyze a large variety of chemical conversions in a highly regio- and stereo-selective manner (Ferraro et al., 2005; Perry, De Los Santos, Alkhalaf, & Challis, 2018; Runda, de Kok et al., 2023; Runda, Kremser et al., 2023).

In addition to catalyzing aromatic dihydroxylations, the involvement of ROs in natural product biosynthesis has received much attention in recent years, as these findings have greatly expanded the scope of their reactivities. For example, ROs were found to catalyze C-C bond formation in the synthesis of prodiginine alkaloids (Hu, Withall, Challis, & Thomson, 2016; Sydor et al., 2011; Withall, Haynes, & Challis, 2015); oxidation of aryl-amines to aryl-nitrates in pyrrolnitrin biosynthesis (van Pée, Salcher, & Lingens, 1980); hydroxylation in the biosynthesis of galbonolides (Kim et al., 2014), promysalin (Li et al., 2011), viomycin (Fei, Yin, Zhang, & Zabriskie, 2007) and saxitoxin (Lukowski et al., 2018); oxidative decarboxylation in the biosynthesis of arcyriarubin (Chang & Brady, 2014); desaturation in the biosynthesis of ambruticin (Guth et al., 2023) and jerangolid (Guth et al., 2023), among other examples (Perry et al., 2018; Runda, de Kok et al., 2023; Runda, Kremser et al., 2023). This large catalytic diversity, combined with high regio- and stereo- selectivity and the ability to functionalize relatively unreactive atoms, gives ROs the potential to be developed into useful bio-catalysts for the pharmaceutical industry or bioremediation applications.

However, to broadly use ROs in a biocatalytic application, several limitations must be overcome. First, ROs are dependent on NAD(P)H cofactors, which are uneconomical to use in an industrial setting. Therefore, the implementation of NAD(P)H cofactor regeneration strategies (Immanuel, Sivasubramanian, Gul, & Dar, 2020; Runda, de Kok et al., 2023; Runda, Kremser et al., 2023; Tian, Boggs et al., 2023; Tian, Garcia, Donnan, & Bridwell-Rabb, 2023; Zhang, Rao, Zhang, Xu, & Yang, 2016) and the exploration of alternative electron sources (e.g. photobiocatalysis (Feyza Özgen et al., 2020; Hu et al., 2022) or electrode immobilization) are essential for the industrial application of ROs. Another obstacle hindering the use of ROs as biocatalysts is that the substrate scope of ROs is currently mostly limited to aromatic hydrocarbons. Similar to cytochrome P450 enzymes (a large and well-studied family of monooxygenases that also rely on an electron transfer chain), ROs share common engineering approaches (Meng et al., 2022). For example, as the oxygenation reaction occurs at the RO-Oxy component, it determines the substrate scope, type of reactivity and stereoselectivity of the RO system. Therefore, substrate binding pockets and substrate access tunnels of the RO-Oxy are prime engineering targets (Heinemann, Armbruster, & Hauer, 2021; Wissner, Schelle, Escobedo-Hinojosa, Vogel, & Hauer, 2021). Recently, flexible loops remote from the active site have been proposed as additional targets of interest for engineering efforts (Brimberry, Garcia, Liu, Tian, & Bridwell-Rabb, 2023).

The discovery of novel RO activities in natural product biosynthetic pathways does not only expand the scope of reactivities, but also provides novel enzyme scaffolds for engineering or as a source of inspiration for the rational engineering of other scaffolds. Furthermore, since ROs are multi-component enzyme systems, another feasible approach is RO-component swapping, where RO components with favorable properties (e.g. well-expressing, highly stable, highly efficient or possessing desired chemoselectivity, substrate specificity, etc.) are combined (Lee, Simurdiak, & Zhao, 2005; Lukowski et al., 2018; Runda, Miao, Kok, & Schmidt, 2024). However, whereas such hybrid systems formed with non-native redox partners as components can have certain advantages (Lee, Simurdiak et al., 2005; Liu, Knapp, Jo, Dill, & Bridwell-Rabb, 2022; Liu, Tian et al., 2022; Lukowski et al., 2018; Runda et al., 2024; Tian, Boggs et al., 2023; Tian, Garcia et al., 2023), the components of different RO systems are not necessarily compatible or efficient in transferring electrons.

In this chapter, cumene dioxygenase (CDO) from *Pseudomonas fluorescens* IP01 is used as a model enzyme to address some of the limitations that one might encounter when purifying and characterizing a typical three-component RO such as CDO. The following methods for studying CDO, parts of which were originally described in Runda, de Kok et al., 2023; Runda, Kremser et al., 2023 and Runda et al., 2024, can serve as a blueprint for working with other RO enzymes and as a starting point for developing enzyme-specific protocols for future research.

2. *In vitro* analysis of Rieske oxygenases

To study this family of enzymes it is important to be able to subject them to *in vitro* experiments. These *in vitro* experiments are paramount for detailed characterization efforts such as the optimization of reaction conditions, determination of kinetic parameters of the enzyme(s) (Bopp, Bernet, Kohler, & Hofstetter, 2022; Pati, Bopp, Kohler, & Hofstetter, 2022), solving crystal structures, and comparing alternative redox partner interactions (Runda et al., 2024). However, the detailed *in vitro* characterization of ROs poses several challenges which are inherent to their structural and functional complexities: ROs consist of multiple components and some of the components exist as (hetero)multimers (Fig. 1). The presence of multiple proteins, each with their own characteristics and activities, makes it difficult to optimize activity assays and isolate the activity, or other behavior of a single component from experimental data. Furthermore, ROs are cofactor and metal ion dependent, and the proper assembly and maintenance of these cofactors are critical for enzyme activity. For example, the Fe-S clusters present in the RO-Fd and RO-Oxy components (and in some cases the RO-Red component as well) can be sensitive to oxygen, leading to their degradation. Thus, the presence of Fe-S clusters in various RO components can complicate expression and handling of these proteins. In addition, the flavin prosthetic group of a RO-Red and the non-heme catalytic iron center embedded in the catalytic α-subunit of the RO-Oxy represent potential sites for electron leakage due to unproductive oxygen activation. This process is referred to as O_2 uncoupling and results in the formation of reactive oxygen species (ROS) (Bopp et al., 2022; Lee, 1999; Runda et al., 2024). These ROS inactivate the enzyme, which is particularly problematic for *in vitro* characterization studies (Runda et al., 2024).

For ROs involved in aromatic hydroxylation reactions, substrates are hydrophobic, resulting in solubility issues (Imbeault, Powlowski, Colbert, Bolin, & Eltis, 2000); or for ROs involved in natural product synthesis, substrates are often not yet known or not accessible (Perry et al., 2018; Runda, de Kok et al., 2023; Runda, Kremser et al., 2023). Another complicating factor is that the large catalytic repertoire of ROs is facilitated by a radical-mediated mechanism, causing the catalytic outcome (at times) to be dependent on the chemical attributes of the substrate (i.e. functional groups, and/or the electronic environment present within the substrate) (Gally, Nestl, & Hauer, 2015; Perry et al., 2018; Resnick, Lee, & Gibson, 1996; Runda, de Kok et al., 2023; Runda, Kremser et al., 2023). This observation suggests that the chemical attributes of the substrate may play a significant role in determining the (apparent) catalytic activity of the enzyme and therefore, the identity of the reaction product. In addition to enzyme stability, the development of functional assays is also challenged with the detection of activity. The aforementioned electron uncoupling causes reliable RO activity assays to be largely dependent on gas chromatography/liquid chromatography-mass spectrometry (GC/LC-MS) techniques measuring the consumption of substrate or the formation of often-(initially)-unknown products; as the depletion of NAD(P)H is not necessarily productive and is therefore not a reliable metric for activity of the RO as a whole.

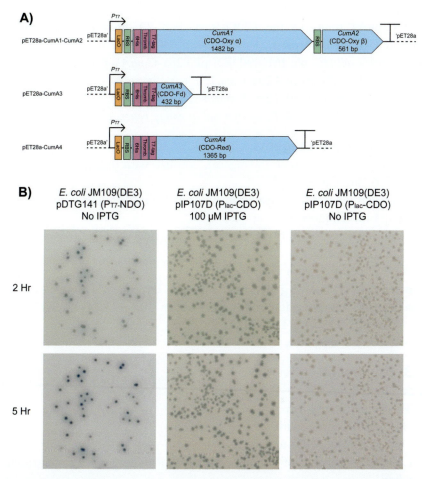

Fig. 2 (A) A schematic overview of the constructs used for overexpression and purification of all three CDO components. Note that only the CDO-Oxy α-subunit bears the His$_6$-tag and that the CDO-Oxy β-subunit is copurified. (B) Colonies of freshly transformed *E. coli* JM109 on a LB agar plate in presence and absence of IPTG expressing NDO (pDTG141) (Suen, 1991) or CDO (pIP107D) (Aoki et al., 1996) after being exposed to indole vapor at room temperature for 2 or 5 h.

3. Expression of Cumene dioxygenase
3.1 Materials
3.1.1 Strains
In the following descriptions, CumA1 and CumA2 are referred to as the α- and β-subunit of CDO-Oxy respectively. CumA3 refers to the CDO-Fd

and CumA4 refers to the CDO-Red. Fig. 2A provides a schematic overview of the constructs.

- *Escherichia coli* (*E. coli*) JM109(DE3) pET28a-*CumA1-CumA2*
- *E. coli* JM109(DE3) pET28a-*CumA3*
- *E. coli* JM109(DE3) pET28a-*CumA4*
- *E. coli JM109(DE3) pET28a (optional as empty vector control)*

3.2 Buffers and reagents

- Kanamycin 1000× stock (50 mg/mL, sterilize with 0.22 μm filter, 1 mL aliquots, store at −20 °C)
- 1 M isopropyl β-d-1-thiogalactopyranoside (IPTG, sterilize with 0.22 μm filter, 1 mL aliquots, store at −20 °C)
- Lysis buffer (50 mM $NaPO_4$ pH 7.2, 300 mM NaCl, 10% glycerol, 30 mM imidazole)
- For 1 L Terrific Broth (TB) medium:
  - o 900 mL TB base (5 g glycerol, 12 g bacto-tryptone, 24 g yeast extract)
  - o 100 mL TB potassium phosphate buffer 10× (23.14 g/L KH_2PO_4 (0.17 M), 125.41 g/L K_2HPO_4 (0.72 M))
  - o Autoclave separately

3.3 Equipment

- Serological pipettes
- Micropipettes and tips
- Shaking incubator (with cooling capabilities)
- Optical density (OD) meter
- Centrifuge (capable of cooling to 4 °C and holding 4 × 1 L centrifuge tubes)

4. Step-by-step method details

Standard sterile inoculation procedures were followed wherein 50 mL of TB media supplemented with kanamycin in a 100 mL baffled Erlenmeyer flask was inoculated from a −80 °C cryo-stock (25% glycerol) for each strain. These inoculated cultures were subsequently incubated overnight at 37 °C with shaking at 200 rpm to facilitate aeration for optimal growth.

The next morning the optical density at 600 nm (OD_{600}) of the overnight cultures was measured and 1 L of TB media supplemented with kanamycin in a 3 L baffled flask was inoculated to an OD_{600} of 0.05

Rieske oxygenase cumene dioxygenase

Cultivation and Protein expression
(Section 2.2.)

Cell harvesting
(Section 3.2.1.)

Cell lysis by Sonication
(Section 3.2.1.)

Purification by IMAC
(Section 3.2.2.)

Desalting by SEC
(Section 3.2.3.)

Concentration by ultrafiltration
(Section 3.2.4.)

Purified protein preparation
(End result)

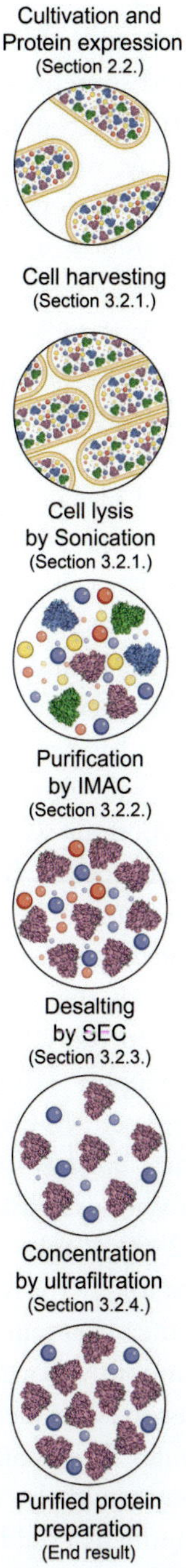

Fig. 3 A schematic overview of key steps in the workflow and

from these overnight cultures. Following inoculation, the cultures were incubated at 37 °C with shaking at 100 rpm for approximately 1.5 h. After this time, OD_{600} measurements were taken to monitor growth and estimate the time at which the OD_{600} would approximate 0.6. Upon reaching this desired OD_{600}, 150 µL of 1 M IPTG was added to induce protein expression and the cultures were transferred to a 20 °C incubator for optimal protein expression.

The following morning, the cultures were transferred to pre-weighed 1 L centrifuge tubes and centrifuged at $3000 \times g$ for 15 min at 4 °C to pellet the cells (Fig. 3). The supernatant was carefully decanted and discarded. The resulting cell pellets were then resuspended in 50 mL of chilled lysis buffer, which was followed by a second round of centrifugation at $3000 \times g$ for 15 min at 4 °C. Following completion of this step, the supernatant was decanted and discarded and the pellets were weighed.

5. General considerations

Besides the typical parameters to optimize protein expression, such as altering the OD_{600} value, or the time, temperature, or IPTG concentration for induction; ROs have

Fig. 3—Cont'd showing the expected state of the sample after each step. Spheres represent various solutes, such as buffers and other additives; differently colored blobs represent different proteins. Note: It is recommended to immediately continue with further purification and activity assays. For CDO-Red the pellet may have a mint-green color instead of yellow.

some additional points to consider, especially due to the presence of Fe-S clusters in the RO-Fd and RO-Oxy components (and sometimes in the two-component RO-Red). Importantly, *E. coli* BL21(DE3) is deficient in Fe-S cluster biogenesis because it lacks the SUF pathway, one of the two Fe-S cluster assembly pathways found in this organism (Corless, Mettert, Kiley, & Antony, 2020). Therefore, the use of Fe-S cluster non-deficient expression strains (such as *E. coli* JM109(DE3) (Parales, Lee et al., 2000; Parales, Resnik et al., 2000; Runda, de Kok et al., 2023; Runda, Kremser et al., 2023) or Suf^{++} strain (Corless et al., 2020)) or strains co-expressing Fe-S cluster biosynthetic gene clusters is recommended (Zheng, Cash, Flint, & Dean, 1998). Additionally, it has been demonstrated that co-expression of Fe-S cluster assembly genes can improve Fe-S loading and overall expression of Fe-S-containing proteins (Corless et al., 2020; Jansing, Mielenbrink, Rosenbach, Metzger, & Span, 2023). To further aid Fe-S assembly and incorporation, some ROs are expressed in the presence of supplemental Fe^{2+} and Fe^{3+} (Liu, Knapp et al., 2022; Liu, Tian et al., 2022; Lukowski et al., 2018; Lukowski, Liu, Bridwell-Rabb, & Narayan, 2020). Alternatively, the (anaerobic) chemical reconstitution of Fe-S Rieske centers has been reported for PrnD (Lee, Simurdiak et al., 2005). Moreover, due to the oxygen sensitivity of Fe-S clusters it can be beneficial to immediately continue with further purification and activity assays instead of storing the cell pellet, lysate, or purified protein. However, if shorter expression times are found to be optimal, storing the cell pellet would be preferred over storing lysate or purified protein.

When all three CDO components are co-expressed, the activity of CDO is easily observed, as the culture turns blue due to the CDO-catalyzed formation of indigo. The formation of indigo upon exposure to indole was originally reported for naphthalene dioxygenase (NDO), which is involved in the degradation of aromatic xenobiotics, but has not been reported for ROs involved in natural product biosynthesis (Ensley & Gibson, 1983; Ensley et al., 1983; Parales, Lee et al., 2000; Parales, Resnik et al., 2000). This property can conveniently be exploited to assess the activity of some ROs *in vivo*. This type of activity measurement can be performed by exposing colonies on an agar plate to indole vapor through the placement of filter paper containing 10% indole dissolved in acetone on a watch glass or in the lids of inverted agar plates (Parales, Lee et al., 2000; Parales, Resnik et al., 2000) (Fig. 2B). This indole-assay could also be used to assess whether an uncharacterized RO-Oxy component could accept electrons from endogenous ferredoxins or reductases, provided that indole is a suitable substrate.

6. Cell lysis and protein purification
6.1 Materials and equipment
6.1.1 Buffers and reagents
- Lysis buffer (50 mM NaPO$_4$ pH 7.2, 300 mM NaCl, 10% glycerol, 30 mM imidazole)
- Wash buffer (50 mM NaPO$_4$ pH 7.2, 300 mM NaCl, 10% glycerol, 100 mM imidazole)
- Elution buffer (50 mM NaPO$_4$ pH 7.2, 300 mM NaCl, 10% glycerol, 400 mM imidazole)
- Desalting buffer (50 mM NaPO$_4$ pH 7.2, 300 mM NaCl, 10% glycerol)
- Ni-NTA agarose (Qiagen, cat: 30210)
- 15 mL unpacked gravity columns (Screening devices, EP .015.08.20.50 P)
- PD-10 desalting columns (Cytiva, cat: 17-0851-01)
- Amicon 10 kDa molecular weight cutoff (MWCO) ultrafiltration column (Merck, cat: UFC901008)
- *Optional: cOmplete EDTA-free Protease inhibitor tablets (Merck, cat: 11873580001)*
- *Optional: lysozyme from chicken egg white (Merck, cat: L6876-10G)*

6.2 Equipment
- Serological pipettes
- Micropipettes and tips
- Ice dispenser
- Sonicator (Branson 450 Analog Sonifier)
- Cold room/fridge
- Rollerbench/rotary shaker
- Centrifuge (capable of cooling to 4 °C, ≥20,000 × *g* and holding 50 mL centrifuge tubes)

7. Step-by-step method details
7.1 Cell lysis by sonication
Note: Keep buffers and biological materials on ice wherever possible from this point onward.

The cell pellets were resuspended in 3 mL of lysis buffer per 1 g wet cell weight (~25% (w/v) cell wet weight). *Optional: 1 mg/g wet cell weight of lysozyme was added; and 0.2 mL/g wet cell weight of EDTA-free protease inhibitor solution (1 tablet in 5 mL lysis buffer) was added.* The resulting cell suspension was then split into 20–30 mL portions in 50 mL centrifuge tubes. To lyse the cells, one of the 50 mL centrifuge tubes containing cell suspension was inserted into ice-water slush and a 13 mm diameter flat-tip sonication probe was inserted to approximately 30% the depth of the liquid (including the probe). The sonicator was then activated for ten 1-minute cycles where the sonicator was ON for 30 s (set at a 50% duty cycle in which cycles have a fixed duration of 1 s) and OFF for 30 s with the "output control setting" on "7". Afterwards the tube, now containing lysate, was transferred back to ice and the process was repeated for each 50 mL centrifuge tube with washed cell suspension. This sonication procedure was followed by centrifugation of all lysate tubes at $17,000 \times g$ for 45 min at 4 °C after which the supernatant, the cell free extract (CFE), was transferred to a new 50 mL centrifuge tube (Fig. 3). 100 µL of CFE was transferred to 1.5 mL microtubes and the 50 mL tubes with the remaining pellet were all stored at −20 °C for sodium dodecyl sulfate – polyacrylamide gel electrophoresis (SDS-PAGE) analysis (Fig. 4).

Note: The freshly clarified CFE of CDO-Red may appear gray-blue due to the stabilized semiquinone radical state of the flavin cofactor. CDO-Red will adopt the typical yellow color of flavoproteins during the following purification steps.

7.2 Purification by immobilized metal ion chromatography (IMAC)

Note: Do not let the Ni-NTA agarose beads run dry.

During the previous centrifugation step, 4 mL (2 mL column volume, CV) of Ni-NTA agarose bead suspension (~50% v/v) per total liter of culture were transferred to a 50 mL centrifuge tube. This solution was briefly pulse-centrifuged to pellet the Ni-NTA agarose beads and the supernatant was decanted and discarded. Then 10 CV of lysis buffer was added to equilibrate the beads. This mixture was briefly pulse centrifuged again and the supernatant discarded. The beads were then resuspended in 1 CV of lysis buffer (to reach the original volume), and 2 mL/L culture of this Ni-NTA agarose bead suspension (~50% v/v) was added into the lysate tubes. *(This step would typically correspond to ~ 1 mL of bead suspension per tube if the cell material was equally split in two portions prior to lysis).*

Note: The following steps were performed in a cold-room (~6 °C).

The beads were incubated with the CFE for 30–60 min with gentle mixing (e.g. on a rollerbench or rotary shaker). During this time, one 15 mL gravity column per protein was positioned above a 50 mL centrifuge tube. After incubation, the lysate-bead suspensions were transferred to the

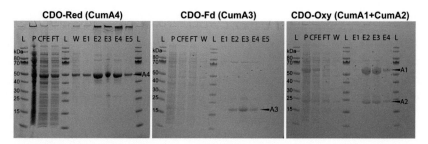

Fig. 4 SDS-PAGE gels allow for various fractions of the CumA1+A2, CumA3 and CumA4 purification process to be analyzed. Proteins of interest are marked with a triangle. Note: The Ni-NTA elution fractions in this image were 0.5 CV instead of 1 CV as described in the protocol.

corresponding gravity columns and the flow-through was collected. After the flow-through fraction was collected, the collection tubes were replaced with 15 mL centrifuge tubes and 10 CV of wash buffer was added to the column. After collection of the wash fraction, the 15 mL centrifuge tubes containing the wash-fraction were replaced with 1.5 mL microtubes. To elute the proteins of interest, 1 CV of elution buffer was added to the column and the elution fraction was collected. This step was then repeated three times, to collect all four elution fractions (Fig. 3). Finally, 12 μL of all fractions was transferred to separate 1.5 mL microtubes and stored at −20 °C to be analyzed by SDS-PAGE (Fig. 4).

Note: Typically, elution fraction 2 will contain the highest amount of His_6-tagged protein. At appreciable concentrations, solutions of fully oxidized RO-Red are typically colored yellow (or orange if containing a Fe-S cluster), RO-Fd brown, and RO-Oxy brown with a slight pink/red-hue compared to RO-Fd.

7.3 Desalting by size exclusion chromatography (SEC)

Note: The following steps were performed in a cold-room (∼6 °C) up to and including the collection of the desalted protein.

To start with desalting of the proteins, the pre-packed PD-10 columns were opened and the storage solution was decanted off. Desalting buffer was then added to fill the columns completely. The flow-through was discarded after the desalting buffer had completely entered the column bed. This series of steps was repeated four more times to completely remove the storage solution and equilibrate the column. Up to 2.5 mL of pooled Ni-NTA fractions containing the protein of interest was added to the column. Typically, this pooled fraction consists of the entire Ni-NTA elution fraction 2, and a pool of fractions 1 and 3, or the samples that typically contain the most amount of protein at an appreciable purity. *Note: if the sample volume is less than 2.5 mL, equilibration buffer should be added to create a total final volume of 2.5 mL.* After the sample completely entered the column bed, the flow-through was discarded and fresh collection tubes were placed under the desalting columns. The proteins of interest were then eluted and collected by addition of 3.5 mL desalting buffer onto the column bed (Fig. 3). After collection of the desalted proteins, the absorbance at 280 nm was measured, compared to the desalting buffer, and used along with sequence-based extinction coefficient predictions to approximate protein concentration.

Note: If the protein concentration is sufficiently high (>100–200 µM), the following concentration step can be skipped.

7.4 Concentration by centrifugal ultrafiltration

To concentrate the desalted proteins, the desalted protein eluates were transferred to the filter units of Amicon 10 kDa MWCO ultrafiltration tubes. The tubes containing the filter units were then centrifuged at $4000 \times g$ for 10 min at 4 °C and the remaining liquid volume was checked. If the desired volume had not been reached yet, the centrifugation step was repeated. If the sample volume was far too low (and thus concentration far too high), desalting buffer was added to approximately dilute the protein solution to the desired concentration. To collect the concentrated protein solutions, slow up- and down-pipetting in the filter unit was used to homogenize the protein samples, which were then transferred into new 1.5 mL tubes (Fig. 3). After collecting the concentrated proteins, the absorbance at 280 nm was measured compared to the desalting buffer and the approximate protein concentration was determined using sequence-based extinction coefficient predictions. A portion of the purified, desalted and concentrated proteins was set aside for activity assays and any excess was aliquoted, flash frozen in liquid nitrogen and stored at −80 °C.

8. General considerations

The proteins will be exposed to buffer in this part of the protocol. Hence, choosing a buffer condition in which the enzyme is stable is crucial. Phosphate buffers are widely used in enzymology, also for work involving ROs; in spite of the successful application of phosphate buffers, non-heme iron containing metalloenzymes can be susceptible to stripping of metal ions. Therefore, buffering systems that minimize metal ion binding – such as Good's buffers (e.g. MES (Wolfe & Lipscomb, 2003), MOPS or CHES) – should be explored at some point in the characterization of ROs. However, Good's buffers containing piperazine rings, e.g. HEPES and PIPES, should be avoided due to possible interference in redox processes and reported formation of radical species (Grady, Chasteen, & Harris, 1988).

The purification of non-heme iron containing metalloenzymes by metal affinity chromatography using a His_6-tag can be problematic due to tag-induced folding or oligomeric state perturbations, metal cofactor stripping or exchange, inhibitory effects of metal ions leached from the column matrix (typically $< 5\,\mu M$) (Block et al., 2009; Kokhan, Marzolf, 2019), or inhibitory effects from imidazole itself (Kokhan et al., 2010). The latter limitations can be avoided by a buffer exchange step (as included in this protocol). The first two obstacles are more fundamental in nature and in these cases, a different purification technique should be chosen, either by using of a different fusion tag (e.g. streptavidin or maltose binding protein (Lukowski et al., 2018)) or a tagless purification approach. However, due to potential oxygen sensitivity and need to expedite the purification process, the use of a tag-based technique is recommended. Nevertheless, the well-studied NDO (Ensley & Gibson, 1983; Ensley et al., 1983; Kyoung, Kauppi, Parales, Gibson, & Ramaswamy, 1997) and other ROs (Dong et al., 2005; Imbeault et al., 2000; Kyoung et al., 1997; Lee, Friemann, Parales, Gibson, & Ramaswamy, 2005), such as nitroarene dioxygenases (NBDO, 2NTDO) (Bopp et al., 2022) have been successfully purified using tagless techniques. Alternatively, oxygen exposure can be limited, as was done for the tagless purification of biphenyl dioxygenase (BPDO), where the majority of the purification procedure was performed in an anaerobic glovebox and the purified protein stored under liquid nitrogen (Imbeault et al., 2000).

The RO-Oxy component is only active as a multimer, as the electron transport required for the catalytic cycle takes place between two α-sub-units of the oxygenase (Ferraro et al., 2005; Runda, de Kok et al., 2023; Runda, Kremser et al., 2023). RO-Oxy components have been found to exist as homotrimeric (α_3) (Nojiri et al., 2005), homohexameric ($\alpha_3\alpha_3$) (Mahto et al., 2021), or heterohexameric ($\alpha_3\beta_3$ or $\alpha_3\alpha'_3$) (Dong et al., 2005; Friemann et al., 2009; Kauppi et al., 1998; Kim et al., 2019) complexes, in which the α-subunits have a catalytic function and the β-subunits have a putative structural function (Runda, de Kok et al., 2023; Runda, Kremser et al., 2023). These different architectural options highlight an important consideration – it is important to consider whether to tag the α-subunit, the α'-, or β-subunit, or both. Published studies on ROs tend to only tag the α-subunit where the α'- or β-subunit is co-purified (Figs. 2 and 4) (Kim et al., 2019; Runda, de Kok et al., 2023; Runda, Kremser et al., 2023; Tsai et al., 2022), or use tagless purification approaches (Dong et al., 2005; Kyoung et al., 1997; Lee, Friemann et al., 2005).

This protocol uses gravity size exclusion chromatography for the desalting (or buffer exchange) step. Whereas this method is time efficient, there is a higher risk of precipitation due to the sudden change in solutes compared to dialysis, which involves a more gradual change in solutes. Despite the potential oxygen sensitivity of RO components, there are reports of dialysis being used successfully in the purification of ROs (Lee, Ang, & Zhao, 2006; Lee, Simurdiak et al., 2005; Lukowski et al., 2018; Lukowski et al., 2020; Ohta, Chakrabarty, Lipscomb, & Solomon, 2008). Moreover, size exclusion chromatography typically results in sample dilution, which must be followed by a concentration step. At sufficiently high concentrations, RO components are visibly colored. This fact can be exploited by collecting only the colored size exclusion eluate, resulting in higher protein concentrations and possibly eliminating or reducing the need for a subsequent concentration step. Alternatively, ultrafiltration can also be used for desalting (or buffer exchange) by repeated dilution with the buffer of choice after concentration (Ensley & Gibson, 1983; Ensley et al., 1983; Lee et al., 2006; Lee, Friemann et al., 2005; Lee, Simurdiak et al., 2005; Lukowski et al., 2018; Lukowski et al., 2020). However, care must be taken not to overconcentrate the proteins. In addition, this method also increases the previously described risk of potentially shocking the proteins of interest by sudden changes in solute concentration and thereby inducing aggregation.

9. Enzymatic activity assay
9.1 Materials and equipment
9.1.1 Buffers and reagents
- MilliQ water (MQ)
- 1 M NaPO$_4$ pH 7.2 (or another buffer)
- 3 M NaCl
- 50 mg/mL Catalase in MQ (Sigma-Aldrich, cat: C9322-1G)
- 100 mM Dithiothreitol in MQ (DTT, freshly prepared)

- 5 mM NAD(P)H in MQ (freshly prepared)
- Indene (or optionally, another suitable substrate)
- Purified and desalted CDO-Oxy, CDO-Fd, CDO-Red
- Dichloromethane (DCM)
- Acetophenone
- MgSO$_4$ (anhydrous)

9.2 Equipment
- Serological pipettes
- Micropipettes and tips
- Rollerbench/rotary shaker
- Microtube centrifuge (capable of cooling to 4 °C)
- Incubator
- 20 mL glass vials with PTFE lined cap
- 1.5 mL glass vials with PTFE lined septum cap
- 1.5 mL microtubes
- Vortex
- GC-MS instrument
  o GC instrument: Shimadzu GC-2010 Plus
  o MS detector: Shimadzu GCMS-QP2010 SE
  o GC column: Agilent HP-5MS (30 m × 0.25 mm, 0.25 μm)
- Chiral Gas chromatography-flame ionization detection (GC-FID) instrument
  o GC instrument: Shimadzu GC-2010
  o GC column: Astec CHIRALDEX G-TA (30 m × 0.25 mm, 0.12 μm)

10. Step-by-step method details
10.1 Enzymatic reaction
Enzymatic reactions with indene as a model substrate to evaluate CDO activity were prepared. The reactions were 1 mL in volume and prepared in 20 mL air-tight sealed glass vials to provide adequate headspace for the supply of atmospheric O$_2$ (Table 1).

The completed reaction mixtures were incubated at 30 °C in a shaking incubator with 120 rpm shaking for 24 h to perform end-point measurements.

10.2 Sample extraction
To extract the analytes from the reactions, 500 μL of the reaction mixture was transferred to a new 1.5 mL microtube and a spatula tip of NaCl was

Table 1 The components of the reaction mixture with their listed stock concentration, final concentration and the volume to add per single reaction. Reaction components were added in descending order.

Component	Stock concentration	Final concentration	Volume to add to reaction
MQ	N/A	N/A	564 µL
NaPO$_4$ buffer pH 7.2	1 M	50 mM	50 µL
NaCl	3 M	200 mM	66 µL
Catalase	50 mg/mL	1 mg/mL	20 µL
DTT	100 mM	1 mM	10 µL
NAD(P)H	100 mM	5 mM	50 µL
Indene	Pure	5 mM	0.583 µL (0.6 µL)
CDO-Oxy	~100 µM	4 µM	40 µL
CDO-Fd	~200 µM	30 µM	150 µL
CDO-Red	~200 µM	10 µM	50 µL

added to saturate the solution. Then, 500 µL of DCM (containing 2 mM of acetophenone as an internal standard) was added to the NaCl-saturated microtubes, which were then vigorously vortexed to mix. To aid phase separation, the tubes were centrifuged at 17,000 × g for 1 min at room temperature and the separated (bottom-) organic phase was transferred to a fresh 1.5 mL microtube. After the organic phase was collected, a spatula tip of anhydrous MgSO$_4$ was added and vigorously vortexed to dry the organic phase. This drying phase was followed by another centrifugation step at 17,000 × g for 5 min at room temperature. The dried organic phase was transferred to a 1.5 mL glass vial, which was subsequently capped, and ready for analysis by GC–MS.

10.3 Non-chiral GC-MS and chiral GC-FID analysis

For all GC analyses, the samples were loaded into the autosampler. First, a solvent blank was injected, followed by the reactions, and finishing with a solvent blank.

For non-chiral GC-MS, the sample injection volume was $2\,\mu L$ at $250\,°C$ with a split ratio of 25. The carrier gas was N_2 with the flow controlled to a linear velocity of $50\,cm/s$. The oven was set to hold at $60\,°C$ for 2 min, then raised to $100\,°C$ over 8 min, and then to $300\,°C$ over 14 min. This final step was followed by holding the temperature at $300\,°C$ for 2 min. The MS source temperature was set at $200\,°C$ with an interface temperature of $250\,°C$ and a scanning range of 50–$350\,m/z$.

For chiral GC-FID, the injection volume was $1\,\mu L$ at $230\,°C$ with a split ratio of 20.5. The carrier gas was N_2 with the flow was controlled to a pressure of 158.7 kPa. The oven was set to hold at $70\,°C$, increase to $140\,°C$ over 3 min, and then hold for 40 min at $140\,°C$. The final steps involved an increase to $170\,°C$ over 15 min and holding at $170\,°C$ for 5 min. The FID temperature was set at $250\,°C$.

The peaks in the chromatograms were compared to (authentic) reference compounds where possible and the mass spectral data was used to aid in identification of enzymatic products. The peak area of analytes was then integrated for quantification against a calibration curve of a reference compound. No authentic reference compound was commercially available for the 1H-indenol product and the synthesis has been reported as challenging due to the formation of an explosive intermediate (Onaran & Seto, 2003). Therefore, a calibration curve of 1-indanone was established and used to evaluate the concentration of 1H-indenol by approximation through the theoretical relative response factor as follows: the theoretical relative response factor of 1H-indenol to 1-indanone was found to be 1.027 based on calculated effective carbon numbers (ECN 1H-indenol and 1-indanone: 8.42 and 8.20 resp.) according to a model by Jorgensen, et al. (Jorgensen, Picel, & Stamoudis, 1990). The obtained RRF value was then used to adjust the integrated peak areas of 1H-indenol. This allowed for approximation of 1H-indenol concentrations based on the linear regression of the 1-indanone calibration curve.

11. General considerations

Since ROs are O_2-dependent enzymes, it is highly recommended to perform the reactions in 20 mL glass vials to provide sufficient headspace. In addition, several additives have been shown to be beneficial to the activity of CDO. For example, the addition of catalase and DTT is recommended to mitigate the effect of electron transfer chain uncoupling, which results in the generation of reactive oxygen species that are harmful to the enzymes. ROs are also dependent on the supply of reducing equivalents from NAD(P)H, but the addition of equimolar amounts of the nicotinamide cofactors is not recommended from an economic standpoint because they are relatively unstable and expensive. Alternatively, small amounts of NAD(P)H (0.4 mM) can be added with 10U or 2 μM of *Bacillus megaterium* glucose dehydrogenase (BmGDH) and 50 mM D-glucose to regenerate NAD(P)H from NAD(P)$^+$ *in situ* (Runda, de Kok et al., 2023; Runda, Kremser et al., 2023). However, the use of this cofactor regeneration system resulted in the formation of more ROS as shown by an 2,2′-Azino-bis(3-ethylbenzothiazoline-6-sulfonic acid) diammonium salt (ABTS) assay (Runda et al., 2024). ROS formation is presumably caused due to increased uncoupling, but ultimately still results in higher activity of the RO system. A disadvantage of cofactor regeneration is that NAD(P)H depletion cannot be measured to assess enzymatic activity, which completely eliminates this option for the development of plate-based screening assays.

Depending on the substrate(s) and product(s) that are being analyzed, different extraction solvents should be considered, such as ethyl acetate (EtOAc), methyl tert-butyl ether (MTBE) or preferably even solvents derived from biowaste or biorenewable sources such as, 2-methyltetrahydrofuran (2-MeTHF) or 1-butanol (1-BuOH).

The analysis technique can also be changed depending on the reaction products and available in-house equipment. In some cases, LC-MS could be used instead of GC-MS. However, if the reaction products are of low molecular weight and/or volatile (as is the case for this protocol), the extract cannot be processed in the same way. To prepare the sample for injection, the organic fraction would have to be dried and the residue redissolved (or diluted into) in a more polar solvent. This step would cause the analytes of interest to evaporate. To avoid this problem, extraction can be skipped altogether, or the extract can be diluted in a more polar organic solvent (at the expense of reduced sensitivity). Detection of analytes in the low m/z range is also more challenging in LC-MS analysis because of the relatively poor signal-to-noise ratio (compared to GC-MS) due to interference from solvent molecules. Therefore, while LC-MS analysis may be a viable alternative technique to GC-MS, it is not recommended for the analysis of CDO-catalyzed aromatic hydrocarbon oxyfunctionalization described in this protocol. In contrast to ROs that primarily oxyfunctionalize aromatic hydrocarbons; ROs that play a role in the biosynthesis of secondary metabolites typically have substrates and associated products that are more polar and are of higher molecular weight as to usually not be volatile. For these ROs, depending on the exact nature of the molecules involved, LC-MS would typically be the analysis technique of choice for direct detection.

12. Summary and conclusions

This chapter summarizes techniques for characterizing a RO that oxyfunctionalizes aromatic hydrocarbons. The protocols described and referenced here can serve as a blueprint for characterizing novel ROs and uncovering novel reactivities, thereby improving the understanding of biosynthetic pathways involving ROs and ultimately expanding the available biocatalytic toolbox. The methodologies for expression, purification, and assay of CDO activity rely on common techniques such as an IPTG-inducible T7 expression system, immobilized metal ion affinity chromatography for purification, a buffer exchange and concentration step based on ultrafiltration and size exclusion chromatography, and an activity assay based on GC-MS. There are several potential improvements to the protocols for CDO characterization; these and several alternative techniques that may prove useful for characterizing other ROs were highlighted. The use of this generally applicable workflow and the proposed solutions to the anticipated limitations of studying ROs *in vitro* thus provide a framework for future RO research and development strategies for this class of enzymes.

Acknowledgments

H.M. acknowledges funding from the China Scholarship Council.

The authors would like to thank M.E. Runda for his input and the fruitful discussions supporting this chapter.

References

Aoki, H., Kimura, T., Habe, H., Yamane, H., Kodama, T., & Omori, T. (1996). Cloning, Nucleotide Sequence, and Characterization of the Genes Encoding Enzymes Involved in the Degradation of Cumene to 2-Hydroxy-6-Oxo-7-Methylocta-2,4-Dienoic Acid in Pseudomonas fluorescens IP01. *Journal of Fermentation and Bioengineering, 81*(3), 187–196. https://doi.org/10.1016/0922-338X(96)82207-0.

Barry, S. M., & Challis, G. L. (2013). Mechanism and catalytic diversity of rieske non-heme iron-dependent oxygenases. *ACS Catalysis, 3*(10), 2362–2370. https://doi.org/10.1021/cs400087p.

Block, H., Maertens, B., Spriestersbach, A., Brinker, N., Kubicek, J., Fabis, R., et al. (2009). *Chapter 27 immobilized-metal affinity chromatography (IMAC). A review. Methods in enzymology, 463*, Academic Press, 439–473. https://doi.org/10.1016/S0076-6879(09)63027-5.

Bopp, C. E., Bernet, N. M., Kohler, H. P. E., & Hofstetter, T. B. (2022). Elucidating the role of O_2 uncoupling in the oxidative biodegradation of organic contaminants by Rieske non-heme iron dioxygenases. *ACS Environmental Au, 2*(5), 428–440. https://doi.org/10.1021/acsenvironau.2c00023.

Brimberry, M., Garcia, A. A., Liu, J., Tian, J., & Bridwell-Rabb, J. (2023). Engineering Rieske oxygenase activity one piece at a time. *Current Opinion in Chemical Biology, 72*, 102227. https://doi.org/10.1016/j.cbpa.2022.102227.

Chang, F. Y., & Brady, S. F. (2014). Characterization of an environmental DNA-derived gene cluster that encodes the Bisindolylmaleimide methylarcyriarubin. *Chembiochem: A European Journal of Chemical Biology, 15*(6), 815–821. https://doi.org/10.1002/cbic.201300756.

Corless, E. I., Mettert, E. L., Kiley, P. J., & Antony, E. (2020). Elevated expression of a functional suf pathway in Escherichia sulfur cluster-containing protein. *Journal of Bacteriology, 202*(3), 1–11. https://jb.asm.org/content/202/3/e00496-19.

Dong, X., Fushinobu, S., Fukuda, E., Terada, T., Nakamura, S., Shimizu, K., et al. (2005). Crystal structure of the terminal oxygenase component of cumene dioxygenase from *Pseudomonas fluorescem* IP01. *Journal of Bacteriology, 187*(7), 2483–2490. https://doi.org/10.1128/JB.187.7.2483-2490.2005.

Ensley, B. D., & Gibson, D. T. (1983). Naphthalene dioxygenase: Purification and properties of a terminal oxygenase component. *Journal of Bacteriology, 155*(2), 505–511. https://doi.org/10.1128/jb.155.2.505-511.1983.

Ensley, B. D., Ratzkin, B. J., Osslund, T. D., Simon, M. J., Wackett, L. P., & Gibson, D. T. (1983). Expression of naphthalene oxidation genes in *Escherichia coli* results in the biosynthesis of indigo. *Science (New York, N. Y.), 222*(4620), 167–169. https://doi.org/10.1126/science.6353574.

Fei, X., Yin, X., Zhang, L., & Zabriskie, T. M. (2007). Roles of VioG and VioQ in the incorporation and modification of the capreomycidine residue in the peptide antibiotic viomycin. *Journal of Natural Products, 70*(4), 618–622. https://doi.org/10.1021/np060605u.

Ferraro, D. J., Gakhar, L., & Ramaswamy, S. (2005). Rieske business: Structure-function of Rieske non-heme oxygenases. *Biochemical and Biophysical Research Communications, 338*(1), 175–190. https://doi.org/10.1016/j.bbrc.2005.08.222.

Feyza Özgen, F. F., Runda, M. E., Burek, B. O., Wied, P., Bloh, J. Z., Kourist, R., et al. (2020). Artificial light-harvesting complexes enable Rieske oxygenase catalyzed hydroxylations in non-photosynthetic cells. *Angewandte Chemie International Edition, 59*(10), 3982–3987. https://doi.org/10.1002/anie.201914519.

Friemann, R., Lee, K., Brown, E. N., Gibson, D. T., Eklund, H., & Ramaswamy, S. (2009). Structures of the multicomponent Rieske non-heme iron toluene 2,3-dioxygenase enzyme system. *Acta Crystallographica Section D: Biological Crystallography, 65*(1), 24–33. https://doi.org/10.1107/S0907444908036524.

Gally, C., Nestl, B. M., & Hauer, B. (2015). Engineering Rieske non-heme iron oxygenases for the asymmetric dihydroxylation of alkenes. *Angewandte Chemie - International Edition, 54*(44), 12952–12956. https://doi.org/10.1002/anie.201506527.

Grady, J. K., Chasteen, N. D., & Harris, D. C. (1988). Radicals from "Good's" buffers. *Analytical Biochemistry, 173*(1), 111–115. https://doi.org/10.1016/0003-2697(88)90167-4.

Guth, F. M., Lindner, F., Rydzek, S., Peil, A., Friedrich, S., Hauer, B., et al. (2023). Rieske oxygenase-catalyzed oxidative late-stage functionalization during complex antifungal polyketide biosynthesis. *ACS Chemical Biology, 18*(12), 2450–2456. https://doi.org/10.1021/acschembio.3c00498.

Heinemann, P. M., Armbruster, D., & Hauer, B. (2021). Active-site loop variations adjust activity and selectivity of the cumene dioxygenase. *Nature Communications, 12*(1), 1095. https://doi.org/10.1038/s41467-021-21328-8.

Hu, D. X., Withall, D. M., Challis, G. L., & Thomson, R. J. (2016). Structure, chemical synthesis, and biosynthesis of prodiginine natural products. *Chemical Reviews, 116*(14), 7818–7853. https://doi.org/10.1021/acs.chemrev.6b00024.

Hu, W. Y., Li, K., Weitz, A., Wen, A., Kim, H., Murray, J. C., et al. (2022). Light-driven oxidative demethylation reaction catalyzed by a Rieske-type non-heme iron enzyme Stc2. *ACS Catalysis, 12*(23), 14559–14570. https://doi.org/10.1021/acscatal.2c04232.

Imbeault, N. Y. R., Powlowski, J. B., Colbert, C. L., Bolin, J. T., & Eltis, L. D. (2000). Steady-state kinetic characterization and crystallization of a polychlorinated biphenyl-transforming dioxygenase. *Journal of Biological Chemistry, 275*(17), 12430–12437. https://doi.org/10.1074/jbc.275.17.12430.

Immanuel, S., Sivasubramanian, R., Gul, R., & Dar, M. A. (2020). Recent progress and perspectives on electrochemical regeneration of reduced nicotinamide adenine dinucleotide (NADH). *Chemistry - An Asian Journal, 15*(24), 4256–4270. https://doi.org/10.1002/asia.202001035.

Jansing, M., Mielenbrink, S., Rosenbach, H., Metzger, S., & Span, I. (2023). Maturation strategy influences expression levels and cofactor occupancy in Fe–S proteins. *Journal of Biological Inorganic Chemistry, 28*(2), 187–204. https://doi.org/10.1007/s00775-022-01972-1.

Jerina, D. M., Daly, J. W., Jeffrey, A. M., & Gibson, D. T. (1971). Cis-1,2-dihydroxy-1,2-dihydronaphthalene: A bacterial metabolite from naphthalene. *Archives of Biochemistry and Biophysics, 142*(1), 394–396. https://doi.org/10.1016/0003-9861(71)90298-0.

Jorgensen, A. D., Picel, K. C., & Stamoudis, V. C. (1990). Prediction of gas chromatography flame ionization detector response factors from molecular structures. *Analytical Chemistry, 62*(7), 683–689. https://doi.org/10.1021/ac00206a007.

Kauppi, B., Lee, K., Carredano, E., Parales, R. E., Gibson, D. T., Eklund, H., et al. (1998). Structure of an aromatic-ring-hydroxylating dioxygenasenaphthalene 1,2-dioxygenase. *Structure (London, England: 1993), 6*(5), 571–586. https://doi.org/10.1016/S0969-2126(98)00059-8.

Kim, H. J., Karki, S., Kwon, S. Y., Park, S. H., Nahm, B. H., Kim, Y. K., et al. (2014). A single module type I polyketide synthase directs de Novo macrolactone biogenesis during galbonolide biosynthesis in *Streptomyces galbus*. *Journal of Biological Chemistry, 289*(50), 34557–34568. https://doi.org/10.1074/jbc.M114.602334.

Kim, J. H., Kim, B. H., Brooks, S., Kang, S. Y., Summers, R. M., & Song, H. K. (2019). Structural and mechanistic insights into caffeine degradation by the bacterial N-demethylase complex. *Journal of Molecular Biology, 431*(19), 3647–3661. https://doi.org/10.1016/j.jmb.2019.08.004.

Kokhan, O., & Marzolf, D. R. (2019). Detection and quantification of transition metal leaching in metal affinity chromatography with hydroxynaphthol blue. *Analytical Biochemistry, 582*, 113347. https://doi.org/10.1016/j.ab.2019.113347.

Kokhan, O., Shinkarev, V. P., & Wraight, C. A. (2010a). Binding of imidazole to the heme of cytochrome c1and inhibition of the bc1 complex from *Rhodobacter sphaeroides*: II. kinetics and mechanism of binding. *Journal of Biological Chemistry, 285*(29), 22522–22531. https://doi.org/10.1074/jbc.M110.128058.

Kweon, O., Kim, S. J., Baek, S., Chae, J. C., Adjei, M. D., Baek, D. H., et al. (2008). A new classification system for bacterial Rieske non-heme iron aromatic ring-hydroxylating oxygenases. *BMC Biochemistry, 9*(1), 11. https://doi.org/10.1186/1471-2091-9-11.

Kyoung, L., Kauppi, B., Parales, R. E., Gibson, D. T., & Ramaswamy, S. (1997). Purification and crystallization of the oxygenase component of naphthalene dioxygenase in native and selenomethionine-derivatized forms. *Biochemical and Biophysical Research Communications, 241*(2), 553–557. https://doi.org/10.1006/bbrc.1997.7863.

Lee, J. K., Ang, E. L., & Zhao, H. (2006). Probing the substrate specificity of aminopyrrolnitrin oxygenase (PrnD) by mutational analysis. *Journal of Bacteriology, 188*(17), 6179–6183. https://doi.org/10.1128/JB.00259-06.

Lee, J., Simurdiak, M., & Zhao, H. (2005). Reconstitution and characterization of aminopyrrolnitrin oxygenase, a Rieske N-oxygenase that catalyzes unusual arylamine oxidation. *Journal of Biological Chemistry, 280*(44), 36719–36728. https://doi.org/10.1074/jbc.M505334200.

Lee, K. (1999). Benzene-induced uncoupling of naphthalene dioxygenase activity and enzyme inactivation by production of hydrogen peroxide. *Journal of Bacteriology, 181*(9), 2719–2725. https://doi.org/10.1128/jb.181.9.2719-2725.1999.

Lee, K., Friemann, R., Parales, J. V., Gibson, D. T., & Ramaswamy, S. (2005). Purification, crystallization and preliminary X-ray diffraction studies of the three components of the toluene 2,3-dioxygenase enzyme system. *Acta Crystallographica Section F: Structural Biology and Crystallization Communications, 61*(7), 669–672. https://doi.org/10.1107/S1744309105017549.

Li, W., Estrada-De Los Santos, P., Matthijs, S., Xie, G. L., Busson, R., Cornelis, P., et al. (2011). Promysalin, a salicylate-containing pseudomonas putida antibiotic, promotes surface colonization and selectively targets other pseudomonas. *Chemistry and Biology, 18*(10), 1320–1330. https://doi.org/10.1016/j.chembiol.2011.08.006.

Liu, J., Knapp, M., Jo, M., Dill, Z., & Bridwell-Rabb, J. (2022). Rieske oxygenase catalyzed C-H bond functionalization reactions in chlorophyll b biosynthesis. *ACS Central Science, 8*(10), 1393–1403. https://doi.org/10.1021/acscentsci.2c00058.

Liu, J., Tian, J., Perry, C., Lukowski, A. L., Doukov, T. I., Narayan, A. R. H., et al. (2022). Design principles for site-selective hydroxylation by a Rieske oxygenase. *Nature Communications, 13*(1), 1–13. https://doi.org/10.1038/s41467-021-27822-3.

Lukowski, A. L., Ellinwood, D. C., Hinze, M. E., Deluca, R. J., Du Bois, J., Hall, S., et al. (2018). C-H hydroxylation in paralytic shellfish toxin biosynthesis. *Journal of the American Chemical Society, 140*(37), 11863–11869. https://doi.org/10.1021/jacs.8b08901.

Lukowski, A. L., Liu, J., Bridwell-Rabb, J., & Narayan, A. R. H. (2020). Structural basis for divergent C–H hydroxylation selectivity in two Rieske oxygenases. *Nature Communications, 11*(1), 1–10. https://doi.org/10.1038/s41467-020-16729-0.

Mahto, J. K., Neetu, N., Waghmode, B., Kuatsjah, E., Sharma, M., Sircar, D., et al. (2021). Molecular insights into substrate recognition and catalysis by phthalate dioxygenase from *Comamonas testosteroni. Journal of Biological Chemistry, 297*(6), 101416. https://doi.org/10.1016/j.jbc.2021.101416.

Meng, S., Ji, Y., Zhu, L., Dhoke, G. V., Davari, M. D., & Schwaneberg, U. (2022). The molecular basis and enzyme engineering strategies for improvement of coupling efficiency in cytochrome P450s. *Biotechnology Advances, 61*, 108051. https://doi.org/10.1016/j.biotechadv.2022.108051.

Nam, J. W., Nojiri, H., Yoshida, T., Habe, H., Yamane, H., & Omori, T. (2001). New classification system for oxygenase components involved in ring-hydroxylating oxygenations. *Bioscience, Biotechnology, and Biochemistry, 65*(2), 254–263. https://doi.org/10.1271/bbb.65.254.

Nojiri, H., Ashikawa, Y., Noguchi, H., Nam, J. W., Urata, M., Fujimoto, Z., et al. (2005). Structure of the terminal oxygenase component of angular dioxygenase, carbazole 1,9a-dioxygenase. *Journal of Molecular Biology, 351*(2), 355–370. https://doi.org/10.1016/j.jmb.2005.05.059.

Ohta, T., Chakrabarty, S., Lipscomb, J. D., & Solomon, E. I. (2008). Near-IR MCD of the nonheme ferrous active site in naphthalene 1,2-dioxygenase: Correlation to crystallography and structural insight into the mechanism of rieske dioxygenases. *Journal of the American Chemical Society, 130*(5), 1601–1610. https://doi.org/10.1021/ja074769o.

Onaran, M. B., & Seto, C. T. (2003). Using a lipase as a high-throughput screening method for measuring the enantiomeric excess of allylic acetates. *Journal of Organic Chemistry, 68*(21), 8136–8141. https://doi.org/10.1021/jo035067u.

Parales, R. E., Parales, J. V., & Gibson, D. T. (1999). Aspartate 205 in the catalytic domain of naphthalene dioxygenase is essential for activity. *Journal of Bacteriology, 181*(6), 1831–1837. https://doi.org/10.1128/jb.181.6.1831-1837.1999.

Parales, R. E., Lee, K., Resnick, S. M., Jiang, H., Lessner, D. J., & Gibson, D. T. (2000). Substrate specificity of naphthalene dioxygenase: Effect of specific amino acids at the active site of the enzyme. *Journal of Bacteriology, 182*(6), 1641–1649. https://doi.org/10.1128/JB.182.6.1641-1649.2000.

Parales, R. E., Resnick, S. M., Yu, C. L., Boyd, D. R., Sharma, N. D., & Gibson, D. T. (2000). Regioselectivity and enantioselectivity of naphthalene dioxygenase during arene cis-dihydroxylation: Control by Phenylalanine 352 in the α subunit. *Journal of Bacteriology, 182*(19), 5495–5504. https://doi.org/10.1128/JB.182.19.5495-5504.2000.

Pati, S. G., Bopp, C. E., Kohler, H. P. E., & Hofstetter, T. B. (2022). Substrate-specific coupling of O_2 activation to hydroxylations of aromatic compounds by Rieske non-heme iron dioxygenases. *ACS Catalysis, 12*(11), 6444–6456. https://doi.org/10.1021/acscatal.2c00383.

Perry, C., De Los Santos, E. L. C., Alkhalaf, L. M., & Challis, G. L. (2018). Rieske non-heme iron-dependent oxygenases catalyse diverse reactions in natural product biosynthesis. *Natural Product Reports, 35*(7), 622–632. https://doi.org/10.1039/c8np00004b.

Resnick, S. M., Lee, K., & Gibson, D. T. (1996). Diverse reactions catalyzed by naphthalene dioxygenase from Pseudomonas sp. strain NCIB 9816. *Journal of Industrial Microbiology and Biotechnology, 17*(5–6), 438–457. https://doi.org/10.1007/bf01574775.

Runda, M. E., de Kok, N. A. W., & Schmidt, S. (2023). Rieske oxygenases and other ferredoxin-dependent enzymes: Electron transfer principles and catalytic capabilities. *Chembiochem: A European Journal of Chemical Biology, 202300078*, e202300078. https://doi.org/10.1002/cbic.202300078.

Runda, M. E., Kremser, B., Özgen, F. F., & Schmidt, S. (2023). An optimized system for the study of Rieske oxygenase-catalyzed hydroxylation reactions in vitro. *ChemCatChem, 15*(16), https://doi.org/10.1002/cctc.202300371.

Runda, M. E., Miao, H., Kok, N. A. W. D., & Schmidt, S. (2024). Developing hybrid systems to address O_2 uncoupling in multi-component Rieske oxygenases. *Journal of Biotechnology, 389*, 22–29. https://doi.org/10.1101/2024.02.16.580709 580709.

Suen, W.-C. (1991). Ph.D. thesis, Gene expression of naphthalene dioxygenase from Pseudomonas sp. NCIB 9816-4 in Escherichia coli. The University of Iowa, Iowa City.

Sydor, P. K., Barry, S. M., Odulate, O. M., Barona-Gomez, F., Haynes, S. W., Corre, C., et al. (2011). Regio- and stereodivergent antibiotic oxidative carbocyclizations catalysed by Rieske oxygenase-like enzymes. *Nature Chemistry, 3*(5), 388–392. https://doi.org/10.1038/nchem.1024.

Tian, J., Boggs, D. G., Donnan, P. H., Barroso, G. T., Garcia, A. A., Dowling, D. P., et al. (2023). The NADH recycling enzymes TsaC and TsaD regenerate reducing equivalents for Rieske oxygenase chemistry. *Journal of Biological Chemistry, 299*(10), 105222. https://doi.org/10.1016/j.jbc.2023.105222.

Tian, J., Garcia, A. A., Donnan, P. H., & Bridwell-Rabb, J. (2023). Leveraging a structural blueprint to rationally engineer the Rieske oxygenase TsaM. *Biochemistry, 62*(11), 1807–1822. https://doi.org/10.1021/ACS.BIOCHEM.3C00150/ASSET/IMAGES/LARGE/BI3C00150_0006.JPEG.

Tsai, P. C., Chakraborty, J., Suzuki-Minakuchi, C., Terada, T., Kotake, T., Matsuzawa, J., et al. (2022). The a- and b-subunit boundary at the stem of the mushroom- and a3b3-type oxygenase component of Rieske non-heme iron oxygenases is the Rieske-type ferredoxin-binding site. *Applied and Environmental Microbiology, 88*(15), e00835-22. https://doi.org/10.1128/aem.00835-22.

van Pée, K.-H., Salcher, O., & Lingens, F. (1980). Formation of pyrrolnitrin and 3-(2-amino-3-chlorophenyl)pyrrole from 7-chlorotryptophan. *Angewandte Chemie International Edition in English, 19*(10), 828–829. https://doi.org/10.1002/anie.198008281.

Wackett, L. P., Kwart, L. D., & Gibson, D. T. (1988). Benzylic monooxygenation catalyzed by toluene dioxygenase from *Pseudomonas putida*. *Biochemistry, 27*(4), 1360–1367. https://doi.org/10.1021/bi00404a041.

Waterhouse, A., Bertoni, M., Bienert, S., Studer, G., Tauriello, G., Gumienny, R., et al. (2018). SWISS-MODEL: Homology modelling of protein structures and complexes. *Nucleic Acids Research, 46*(W1), W296–W303. https://doi.org/10.1093/nar/gky427.

Wissner, J. L., Schelle, J. T., Escobedo-Hinojosa, W., Vogel, A., & Hauer, B. (2021). Semirational engineering of toluene dioxygenase from *Pseudomonas putida* F1 towards oxyfunctionalization of bicyclic aromatics. *Advanced Synthesis and Catalysis, 363*(21), 4905–4914. https://doi.org/10.1002/adsc.202100296.

Withall, D. M., Haynes, S. W., & Challis, G. L. (2015). Stereochemistry and mechanism of undecylprodigiosin oxidative carbocyclization to streptorubin B by the Rieske oxygenase RedG. *Journal of the American Chemical Society, 137*(24), 7889–7897. https://doi.org/10.1021/jacs.5b03994.

Wolfe, M. D., & Lipscomb, J. D. (2003). Hydrogen peroxide-coupled cis-diol formation catalyzed by naphthalene 1,2-dioxygenase. *Journal of Biological Chemistry, 278*(2), 829–835. https://doi.org/10.1074/jbc.M209604200.

Yeh, W. K., Gibson, D. T., & Liu, T. N. (1977). Toluene dioxygenase: A multicomponent enzyme system. *Biochemical and Biophysical Research Communications, 78*(1), 401–410. https://doi.org/10.1016/0006-291X(77)91268-2.

Zhang, X., Rao, Z., Zhang, L., Xu, M., & Yang, T. (2016). Efficient 9α-hydroxy-4-androstene-3,17-dione production by engineered *Bacillus subtilis* co-expressing *Mycobacterium neoaurum* 3-ketosteroid 9α-hydroxylase and *B. subtilis* glucose 1-dehydrogenase with NADH regeneration. *SpringerPlus, 5*(1), 1–8. https://doi.org/10.1186/s40064-016-2871-4.

Zheng, L., Cash, V. L., Flint, D. H., & Dean, D. R. (1998). Assembly of iron-sulfur clusters. Identification of an iscSUA-hscBA-fdx gene cluster from *Azotobacter vinelandii*. *Journal of Biological Chemistry, 273*(21), 13264–13272. https://doi.org/10.1074/jbc.273.21.13264.

SECTION 2

Leveraging mononuclear non-heme iron enzymes for biocatalysis

CHAPTER NINE

Radical-relay C(sp³)–H azidation catalyzed by an engineered nonheme iron enzyme

Qun Zhao[a,*], Jinyan Rui[b], and Xiongyi Huang[b,*]

[a]School of Biotechnology and Key Laboratory of Industrial Biotechnology of Ministry of Education, Jiangnan University, Wuxi, P.R. China
[b]Department of Chemistry, Johns Hopkins University, Baltimore, MD, United States
*Corresponding authors. e-mail address: qunzhao@jiangnan.edu.cn; xiongyi@jhu.edu

Contents

1. Introduction	196
2. Materials	198
2.1 Cloning	198
2.2 Enzyme expression in *E. coli*	201
2.3 Whole-cell reaction	201
2.4 GCMS (gas chromatography–mass spectrometry) and normal phase HPLC (high performance liquid chromatography) analysis	201
3. Protocols	202
3.1 Cloning for a site-saturated mutagenesis screening library	202
3.2 High-throughput experimentation in 96-well plates	205
3.3 Analytical scale reactions to validate the screening hits	207
3.4 Preparative-scale reactions	209
4. Summary	211
Acknowledgments	211
References	211

Abstract

Nonheme iron enzymes are versatile biocatalysts for a broad range of unique and powerful transformations, such as hydroxylation, chlorination, and epimerization as well as cyclization/ring-opening of organic molecules. Beyond their native biological functions, these enzymes are robust for engineering due to their structural diversity and high evolvability. Based on enzyme promiscuity and directed evolution as well as inspired by synthetic organic chemistry, nonheme iron enzymes can be repurposed to catalyze reactions previously only accessible with synthetic catalysts. To this end, our group has engineered a series of nonheme iron enzymes to employ non-natural radical-relay mechanisms for new-to-nature radical transformations. In particular, we have demonstrated that a nonheme iron enzyme, (4-hydroxyphenyl) pyruvate dioxygenase from streptomyces avermitilis (*Sav*HppD), can be repurposed

---

Methods in Enzymology, Volume 703
ISSN 0076-6879, https://doi.org/10.1016/bs.mie.2024.07.003
Copyright © 2024 Elsevier Inc. All rights are reserved, including those for text and data mining, AI training, and similar technologies

to enable abiological radical-relay process to access C(sp^3)-H azidation products. This represents the first known instance of enzymatic radical relay azidation reactions. In this chapter, we describe the detailed experimental protocol to convert promiscuous nonheme iron enzymes into efficient and selective biocatalyst for radical relay azidation reactions. One round of directed evolution is described in detail, which includes the generation and handling of site-saturation mutagenesis, protein expression and whole-cell reactions screening in a 96-well plate. These protocol details might be useful to engineer various nonheme iron enzymes for other applications.

1. Introduction

The nonheme iron enzymes are ubiquitous in nature and involved in a wide range of fundamental biological processes (Solomon et al., 2000), including DNA repair (Falnes, Klungland, & Alseth, 2007), gene regulation (Mahon, Hirota, & Semenza, 2001), O$_2$ sensing (Mahon et al., 2001) and biosynthesis and metabolism (Baldwin & Abraham, 1988). By activating molecular oxygen, nonheme iron enzymes catalyze an astoundingly broad array of oxidation reactions, including chemically challenging transformations such as hydroxylation, halogenation, desaturation, and carbon–carbon bond formation (Abu-Omar, Loaiza, & Hontzeas, 2005; Ali, Warwicker, & de Visser, 2023; Blasiak & Drennan, 2009; Bollinger et al., 2015; Kovaleva, Neibergall, Chakrabarty, & Lipscomb, 2007; Krebs, Galonić Fujimori, Walsh, & Bollinger, 2007; Liao et al., 2018; Ryle & Hausinger, 2002; Song, Naowarojna, Cheng, Lopez, & Liu, 2019; Timmins & De Visser, 2018). Beyond their native biological functions, nonheme iron enzymes have emerged as a potent biocatalytic platform for abiotic synthetic transformations. Pioneering research by the Bollinger Lab has showcased the abilities of nonheme iron halogenases as radical transfer enzymes for incorporating non–native functionalities like azide and nitrate. Drawing on the resemblance between the iron-oxo intermediates in nonheme iron enzymes and iron-nitrene intermediates used in synthetic chemistry, both the Fasan and Arnold groups have successfully repurposed these enzymes for nitrene transfer reactions (Goldberg, Knight, Zhang, & Arnold, 2019; Vila, Steck, Giordano, Carrera, & Fasan, 2020). Furthermore, the Chang Group has also explored the potential of nonheme iron enzymes for nitrene chemistry by demonstrating an azide-to-nitrile transformation (Davidson, McNamee, Fan, Guo, & Chang, 2019). These investigations underscore the versatility of nonheme iron enzymes in catalyzing synthetically valuable reactions that are not previously present in nature.

Radical-Relay C(sp3)–H Azidation Catalyzed by an Engineered Nonheme Iron Enzyme 197

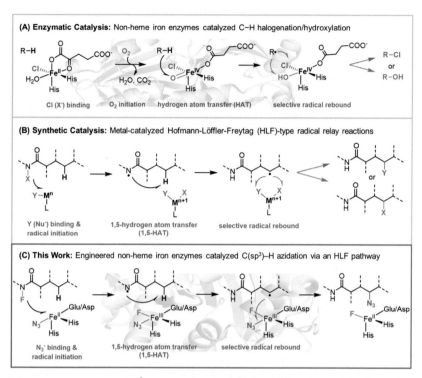

Fig. 1 Enantioselective C(sp^3)-H azidation catalyzed by nonheme iron enzymes via a radical relay pathway. (A) Representative native C–H halogenation/hydroxylation catalyzed by nonheme iron enzymes. (B) Metal-catalyzed Hofmann–Löffler–Freytag (HLF)-type radical relay reactions. (C) Repurposing nonheme iron enzymes to catalyze abiological radical-relay C(sp^3)-H azidation via an HLF pathway.

To this end, our group has recently expanded the reaction space of nonheme iron enzymes by introducing abiological radical relay reaction mechanisms into metalloenzymatic catalysis. By drawing mechanistic similarities between nonheme iron dependent hydroxynases/halogenases and the transition metal catalyzed Hofmann–Löffler–Freytag (HLF)-type reactions (Fig. 1A and B), nonheme iron dependent dioxygenase, *Sav*HPPD, has been repurposed to catalyze radical relay azidation reactions via an HLF pathway (Rui et al., 2022). The key mechanistic features of this new reaction mode involve the initial activation of an *N*-fluoroamide substrate for *N*-centered radical generation, followed by an intramolecular 1,5-hydrogen atom transfer to generate the carbon-centered radical. The final step involves the interception of the carbon-centered radical by the iron(III) complex for

C—N bond formation (Fig. 1C). Following this work, our group and the Yang group has also successfully developed a biocatalytic system for asymmetric fluorine atom transfer reactions via engineered nonheme enzymes (Zhao et al., 2024; Zhao, Chen, Rui, & Huang, 2024; Zhao, Chen, Soler, et al., 2024). In this chapter, we describe the detailed experimental protocol for the directed evolution of the nonheme iron enzyme *Sav*HPPD for radical relay azidation reactions. The generation and handling of site-saturation mutagenesis (SSM), protein expression and whole-cell reactions screening in a 96-well plate are provided in detail.

2. Materials
2.1 Cloning
- Eppendorf Research Plus single channel mechanical pipettes, variable volumes
- TempAssure 0.2 mL tubes for PCR (polymerase chain reaction) applications, flat caps (USA Scientific, Catalog number: 1402-8100)
- TOPQSC XK-400 Palm-Series Mini-Centrifuge
- Bio-Rad T100 Thermal Cycler
- Bio-Rad PowerPac Basic power supply
- Bio-Rad Mini-Sub Cell GT
- VWR Blue Light Transilluminator
- Bio-Rad MicroPulser electroporator
- Fisherbrand electroporation cuvettes-2 mm gap
- Gene of interest (GOI) cloned into pET-22b(+) vector between *Nde*I and *Xho*I with a C-terminal 6×His tag
- Primers (005, 006, 007, 008, NDT, VHG, TGG, and rev primers; purchased from GENEWIZ)
- Forward primer 005: for the amplification of GOI: 5′-GAAATAATT TTGTTTAACTTTA AGAAGGAGATATACATATG-3′
- Reverse primer 006 for the amplification of GOI: 5′-GCCGGATCT CAGTGGTGGTGGT GGTGGTGCTCGAG-3′
- Reverse primer 007 for the amplification of the backbone: reverse complement strand of 005
- Forward primer 008 for the amplification of the backbone: reverse complement strand of 006
- Forward (NDT, VHG, TGG) and reverse primers designed for SSM using the "22-c trick" method (Kille et al., 2013).

- DNA sequence of wt SavHppD. Further specification: The provided sequence encompasses the ribosome binding site of the pET-22b(+) vector (AAGGAG, highlighted in bold blue). The start codon ATG is highlighted in bold black. This construct features a C-terminus 6xHis tag (highlighted in bold green) that is attached to the protein with a leucine-glutamate linker (underscored in green). The TAA stop codon is highlighted in bold red.

AAGGAGATATACAT**ATG**ACGCAGACTACACATCACACG
CCCGACACGGCACGTCAGGCAGATCCATTTCCAGTGAAG
GGTATGGATGCTGTTGTGTTCGCTGTAGGTAATGCAAAA
CAGGCGGCGCACTATTACAGCACAGCCTTTGGGATGCAG
TTGGTGGCCTACAGTGGTCCCGAAAATGGGAGCCGTGAG
ACCGCCTCGTATGTCTTGACCAACGGATCAGCGCGTTTC
GTCCTGACAAGCGTTATTAAGCCGGCCACGCCATGGGGA
CATTTCCTTGCAGATCACGTTGCAGAGCACGGAGATGGA
GTTGTAGACCTGGCTATTGAAGTCCCCGACGCCCGTGCC
GCTCACGCCTACGCTATCGAACACGGTGCCCGCTCCGTG
GCGGAACCGTACGAATTAAAGGACGAGCACGGCACGGTC
GTTTTGGCTGCCATCGCCACCTACGGCAAAACGCGCCAC
ACACTGGTAGACCGTACGGGCTACGACGGGCCATACTTG
CCGGGGTACGTAGCTGCCGCCCCTATTGTCGAGCCCCCT
GCGCACCGCACCTTCCAAGCTATTGACCACTGTGTAGGT
AATGTGGAATTAGGACGCATGAACGAATGGGTGGGCTTC
TATAATAAGGTTATGGGGTTCACCAACATGAAAGAGTTTG
TAGGGGATGACATTGCAACAGAATATTCGGCCCTGATGT
CAAAAGTGGTCGCTGATGGGACCCTTAAAGTAAAATTTC
CCATTAACGAACCCGCTTTAGCAAAGAAGAAATCTCAAAT
TGATGAATACTTAGAATTTTACGGAGGAGCGGGAGTCCAA
CATATCGCTTTAAACACGGGCGACATCGTGGAGACGGTC
CGTACCATGCGTGCAGCTGGGGTACAATTCCTGGACACT
CCCGATTCATACTATGACACGCTTGGTGAGTGGGTTGGC
GATACTCGTGTTCCGGTCGACACTCTTCGTGAGCTGAAA
ATCTTGGCGGATCGCGACGAGGATGGATACTTATTACAA
ATTTTTACTAAACCAGTGCAGGACCGTCCTACCGTTTTC
TTCGAAATTATTGAGCGTCATGGGAGCATGGGGTTTGGT
AAGGGGAATTTCAAGGCCCTTTTTGAGGCAATCGAGCGT
GAGCAAGAGAAACGCGGGAATTT<u>ACTCGAG</u>**CACCACCAC**
CACCACCACTAA

- Example primers designed for SSM at the V189X site using "wt *Sav*HppD" as the parent:
- V189X_fwd_NDT: 5′-CACCTTCCAAGCTATTGACCACTGT**N-DT**GGTAATG-3′
- V189X_fwd_VHG: 5′-CACCTTCCAAGCTATTGACCACTGT**V-HG**GGTAATG-3′
- V189X_fwd_TGG: 5′-CACCTTCCAAGCTATTGACCACTGT**TG** **G**GGTAATG-3′
- V189X_rev: 5′-CAGTGGTCAATAGCTTGGAAGGTG-3′
- Backbone (ca. 100 ng/μL as determined by NanoDrop) of the pET-22b (+) vector; prepared by long-range PCR using pET-22b(+) as the template and 007 and 008 as primers
- Phusion High-Fidelity DNA Polymerase, DMSO (dimethyl sulfoxide), Phusion 5× HF (High-Fidelity) Buffer, doxyribonucleotide triphosphate solutions (dNTPs), *Dpn*I, from New England Biolabs (NEB)
- ddI water (autoclaved MilliQ water)
- 50× TAE buffer (50 mM EDTA, 2 M Tris, 1 M acetic acid)
- 2% agarose gel (10 g agarose (low electroendosmosis, GOLDBIO, Catalog number: A-201-500) in 500 mL 1× TAE buffer, microwave carefully until transparent)
- 100 bp and 1 kb DNA ladders (NEB, Catalog number: N3231S, N3232S)
- DNA loading dye (6×), no SDS (NEB, Catalog number: B7025S) with SYBR Gold nucleic acid gel stain (10,000×) (Invitrogen, Catalog number: S11494), kept at −20 °C and avoid light
- Monarch PCR and DNA Gel Extraction Kit (NEB)
- Gibson Mastermix (Gibson et al., 2009) [320 μL 5× isothermal buffer (25% PEG-8000, 500 mM Tris–HCl pH 7.5, 50 mM $MgCl_2$, 50 mM dithiothreitol, 1 mM each of the dNTPs, and 5 mM NAD)], 0.64 μL of 10 U/μL T5 exonuclease, 20 μL of 2 U/μL Phusion High-Fidelity DNA Polymerase, 160 μL of 40 U/μL Taq DNA ligase from NEB
- SOC medium (2% tryptone, 0.5% yeast extract, 10 mM NaCl, 2.5 mM KCl, 10 mM $MgCl_2$, 10 mM $MgSO_4$, and 20 mM glucose, autoclaved)
- Electrocompetent *Escherichia coli* strain E. cloni BL21(DE3) cells (Lucigen)
- LBamp agar plate [15 g agar (molecular genetics grade, Fisher BioReagents, Catalog number: BP1423-500) in Luria-Bertani medium with 0.1 mg/mL ampicillin sodium salt, Chem-Impex Int'l. Inc. Catalog number: 00516]
- New Brunswick Innova 44 R shakers
- Microcentrifuge tubes (1.5 mL)
- New Brunswick U535 ultra-low temperature freezer

2.2 Enzyme expression in *E. coli*
- Eppendorf Xplore plus 12-channel pipettes, variable volumes
- Toothpicks (autoclaved)
- EasyApp microporous film rolls, sterile (USA Scientific, Catalog number: 2977-6282)
- Fisherbrand disposable sterile plastic culture tubes (Catalog number: FB149566B)
- Erlenmeyer flasks, variable volumes
- LB_{amp} medium (Lennox L Broth with 0.1 mg/mL ampicillin)
- TB_{amp} medium (Terrific Broth with 0.1 mg/mL ampicillin)
- 1 M isopropyl β-D-1-thiogalactopyranoside (IPTG, 1000×) in ddI H_2O (sterilized by membrane filtration through a 0.22-μm sterile cellulose acetate)

2.3 Whole-cell reaction
- Eppendorf tabletop centrifuge 5810R
- Eppendorf 5424R centrifuge
- Fisherbrand microplate shaker
- Coy Lab vinyl anaerobic chamber
- falcon tubes (50 and 15 mL)
- Biotage Isolera One Flash Chromatography System
- SiliaFlash P60, 40–63 μm (230–400 mesh), 60 Å (SiliCycle, Catalog number: R12030B)
- Kpi buffer (34.8 mM K_2HPO_4, 15.2 mM KH_2PO_4, pH 7.4)
- DME (1,2-dimethoxyethane), ReagentPlus, ≥99%, inhibitor-free (Sigma-Aldrich, Catalog number: E27408)
- 400 mM *N*-fluoroamide substrate in DME
- 100 mM ferrous ammonium sulfate in ddI H_2O
- 1 M sodium azide in ddI H_2O

2.4 GCMS (gas chromatography–mass spectrometry) and normal phase HPLC (high performance liquid chromatography) analysis
- Agilent 5977B GC/MSD
- HP-5MS UI column (30.0 m×0.25 mm)
- Agilent 1260-series HPLC
- CHIRALPAK IC column (4.6 mm×250 mm, 5 mic)
- Vortex-Genie 2 Mixer (Fisherbrand, Catalog number: 12-812)
- 2-Propanol (HPLC grade, Sigma-Aldrich, Catalog number: 650447)

- Hexanes (98.5% hexane, mixture of isomers, HPLC grade, Sigma-Aldrich, Catalog number: 293253)
- Extraction solution (0.5 mM 1,3,5-trimethoxybenzene in EtOAc (ethyl acetate)/hexanes: 3/2)

3. Protocols

The primary mutagenesis technique used in this directed evolution campaign is SSM, which has proven to be a reliable and efficient method for generating mutant libraries aimed for enzyme function improvement (Reetz & Carballeira, 2007). In SSM, one or a few residues of the target enzyme are selected for random mutagenesis. Random mutations are often introduced using primers containing degenerate codons such as NNK, NDT, and VHG. In this work, the NNK codon was chosen for its ease of ordering and its coverage of all 20 canonical amino acids with only moderate codon redundancy. The selection of residues for random mutagenesis typically relies on structural or modeling data. In this study, the residues chosen for SSM were primarily those in proximity to the iron center and/or crucial for binding the native substrate of SavHPPD. The hypothesis was that these residues would likely need modification to better accommodate the non-natural radical relay chemistry, given the significant differences between radical-relay reactions and the native reaction of SavHPPD. For all selected residues, we adopted a sequential SSM strategy. In each round of directed evolution, single-site SSM and screening were performed for one or a few specific sites, and the best-performing mutant identified was used as the parent for screening at the next site(s). This strategy benefits from rapid screening time since only single-site SSM libraries were assessed in each round of engineering. However, it does not fully account for cooperativity among all chosen residues, nor does it consider how the order of residue selection for directed evolution impacts the final outcome. In this work, the order of residues for evolution was chosen based on their proximity to the iron center.

3.1 Cloning for a site-saturated mutagenesis screening library

1. **Prepare the primer mix:**
   - Dilute all primers to 25 μM with ddI H$_2$O to get:
     V189X_fwd_NDT (25 μM), V189X_fwd_VHG (25 μM), V189X_fwd_TGG (25 μM), V189X_rev (25 μM).

- Fwd_primers mix: mix V189X forward primers (NDT/VHG/TGG: 120 μL/90 μL/10 μL) in a 1.5 mL microcentrifuge tube, vortex and mix it well to get the Fwd-primers mix.

2. **Set up the PCR ice (for fragment 1 and 2):**
   - Step 1 (PCR1): Transfer 34 μL ddI H_2O, 10 μL 5× Phusion HF buffer (NEB), 1.5 μL DMSO, 1 μL 10 mM dNTP mix, 1 μL DNA template (30–150 ng/μL), 1 μL Fwd_primers mix, and 1 μL 006 primer to a PCR tube, mix it by tapping gently and quickly spin it down.
   - Step 2 (PCR2): Transfer 34 μL ddI H_2O, 10 μL 5× HF buffer, 1.5 μL DMSO, 1 μL 10 mM dNTPs, 1 μL DNA template (30–150 ng/μL), 1 μL Rev_prime, and 1 μL 005 primer to a PCR tube, mix it by tapping gently and quickly spin it down.
   - Step 3: Add 0.5 μL Phusion DNA polymerase to PCR1 and PCR2 separately, mix it by tapping gently and quickly spin it down.

3. **Set up the thermoCycler's parameters (for fragment 1 and 2):**
   Initial denaturation step (98 °C, 30 s), denaturation step (98 °C, 10 s), annealing step (55 °C, 30 s), extension step (72 °C, 30 s), cycles (25–35 cycles), final extension step (72 °C, 10 min), and hold at 4 °C. (For Phusion DNA polymerase, the extension rate is around 1000 bp/30 s. Calculate the extension time based on the length of the DNA and extension rate of Phusion DNA polymerase.)

4. **Run PCR (for fragment 1 and 2)**
   Place PCR tubes in the thermoCycler to start the PCR program: Get fragment 1 from PCR1 and fragment 2 from PCR2.

5. **DNA gel electrophoresis (for fragment 1 and 2):**
   - Add 2% Agarose gel in 1× TAE buffer.
   - When the PCR completes, add 0.1% of SYBE gold Nucleic Acid Gel Stain (10,000×) into DNA loading dye (6×) in each PCR tube.
   - Load PCR products (PCR1 and PCR2) to the agarose gel separately.
   - Load DNA ladder (100 bp) to the agarose gel as a reference.
   - Start DNA gel electrophoresis (130 V, 20 min).
   - Cut out the mix of PCR product and agarose gel detected by a blue light transilluminator. Combine the PCR1 mix and PCR2 mix.
   - Isolate the combined PCR products (fragment 1 and fragment 2) from the above combination (PCR1 mix and PCR2 mix) using the NEBs Monarch DNA Gel Extraction Kit; usually 10 μL elution buffer is used for each PCR DNA to ensure a relatively high final concentration of DNA products.

- Measure the DNA concentration using Nanodrop and adjust the concentration to $100\,ng/\mu L$ with elution buffer, then, keep the combined PCR products at $-20\,°C$ for storage.

6. **Set up the PCR on ice (for** GOI**):**
   - PCR3: Transfer $34\,\mu L$ ddI H_2O, $10\,\mu L$ $5\times$ HF buffer, $1.5\,\mu L$ DMSO, $1\,\mu L$ $10\,mM$ dNTP mix (dATP, dGTP, dTTP, dCTP), $1\,\mu L$ DNA template (the combined PCR products from step 5), $1\,\mu L$ 005 and $1\,\mu L$ 006 primer to a PCR tube, then, add $0.5\,\mu L$ Phusion DNA polymerase, mix it by tapping gently and quickly spin it down.

7. **Set up the thermoCycler's parameters (for** GOI**):**
   Initial denaturation step ($98\,°C$, $30\,s$), Denaturation step ($98\,°C$, $10\,s$), Annealing step ($55\,°C$, $30\,s$), Extension step ($72\,°C$, $30\,s$), cycles (25–35 cycles), Final extension step ($72\,°C$, $10\,min$), and hold at $4\,°C$. (For Phusion DNA polymerase, the extension rate is around $1000\,bp/30\,s$. Calculate the extension time based on the length of the DNA and extension rate of Phusion DNA polymerase).

8. **DNA gel electrophoresis (for** GOI**):**
   - Add 2% Agarose gel in $1\times$ TAE buffer.
   - When the PCR completes, add 0.1% of SYBE gold Nucleic Acid Gel Stain ($10,000\times$) into DNA loading dye ($6\times$) in each PCR tube.
   - Load PCR products (PCR3) to the agarose gel.
   - Load DNA ladder ($100\,bp$) to the agarose gel as a reference.
   - Start DNA gel electrophoresis ($130\,mV$, $20\,min$).
   - Cut out the mix of PCR product and agarose gel detected by a blue light transilluminator.
   - Isolate the PCR3 products from the above PCR3 mix using the NEB's Monarch DNA Gel Extraction Kit; usually $10\,\mu L$ elution buffer is used for each PCR DNA to ensure a relatively high final concentration of DNA products.
   - Measure the DNA concentration using Nanodrop and adjust the concentration to $100\,ng/\mu L$ with elution buffer, then, keep the PCR3 product at $-20\,°C$ for storage.

9. **Gibson assembly (for plasmid):**
   - Add $1.25\,\mu L$ backbone and $1.25\,\mu L$ PCR3 product into a PCR tube containing $7.5\,\mu L$ of Gibson Mastermix.
   - Incubate the above Gibson mixture at $50\,°C$ for $1\,h$ in the ThermoCycler.
   - Keep the resulting Gibson products on ice for transformation.

10. Transformation:

- Thaw the SOC medium on ice and cool down the electroporation cuvette on ice.
- Thaw a tube of electrocompetent *E. coli* cells (50 μL) from the −80 °C freezer on ice.
- Add the Gibson product (1 μL) into electrocompetent *E. coli* cells.
- Transfer electrocompetent *E. coli* cells with Gibson products into an electroporation cuvette.
- Wipe dry electrodes of the cuvette and do electroporation in "Bacteria" mode (1.8 kV).
- Quickly add 500 μL SOC medium into the cuvette and incubate at 37 °C, 240 rpm for 45 min
- Inoculate 50–100 μL SOC culture on an LBamp Agar plate and incubate at 37 °C for 10–12 h.

11. Parent controls are parallelly done in the screening 96-well plates.

- Freshly transformed *E. coli* cells with plasmids encoding parent enzymes are used as a control.

3.2 High-throughput experimentation in 96-well plates

To develop enzymes for non-natural reactions, library screening is primarily conducted using chromatography-based analytical methods such as HPLC and GCMS. In this work, because the inorganic azide cannot undergo copper-catalyzed azide-alkyne cyclization (CuAAC), we developed a high-throughput screening (HTS) platform based on CuAAC for the rapid assessment of enzymatic azidation activity. The following protocol outlines the procedures to perform this HTS assay for each library assessment.

1. The overnight culture growth in a 96-well plate:

- Transfer 400 μL LB$_{amp}$ medium to each well of a 96-well plate using an Eppendorf Xplorer 12-channel pipette.
- Pick the colony of parent with autoclaved toothpicks and add to different wells (A1, B2, C3, D4, E5, F6, G7, and H8) for parent controls.
- Pick single colonies from the agar plate with the SSM screening library using autoclaved toothpicks and add every single colony to each well except wells with parents.
- Remove all the toothpicks. (do carefully to avoid cross contaminations)

- Cover the 96-well plate with a microporous film.
- Incubate starter cultures in the 96-well plate at 37 °C, 240 rpm for 16 h in a New Brunswick Innova 44R shaker.

2. **Protein expression in 96-well plate:**
- Transfer 1 mL TBamp medium to each well of a new sterilized 96-well plate.
- Transfer 50 µL LBamp overnight culture to each well of the 96-well plate containing TBamp medium.
- Cover the 96-well plate with a microporous film.
- Incubate the 96-well plate at 37 °C, 240 rpm for 3 h in a New Brunswick Innova 44R shaker until the OD600 is around 2.0.
- Glycerol stock storage: Transfer 100 µL sterilized glycerol (50% in ddI H_2O) into each well of a new 96-well microplate for glycerol stock storage. Transfer 100 µL LBamp culture from the starter culture plate to each well of this microplate. Cover the lid and mix it by gentle swirling. Keep the culture glycerol stock at −80 °C.
- After 3 h, put the 96-well plate with expression cultures on ice for 30 min
- Thaw the stock of IPTG solution (100 mM) from the −20 °C freezer and transfer 10 µL of this stock to each well of the 96-well plate containing expression cultures using an Eppendorf Xplorer 12-channel pipette.
- Gently shake the plate to ensure good mixing.
- Cover the 96-well plate with microporous films.
- Incubate the expression cultures at 20.5 °C, 230 rpm for 24 h in a New Brunswick Innova 44R shaker.

3. **Set up whole-cell screening reactions in 96-well plate:**
- Cool down the Eppendorf tabletop centrifuge 5810R to 4 °C.
- Spin down the expression cultures at 4 °C, 4000×g for 3 min, then, keep the cell pellets and discard the supernatant.
- Transfer 380 µL Kpi buffer into each well and shake the plate on a Fisher Scientific microplate shaker at 680 rpm for resuspension.
- Transfer the 96-well plate, substrate stock, sodium azide stock (1 M in ddI H_2O) and Fe(II) stock (ferrous ammonium sulfate, 100 mM in ddI H_2O) to a Coy anaerobic chamber.
- Add 10 µL Fe(II) stock solution into each well using an Eppendorf Xplorer 12-channel pipette, shake the plate on a Fisher Scientific microplate shaker at 680 rpm for 2 min
- Add 10 µL sodium azide stock solution into each well using an Eppendorf Xplorer 12-channel pipette, shake the plate on a Fisher Scientific microplate shaker at 680 rpm for 2 min

- Add 10 μL substrate stock solution (400 mmol in DME) into each well using an Eppendorf Xplorer 12-channel pipette.
- Cover the plate with a resealable aluminum foil.
- Shake the plate on the Fisher Scientific microplate shaker at 680 rpm for 24 h.

4. High-throughput fluorescent detection:
- 96-well black fluorescence plate (Caplugs Evergreen)
- TECAN Spark microplate reader.
- CuSO4 (Copper(II) sulfate), anhydrous, powder, ≥99.99% trace metals basis, (Sigma-Aldrich, Product number: 451657)
- BTTAA (2-(4-((Bis((1-(tert-butyl)-1H-1,2,3-triazol-4-yl)methyl)amino)methyl)-1H-1,2,3-triazol-1-yl)acetic acid), anhydrous, ≥95%, (Sigma-Aldrich, Product number: 906328).
- Ascorbic acid (L-threoascorbic acid), reagent grade, crystalline, (Sigma-Aldrich, Product number: A7506).
- Fluorogenic alkyne probe (4-ethynyl-N-ethyl-1,8-naphthalimide, home made, NMR pure).

DMF (N,N-dimethylformamide), anhydrous, ≥99.8%, (Sigma-Aldrich, Product number: 227056).

Procedure:
- After the reaction, 400 μL of DMF was added to each well and the plate was incubated for 1 h at room temperature.
- The plate was then centrifuged to remove the insolubles. From each well, 5 μL of the supernatant was transferred to a 96-well black fluorescence plate.
- Add 195 μL of 25% aqueous solution of DMF containing 77 μM CuSO$_4$, 154 μM BTTAA ligand (Click Chemistry Tools), 5.1 mM ascorbic acid, 25.6 mM KPi (pH 7.4), and 103 μM of fluorogenic alkyne probe.
- The fluorescence plate was incubated and the formation of the fluorescent triazole product was monitored by a TECAN Spark microplate reader outfitted with a plate stacker (excitation wavelength, 357 nm: emission wavelength 462 nm; bandwidth, 20 nm).

3.3 Analytical scale reactions to validate the screening hits

To validate the beneficial mutation(s) identified during library screening, it is necessary to perform analytical-scale reactions using enzymes obtained from larger volumes of cell cultures. This process helps eliminate false positives in the initial screening.

1. Grow the overnight culture in culture tubes:
- Transfer 4.5 mL LB$_{amp}$ medium into a culture tube.

- Inoculate bacteria glycerol stock in LB_{amp} using sterile toothpicks.
- Cap the culture tube loosely for ventilation.
- Incubate the starter culture at 37 °C, 240 rpm for 12–14 h in an Eppendorf Innova 44R shaker.

2. **Protein expression in 250 mL Erlenmeyer flasks:**
   - To validate hits from HTS, a copy of bacterial glycerol stock is usually saved. The hits are also sequenced by Sanger sequencing.
   - Transfer 50 mL TBamp medium into a 250 mL Erlenmeyer flask.
   - Inoculate 1.0 mL of the overnight culture into the TBamp medium.
   - Incubate at 37 °C, 230 rpm in an Eppendorf Innova 44R shaker until OD600 reaches ca. 2.0.
   - Thaw the IPTG (1000×) stock from the −20 °C freezer.
   - After 2 h, cool down the Erlenmeyer flask on ice for 20 min
   - Transfer 50 μL IPTG stock solution to each flask.
   - Incubate the expression culture at 20.5 °C, 230 rpm for 24 h in an Eppendorf Innova 44R shaker.

3. **Set up validation reactions with the whole *E. coli* cells in 2 mL sample vials:**
   - Analytical scale reactions are carried out in triplicate and the reaction with parent is carried out as a control.
   - Set the temperature of Eppendorf 5810R tabletop centrifuge to 4 °C before use.
   - Transfer the expression cultures into 50 mL falcon tubes.
   - Spin down the cell pellets at 4000×g at 4 °C for 5 min. Keep the pellets and discard the supernatant.
   - Add the Kpi buffer to normalize the OD600 to 40.
   - Resuspend the pellets in Kpi buffer by gently pipetting.
   - Transfer 380 μL of the cell suspension into the 2 mL sample vials.
   - Transfer the sample vials, Fe(II) stock solution, and substrate stock solution into a Coy anaerobic chamber.
   - Add 10 μL Fe(II) stock solution (100 mM in ddI H_2O) into each vial, quickly cap the vials, then, fix and shake the vials on a Fisher Scientific microplate shaker at 680 rpm for 2 min
   - Add 10 μL substrate stock solution (400 mM in DME) into each vial, quickly cap the vials, then, fix and shake the vials on a Fisher Scientific microplate shaker at 680 rpm for 24 h.

4. **Determine the enzyme concentration in the whole-cell:**
   - Prepare the denaturing buffer (laemmli buffer/BME: 20/1, BME: β-mercaptoethanol).

- Have a known concentration of the protein of interest as a standard (~150 µg/mL) to make a calibration curve with 2×, 4×, 6× and 8× dilution.
- Aliquot 20 µL of whole-cell samples.
- Mix aliquot samples in 3 with 20 µL of denaturing buffer from step 1 and incubate in heat block at 95 °C for 10 min
- Load the whole-cell and standard samples into wells of the SDS–PAGE (sodium dodecyl sulfate polyacrylamide gel electrophoresis) protein gel.
- Run gel electrophoresis at 130 V for 45 min, adjust time and voltage as needed.
- Afterwards the protein bands can be visualized in gel imager with 280 nm UV detection.
- Extrapolate the concentration by using the imageJ software.

5. Reactions' work-up:
- Add 800 µL extraction solution into each well to quench the reaction.
- Transfer the mixture into 1.5 mL centrifuge tubes.
- Load the tubes in an Eppendorf 5424R centrifuge to separate the organic and the aqueous layers at $14,000 \times g$ for 5 min
- Transfer 300 µL of the organic layer into a 0.5 mL insert and place these inserts in sample vials for GCMS and HPLC analysis.
- Load sample vials in GCMS [Agilent 5977B GC/MSD system, HP-5MS UI column (30.0 m×0.25 mm) with the following oven temperature setting (helium flow: 1 mL/min): Initial: 110 °C (hold 0 min); Ramp 1: 110–160 °C (20 °C/min, hold 0 min); Ramp 2: 160–225 °C (15 °C/min, hold 0 min); Ramp 3: 225–270 °C (30 °C/min, hold 4 min)].
- Load the sample vials in HPLC. Determination of enantioselectivity by using normal phase chiral HPLC [Agilent 1260 series, CHIRALPAK IC (4.6×250 mm, 5 mic), isocratic elution: 1% iPrOH/hexanes, 0.5 mL/min].
- Identify the hits by both GCMS and HPLC analysis.

3.4 Preparative-scale reactions

To enhance the synthetic utility of this enzymatic system, the following section outlines the procedures for conducting the reaction on a preparative scale to obtain the azidation product at a 100 mg scale.

1. Grow the overnight culture in a culture tube:
- Transfer 4.5 mL LB_{amp} medium into a culture tube.

- Inoculate the bacteria glycerol stock in LB_{amp} using a sterile toothpick.
- Cap the culture tube loosely for ventilation.
- Incubate the starter culture at 37 °C, 240 rpm for 16 h in an Eppendorf Innova 44R shaker.

2. **Protein expression in 1 L Erlenmeyer flasks:**
   - Transfer 200 mL TBamp medium into a 1 L Erlenmeyer flask.
   - Inoculate 2.0 mL of the overnight culture into the TBamp medium.
   - Incubate at 37 °C, 240 rpm in an Eppendorf Innova 44R shaker until OD600 reaches ca. 0.7.
   - Thaw the IPTG (1000×) stock from the −20 °C freezer.
   - After 2 h (OD600 reaches ca. 0.7), cool down the Erlenmeyer flask on ice for 20 min
   - Transfer 200 μL IPTG stock solution into the Erlenmeyer flask.
   - Incubate the expression culture at 20.5 °C, 230 rpm for 24 h in an Eppendorf Innova 44R shaker.

3. **Set up the preparative-scale reaction with the whole *E. coli* cells in a 40 mL vial:**
   - Set the temperature of Eppendorf 5810R tabletop centrifuge to 4 °C before use.
   - Transfer the expression cultures into 50 mL falcon tubes.
   - Spin down the cell pellets at 4000×g at 4 °C for 15 min. Keep the pellets and discard the supernatant.
   - Add the Kpi buffer to normalize the OD600 to 20.
   - Resuspend the pellets in Kpi buffer by gently pipetting.
   - Transfer 15 mL of the SavHppD Az1 whole-cell suspension into the 40 mL vial.
   - Place the vial on ice and bubble with Ar for 15 min
   - Transfer the cell suspension vial, Fe(II) stock solution, and substrate stock solution into a Coy anaerobic chamber.
   - Add 0.5 mL Fe(II) stock solution (100 mM in ddI H_2O) into the vial, quickly cap the vial, then, fix and shake the vial on a Fisher Scientific microplate shaker at 680 rpm for 2 min
   - Add 1 mL sodium azide stock solution (1 M in ddI H_2O) into each well using an Eppendorf Xplorer 12-channel pipette, shake the plate on a Fisher Scientific microplate shaker at 680 rpm for 2 min
   - Add 0.5 mL substrate stock solution (1.5 M in DME) into the vial, quickly cap the vial, then, fix and shake the vial on a Fisher Scientific microplate shaker at 680 rpm for 48 h.

4. Reaction work-up:

- Transfer the reaction solution into a 50 mL falcon tube and mixed with 30 mL EtOAc/hexanes (3/2) via vortexing (30 s for three times).
- Centrifuge the mixture at $10,500 \times g$ for 5 min to separate the organic and aqueous layers.
- Do two additional rounds of extraction.
- Dry the combined organic extracts over anhydrous $MgSO_4$.
- Filter the combined organic extracts through a celite plug to remove Na_2SO_4.
- Concentrate the organic phase via rotary evaporation.
- Purify the azidation product by flash chromatography using a Biotage Isolera.

4. Summary

By drawing mechanistic similarities between abiological reactions and natural biological processes, an array of natural enzymes has been successfully engineered to catalyze reactions previously only accessible in synthetic organic chemistry. Directed evolution has been demonstrated powerfully to enable engineered enzymes to access abiological reactions with high activity and selectivity. In this chapter, we described detailed methods to perform radical-relay $C(sp^3)$–H azidation reactions catalyzed by nonheme iron enzymes. Protocols for the high-throughput enzyme engineering and activity screening in 96-well plates, analytical and large-scale reactions were provided. These protocols with details may stimulate further expansion of extra catalytic functions of nonheme iron enzymes.

Acknowledgments

We acknowledge the Johns Hopkins University and National Institute for General Medical Sciences (R00GM129419 & R35GM147639) for financial support.

References

Abu-Omar, M. M., Loaiza, A., & Hontzeas, N. (2005). Reaction mechanisms of mononuclear non-heme iron oxygenases. *Chemical Reviews, 105*(6), 2227–2252.

Ali, H. S., Warwicker, J., & de Visser, S. P. (2023). How does the nonheme iron enzyme Napi react through l-arginine desaturation rather than hydroxylation? A quantum mechanics/molecular mechanics study. *ACS Catalysis, 13*(16), 10705–10721.

Baldwin, J., & Abraham, E. (1988). The biosynthesis of penicillins and cephalosporins. *Natural Product Reports, 5*(2), 129–145.

Blasiak, L. C., & Drennan, C. L. (2009). Structural perspective on enzymatic halogenation. *Accounts of Chemical Research, 42*(1), 147–155.

Bollinger, J. M., Jr., Chang, W.-C., Matthews, M. L., Martinie, R. J., Boal, A. K., & Krebs, C. (2015). Mechanisms of 2-oxoglutarate-dependent oxygenases: The hydroxylation paradigm and beyond. In C. Schofield, & R. Hausinger (Eds.). *2-Oxoglutarate-dependent oxygenases* (pp. 487)The Royal Society of Chemistry.

Davidson, M., McNamee, M., Fan, R., Guo, Y., & Chang, W.-C. (2019). Repurposing nonheme iron hydroxylases to enable catalytic nitrile installation through an azido group assistance. *Journal of the American Chemical Society, 141*(8), 3419–3423.

Falnes, P., Klungland, A., & Alseth, I. (2007). Repair of methyl lesions in DNA and RNA by oxidative demethylation. *Neuroscience, 145*(4), 1222–1232.

Gibson, D. G., Young, L., Chuang, R.-Y., Venter, J. C., Hutchison, C. A., & Smith, H. O. (2009). Enzymatic assembly of DNA molecules up to several hundred kilobases. *Nature Methods, 6*(5), 343–345.

Goldberg, N. W., Knight, A. M., Zhang, R. K., & Arnold, F. H. (2019). Nitrene transfer catalyzed by a non-heme iron enzyme and enhanced by non-native small-molecule ligands. *The Journal of the American Chemical Society, 141*(50), 19585–19588.

Kille, S., Acevedo-Rocha, C. G., Parra, L. P., Zhang, Z.-G., Opperman, D. J., Reetz, M. T., et al. (2013). Reducing codon redundancy and screening effort of combinatorial protein libraries created by saturation mutagenesis. *ACS Synthetic Biology, 2*(2), 83–92.

Kovaleva, E., Neibergall, M., Chakrabarty, S., & Lipscomb, J. D. (2007). Finding intermediates in the O_2 activation pathways of non-heme iron oxygenases. *Accounts of Chemical Research, 40*(7), 475–483.

Krebs, C., Galonić Fujimori, D., Walsh, C. T., & Bollinger, J. M., Jr. (2007). Non-heme Fe (IV)–oxo intermediates. *Accounts of Chemical Research, 40*(7), 484–492.

Liao, H. J., Li, J., Huang, J. L., Davidson, M., Kurnikov, I., Lin, T. S., et al. (2018). Insights into the desaturation of cyclopeptin and its C3 epimer catalyzed by a non-heme iron enzyme: Structural characterization and mechanism Elucidation. *Angewandte Chemie International Edition, 57*(7), 1831–1835.

Mahon, P. C., Hirota, K., & Semenza, G. L. (2001). FIH-1: A novel protein that interacts with HIF-1α and VHL to mediate repression of HIF-1 transcriptional activity. *Genes & Development, 15*(20), 2675–2686.

Reetz, M. T., & Carballeira, J. D. (2007). Iterative saturation mutagenesis (ISM) for rapid directed evolution of functional enzymes. *Nature Protocols, 2*(4), 891–903.

Rui, J., Zhao, Q., Huls, A. J., Soler, J., Paris, J. C., Chen, Z., et al. (2022). Directed evolution of nonheme iron enzymes to access abiological radical-relay $C(sp^3)$–H azidation. *Science (New York, N.Y.), 376*(6595), 869–874.

Ryle, M. J., & Hausinger, R. P. (2002). Non-heme iron oxygenases. *Current Opinion in Chemical Biology, 6*(2), 193–201.

Solomon, E. I., Brunold, T. C., Davis, M. I., Kemsley, J. N., Lee, S.-K., Lehnert, N., et al. (2000). Geometric and electronic structure/function correlations in non-heme iron enzymes. *Chemical Reviews, 100*(1), 235–350.

Song, H., Naowarojna, N., Cheng, R., Lopez, J., & Liu, P. (2019). Non-heme iron enzyme-catalyzed complex transformations: Endoperoxidation, cyclopropanation, orthoester, oxidative C–C and C–S bond formation reactions in natural product biosynthesis. *Advances in Protein Chemistry and Structural Biology, 117*, 1–61.

Timmins, A., & De Visser, S. P. (2018). A comparative review on the catalytic mechanism of nonheme iron hydroxylases and halogenases. *Catalysts, 8*(8), 314.

Vila, M. A., Steck, V., Giordano, S. R., Carrera, I., & Fasan, R. (2020). C–H amination via nitrene transfer catalyzed by mononuclear non-heme iron-dependent enzymes. *Chembiochem: a European Journal of Chemical Biology, 21*(14), 1981–1987.

Zhao, L.-P., Mai, B. K., Cheng, L., Gao, F., Zhao, Y., Guo, R., et al. (2024). Biocatalytic enantioselective $C(sp^3)$–H fluorination enabled by directed evolution of non-haem iron enzymes. *Nature Synthesis*. https://doi.org/10.1038/s44160-024-00536-2.

Zhao, Q., Chen, Z., Rui, J., & Huang, X. (2024). Radical fluorine transfer catalysed by an engineered nonheme iron enzyme. *Methods in Enzymology, 696*(14), 231–237.

Zhao, Q., Chen, Z., Soler, J., Chen, X., Rui, J., Ji, N. T., et al. (2024). Engineering non-haem iron enzymes for enantioselective $C(sp^3)$–F bond formation via radical fluorine transfer. *Nature Synthesis*. https://doi.org/10.1038/s44160-024-00507-7.

CHAPTER TEN

Purification and characterization of a Rieske oxygenase and its NADH-regenerating partner proteins

Gage T. Barroso[1], Alejandro Arcadio Garcia[1], Madison Knapp, David G. Boggs, and Jennifer Bridwell-Rabb*

Department of Chemistry, University of Michigan, Ann Arbor, MI, United States
*Corresponding author. e-mail address: jebridwe@umich.edu

Contents

1. Introduction	216
2. Considerations for assembling a Rieske oxygenase pathway *in vitro*	219
3. Protein constructs for recombinant expression and purification	220
3.1 Assembly of needed constructs for protein isolation	220
3.2 Transformation protocol for the TsaMBCD pathway encoding genes	221
4. Recombinant expression and purification of the TsaM, TsaC, TsaD, and VanB	221
4.1 Recombinant expression of the TsaM, TsaC, TsaD, and VanB encoding genes	221
4.2 Purification of TsaM, VanB, and TsaC	223
4.3 Purification of the NAD$^+$-dependent aldehyde dehydrogenase TsaD	225
5. Methods for assessing the quality of the purified TsaMBCD pathway proteins	226
5.1 Biochemical analysis of purified proteins	226
5.2 Quantification of the iron content in TsaM and VanB	227
6. Enzymatic assays for the TsaMBCD pathway	228
6.1 Liquid chromatography mass spectrometry (LC-MS) methods for activity assays	228
6.2 Separation of TsaMBCD pathway intermediates using LC-MS	230
6.3 Identification of the optimal conditions for measuring the activity of TsaM	231
6.4 Total turnover number (TTN) determination using LC-MS	232
6.5 Spectroscopic assay for analysis of NAD(P)H consumption and production	234
7. Crystallization of the short-chain dehydrogenase/reductase (SDR) enzyme TsaC	236
8. Conclusions	238
Acknowledgments	239
References	239

---

[1] These authors contributed equally to this work.

Methods in Enzymology, Volume 703
ISSN 0076-6879, https://doi.org/10.1016/bs.mie.2024.05.015
Copyright © 2024 Jennifer D Bridwell-Rabb. Published by Elsevier Inc. All rights are reserved.

Abstract

The Rieske non-heme iron oxygenases (Rieske oxygenases) comprise a class of metalloenzymes that are involved in the biosynthesis of complex natural products and the biodegradation of aromatic pollutants. Despite this desirable catalytic repertoire, industrial implementation of Rieske oxygenases has been hindered by the multicomponent nature of these enzymes and their requirement for expensive reducing equivalents in the form of a reduced nicotinamide adenine dinucleotide cosubstrate (NAD(P)H). Fortunately, however, some Rieske oxygenases co-occur with accessory proteins, that through a downstream reaction, recycle the needed NAD(P)H for catalysis. As these pathways and accessory proteins are attractive for bioremediation applications and enzyme engineering campaigns, herein, we describe methods for assembling Rieske oxygenase pathways *in vitro*. Further, using the TsaMBCD pathway as a model system, in this chapter, we provide enzymatic, spectroscopic, and crystallographic methods that can be adapted to explore both Rieske oxygenases and their co-occurring accessory proteins.

1. Introduction

The annotated members of the Rieske oxygenase enzyme family are characterized by the presence of a [2Fe-2S] Rieske cluster and a mononuclear iron center. The Rieske cluster is involved in electron transfer, and the mononuclear iron center is credited with the reductive activation of molecular oxygen (O_2) and the ensuing oxidation of a substrate (Barry & Challis, 2013; Ferraro, Gakhar, & Ramaswamy, 2005; Knapp, Mendoza, & Bridwell-Rabb, 2021; Kovaleva & Lipscomb, 2008; Perry, de Los Santos, Alkhalaf, & Challis, 2018). Using these metallocenters, Rieske oxygenases catalyze reactions that produce commercially and medically relevant molecules and reactions that degrade carbon–rich compounds found in the environment (Barry & Challis, 2013; Ferraro et al., 2005; Knapp et al., 2021; Kovaleva & Lipscomb, 2008; Perry et al., 2018). This extensive catalytic repertoire makes these enzymes attractive for implementation into a wide range of chemical processes. Indeed, several foundational studies have shown promise in their ability to harness Rieske oxygenase chemistry for biocatalytic applications (Ballard, Courtis, Shirley, & Taylor, 1983; Buckland et al., 1999; Ensley et al., 1983; Fabara & Fraaije, 2020; Newman, Garcia, Hudlicky, & Selifonov, 2004; Reddy, Lee, Neeper, Greasham, & Zhang, 1999). However, several bottlenecks still limit industrial use of this metalloenzyme chemistry. In particular, the need for a partner reductase protein and a reduced nicotinamide adenine dinucleotide cosubstrate (NAD(P)H) both pose challenges to realizing the potential of this enzyme class (Brimberry, Garcia, Liu, Tian, & Bridwell-Rabb, 2023; Runda, de Kok, & Schmidt, 2023). As such, pioneering studies have explored the viability of using photoreduction systems or non-native NAD (P)H recycling systems as sources of electrons for Rieske oxygenase chemistry (Feyza Ozgen et al., 2020; Hu et al., 2022; Lanfranchi, Trajkovic, Barta, de Vries, & Janssen, 2019).

As a complement to these approaches, it is also prudent to investigate the Rieske oxygenase pathways that contain NAD(P)H recycling enzymes (Kincannon et al., 2022; Tian, Boggs, et al., 2023). One such pathway, of interest to our laboratory, is the TsaMBCD pathway (Tian, Boggs, et al., 2023). This pathway involves the Rieske oxygenase TsaM, which works in cooperation with TsaB, TsaC, and TsaD to initiate degradation of the xenobiotic compounds 4-methylbenzenesulfonate and 4-methylbenzoate (Junker et al., 1996; Junker, Kiewitz, & Cook, 1997; Locher, Leisinger, & Cook, 1991). TsaB is an NADH-dependent reductase that supplies two electrons to TsaM to catalyze a monooxygenation reaction that creates either 4-(hydroxymethyl)benzenesulfonate or 4-(hydroxymethyl)benzoate. TsaC and TsaD are both annotated NAD^+-dependent enzymes that produce a molar equivalent of NADH through the iterative conversion of the TsaM-created products into 4-sulfobenzoate or 1,4-benzenedicarboxylate. This design makes the TsaMBCD pathway self-sufficient: the reducing equivalents consumed by TsaM and TsaB in the first step are regenerated by the downstream enzymes in the pathway (Fig. 1). This pathway blueprint is attractive for enzyme engineering because it eliminates the need for an external supply of reducing equivalents, which is an unfeasible expense that typically impedes the industrial use of NAD(P)H-dependent enzymes (Hollmann, Opperman, & Paul, 2021; Lanfranchi et al., 2019; Runda et al., 2023; van der Donk & Zhao, 2003). It is also a design that allows a wide range of bacteria to use aromatic pollutants and other compounds deposited in the environment as sources of carbon and energy. For the TsaMBCD pathway, one or two subsequent steps transform 4-sulfobenzoate or 1,4-benzenedicarboxylate into protocatechuate which, like several other dioxygenated compounds, can be converted into a citric acid cycle intermediate (Fig. 1) (Junker et al., 1996; Junker et al., 1997; Locher et al., 1991; Sgro et al., 2023).

Prior studies have illustrated that this practical design is not unique to the TsaMBCD pathway (Dhindwal et al., 2011; Eaton & Chapman, 1992; Furukawa, Hirose, Suyama, Zaiki, & Hayashida, 1993; Irie, Doi, Yorifuji, Takagi, & Yano, 1987; Kincannon et al., 2022; Li, Wang, & Feng, 2013;

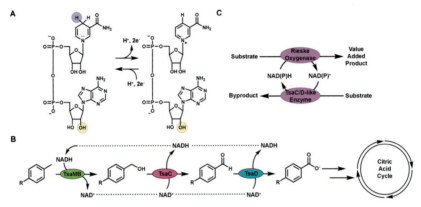

Fig. 1 The nicotinamide adenine dinucleotide (NAD(P)H and NAD(P)$^+$) redox pair play a central role in biochemical pathways. (A) NAD(P)H, a two-electron carrier, is oxidized to NAD(P)$^+$ through the loss of a hydride (purple). Likewise, NAD(P)$^+$ is reduced to NAD(P)H in the reverse reaction. NADPH and NADP$^+$ differ from NADH by the presence of a phosphate moiety at the 2′ position of the ribose (yellow). (B) One biochemical pathway highlighted here is the TsaMBCD pathway, which metabolizes 4-methylbenzenesulfonate (R = SO$_3^-$) and 4-methylbenzoate (R = CO$_2^-$) substrates. The first step of this pathway requires the oxidation of NADH to NAD$^+$. The NADH is regenerated in either the TsaC or TsaD catalyzed reactions from NAD$^+$. This pathway of enzymes catalyzes the initial steps needed to funnel these molecules into the citric acid cycle. (C) A general scheme that shows how NAD(P)H-dependent enzymatic systems can use auxiliary enzymes, such as TsaC or TsaD, to regenerate NAD(P)H at the expense of a sacrificial substrate.

Neidle et al., 1991; Neidle et al., 1992; Tian, Boggs, et al., 2023; Zylstra, McCombie, Gibson, & Finette, 1988). Rather, there are a variety of Rieske oxygenases that naturally co-occur with an NAD(P)H recycling enzyme like TsaC (Dhindwal et al., 2011; Eaton & Chapman, 1992; Furukawa et al., 1993; Irie et al., 1987; Kincannon et al., 2022; Li et al., 2013; Neidle et al., 1991; Neidle et al., 1992; Tian, Boggs, et al., 2023; Zylstra et al., 1988). As these types of pathways are ideal for bioremediation, an investigation into how to capitalize on the use of these pathways in engineered microorganisms is paramount (Tian, Boggs, et al., 2023). Likewise, as TsaC- and TsaD-like enzymes can be used to reduce "sacrificial substrates" and provide NAD(P)H for Rieske oxygenase catalysis, investigation of these accessory proteins is also imperative (Wang et al., 2017). Here, drawing on prior work from our laboratory, we describe how to isolate and reconstitute the TsaMBCD pathway *in vitro* (Tian, Boggs, et al., 2023; Tian, Garcia, Donnan, & Bridwell-Rabb, 2023; Tian, Liu, et al., 2023). These methods can likely be tailored to the reconstitution of other pathways of interest or used to isolate and characterize homologs of the accessory NAD(P)H recycling enzymes. Therefore, in this chapter, we highlight important considerations for performing these types of pathway experiments and we provide methods to (1) isolate the TsaMBCD pathway enzymes, (2) probe enzyme turnover and NAD(P)H production, and (3) structurally characterize the NADH-regenerating enzyme TsaC.

2. Considerations for assembling a Rieske oxygenase pathway *in vitro*

As described in Section 1, a key ingredient for any Rieske oxygenase catalyzed reaction, is a supply of reducing equivalents. Canonical electron donating partners for Rieske oxygenases are classified into three major and two minor sub-classes (Kweon et al., 2008). These classes are assigned based on the identities of the embedded electron carrying cofactors and based on the one- or two-protein component architecture of the reductase (Kweon et al., 2008). Despite the plethora of available reductase options, current data shows that Rieske oxygenase chemistry can be supported by electron donation from non-native proteins that belong to the same, different, or even non-Rieske type classes of reductases (Lee & Zhao, 2006; Lee, Simurdiak, & Zhao, 2005; Tian, Boggs, et al., 2023; Tian, Garcia, et al., 2023; Tian, Liu, et al., 2023). This versatility suggests that despite the previously described requirement for the reductase proteins to associate at the Rieske oxygenase subunit interfaces, there is flexibility with respect to the identity of the electron donor that docks at this position (Ashikawa et al., 2006). The resultant promiscuity of the annotated electron donors turns out to be important for *in vitro* enzyme assays because, for many Rieske oxygenases, the required reductase component is either not annotated, or does not support high levels of activity (Liu, Knapp, Jo, Dill, & Bridwell-Rabb, 2022; Lukowski et al., 2018; Tian, Boggs, et al., 2023; Tian, Garcia, et al., 2023; Tian, Liu, et al., 2023). Indeed, for TsaM, the native annotated reductase (TsaB) can be purified, but VanB, an alternative electron donor from the same reductase class, supports markedly higher levels of activity. Therefore, VanB has been typically employed for *in vitro* TsaM-based experiments in our laboratory (Tian, Boggs, et al., 2023; Tian, Garcia, et al., 2023; Tian, Liu, et al., 2023).

Analogous to the ability of different reductase proteins to functionally replace the annotated partner of a Rieske oxygenase, it has also been shown that the Rieske oxygenase VanA can persist when coupled to its native reductase as well as a supplementary NAD(P)H recycling system (Lanfranchi et al., 2019). This result suggests that Rieske oxygenase reductase proteins also do not have strong preferences regarding their source of NAD(P)H. Therefore, in designing experiments that involve Rieske oxygenases, it is important to broadly screen the available reductase options, and to consider whether an NAD(P)H regeneration system can be adapted, integrated, or coupled with the Rieske oxygenase of interest. The final choices should be selected based upon the desired application and needed levels of activity.

3. Protein constructs for recombinant expression and purification

3.1 Assembly of needed constructs for protein isolation

Once a pathway of interest is chosen, the genes encoding each of the different enzymes need to be obtained using genetic cloning or gene synthesis techniques. In either case, the gene of interest should be inserted into a vector for recombinant protein purification. Subsequently, each of the pathway proteins needs to be expressed and isolated. Previous studies indicate that the isolation of these proteins should involve the use of codon-optimized and synthesized *Comamonas testosteroni* TsaM and TsaC encoding genes that are sub-cloned into pET-28a(+)-tobacco etch virus (TEV) vectors (Tian, Boggs, et al., 2023; Tian, Garcia, et al., 2023; Tian, Liu, et al., 2023). These vectors carry kanamycin resistance and should be assembled such that the encoded proteins contain N-terminal His$_6$-tags and TEV protease tag-cleavage sites. *C. testosteroni* TsaD should instead be codon-optimized, synthesized, and sub-cloned into a pMCSG9 vector that contains an N-terminal His$_6$-maltose binding protein (MBP) affinity tag followed by a TEV protease cleavage site. Last, *Pseudomonas* sp. VanB, which was originally identified as a robust non-native reductase for the paralytic shellfish toxin biosynthetic enzymes, has been shown to be purifiable when codon-optimized and subcloned into a pMCSG7-TEV vector (Lukowski et al., 2018). Both of these latter pMCSG vectors contain an ampicillin, rather than a kanamycin, resistance marker for selection. The suggested working concentrations with these antibiotics are 100 µg/mL (ampicillin) and 50 µg/mL (kanamycin). For the TsaMBCD pathway, the following previously described methods have been used to recombinantly express and purify each of the individual enzymes (Tian, Boggs, et al., 2023; Tian, Garcia, et al., 2023; Tian, Liu, et al., 2023).

3.2 Transformation protocol for the TsaMBCD pathway encoding genes

1. Transform 20–100 ng of the plasmids into one aliquot of C41(DE3) (TsaM, TsaC, and TsaD) or BL21(DE3) (VanB) chemically competent *Escherichia coli* cells. These cells should be kept at −80 °C and allowed to thaw on ice for five minutes prior to transfection with a plasmid. Additionally, no more than 5 μL of plasmid should be introduced into an aliquot of 50 μL of competent cells. This transfer volume is kept low to prevent dilution of the competent cell buffer, which typically contains calcium chloride and additional ions that render the *E. coli* cellular membrane more permeable to transfection with a DNA plasmid (Dagert & Ehrlich, 1979; Mandel & Higa, 1970).
2. Following incubation of the resultant transformed cell mixture on ice for 30 min, use standard protocols to heat-shock the transformed cell mixture, an important step that facilitates the transfer of the desired DNA plasmid of interest into the *E. coli* cell (Dagert & Ehrlich, 1979; Mandel & Higa, 1970).
3. Add 900 μL of autoclaved Lysogeny Broth–Miller formulation (LB) media to the mixture and incubate at 37 °C with shaking at 200 rpm for 1 h. At the same time, pre-heat the required ampicillin (VanB and TsaD) and kanamycin (TsaM and TsaC) containing LB agar plates to 37 °C.
4. Centrifuge the mixtures at 5000 rpm for one minute to pellet the cells.
5. Decant approximately 700–800 μL of LB from the tube in a sterile environment.
6. Resuspend the cells with the remaining volume and transfer to the correct antibiotic containing plate. Allow the added liquid to soak in for about a minute before inverting and incubating the plate at 37 °C for 16–18 h.

4. Recombinant expression and purification of the TsaM, TsaC, TsaD, and VanB

4.1 Recombinant expression of the TsaM, TsaC, TsaD, and VanB encoding genes

1. Remove a single colony from the agar plate that contains the desired transformant and place into a 50 mL starter culture of antibiotic-treated

LB media in a 125 mL culture flask. Loosely cover the flask with foil and incubate the starter culture at 37 °C with shaking (200 rpm) for 14–16 h.

2. Add 5–10 mL of the starter culture into four 1 L cultures of LB media that have been inoculated with the appropriate antibiotic. For this step, the number of 1 L growths can change depending on the needed protein yield. For reference, the typical approximate yield of purified protein is 34 mg (TsaM), 19 mg (VanB), 22 mg (TsaC), and 1 mg (TsaD), from a 1 L growth (Tian, Boggs, et al., 2023; Tian, Garcia, et al., 2023; Tian, Liu, et al., 2023). It should be noted for this step that all culture flasks need to be sterilized prior to culture growth.

3. Incubate the large growths at 37 °C with shaking (200 rpm) until the optical density at 600 nm (OD_{600}) reaches a value of 0.6–0.8. For *E. coli* cells, it is typically stated that the OD_{600} should double every 20 min in a healthy, nutrient-replete, aerobic growth (Gibson, Wilson, Feil, & Eyre-Walker, 2018). However, in our laboratory, for the TsaMBCD pathway enzymes, we estimate that this step instead takes an average of 30–45 min.

4. Once the OD_{600} reaches 0.6–0.8, induce the growths using the following methods:
   a. For TsaM, to each flask, first add 2 mL of a sterile filtered 100 mg/mL ferrous ammonium sulfate hexahydrate ($(NH_4)_2Fe(SO_4)_2{\cdot}6H_2O$) solution, swirl the flask by hand, and then repeat again with an additional 2 mL of the $(NH_4)_2Fe(SO_4)_2{\cdot}6H_2O$ solution. Second, add 2 mL of a sterile filtered 100 mg/mL ammonium ferric citrate solution to the flask, and again, swirl the flask by hand. Third, add 100 μL of 1 M of isopropyl β-d-1-thiogalactopyranoside (IPTG) to the cultures to induce protein expression.
   b. For TsaC and TsaD, add 500 μL of 1 M IPTG to each culture flask.
   c. For VanB, add 100 μL of 1 M IPTG to each flask of *E. coli* cells.

5. Once IPTG has been added to the culture flasks, use the following methods to complete this stage of recombinant expression.
   a. For TsaM and VanB, incubate the cultures at 18 °C and shake at 160 rpm for 16–18 h.
   b. For TsaC and TsaD, incubate the culture flasks at 25 °C and shake at 160 rpm for 18 h.

6. Collect the growths *via* centrifugation at 4 °C and 5000 rpm for 20 min. Transfer the harvested cells into low-temperature safe containers, freeze with liquid nitrogen, and store at −80 °C until purification can be accomplished. As previously reported, the typical amount of cell pellet

from each recombinant expression experiment is 5–6 g (TsaM), 6 g (TsaC), 4 g (TsaD), and 6 g (VanB) from a 1 L growth (Lukowski, Liu, Bridwell-Rabb, & Narayan, 2020; Tian, Boggs, et al., 2023; Tian, Liu, et al., 2023).

4.2 Purification of TsaM, VanB, and TsaC

1. Thaw the cell pellet from two 1 L growths (10–15 g) of TsaM, VanB, or TsaC at room temperature. This step typically takes 20–30 min.
2. Resuspend the thawed pellets with 5–10 mL of buffer A (50 mM Tris (hydroxymethyl)aminomethane hydrochloride (Tris-HCl) pH 7.5, 200 mM NaCl, 10 mM imidazole, 10 mM 2-mercaptoethanol) per gram of cell pellet *via* stirring in an ice bath for 20 min.
3. Sonicate the resuspension on ice for 10 min at 30% amplitude using pulsing cycles of sonication (5 s of pulsation followed by 10 s of rest).
4. Pellet the lysate using centrifugation at 12,000 rpm and 4 °C. Complete clarification of the supernatant requires 60 min of centrifugation.
5. Decant the clarified solution into a new tube or flask and load onto a pre-equilibrated 5 mL HisTrap HP Column (Cytiva). This column should be run using the following method:
   a. First, wash the column with 5 column volumes of buffer A to facilitate the removal of non His$_6$-tagged protein impurities from the column.
   b. Second, remove additional contaminant proteins from the column matrix through use of a two-step series of 10% buffer B (50 mM Tris-HCl pH 7.5, 200 mM NaCl, 200 mM imidazole, 10 mM 2-mercaptoethanol) and 20% buffer B. Run each step in this series for 5 column volumes.
   c. Last, using 100% buffer B, elute the protein of interest from the column. The high levels of imidazole in this buffer should be enough to out-compete the interaction of the His$_6$-tagged protein with the column matrix.
6. Identify the protein-containing fractions from the HisTrap HP column-purification step using sodium dodecyl sulfate-polyacrylamide gel electrophoresis (SDS-PAGE). Once identified, pool together eluted fractions of TsaC and concentrate to less than 5 mL of volume using a 10 kDa molecular weight cutoff (MWCO) centrifugal filter. Likewise, pool together the eluted fractions of TsaM and concentrate to less than 5 mL of volume using a 30 kDa MWCO centrifugal filter.

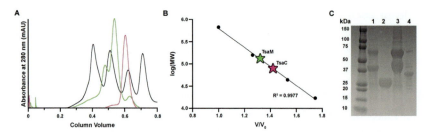

Fig. 2 The individual enzymes from the TsaMBCD pathway can be recombinantly expressed and purified for analysis. (A) Size-exclusion chromatography chromatograms of TsaM (green) and TsaC (pink) overlayed with a Gel Filtration Standard (Bio-Rad, black). (B) The linear relationship between elution volume and logarithmic molecular weight of the standard can be used to approximate the molecular weight of the purified protein. This linear analysis was performed in GraphPad Prism with an $R^2 = 0.9977$. The elution volumes of TsaM and TsaC are plotted against the gel filtration standard. The migration times of TsaM and TsaC through the column suggest corresponding molecular weights of 130.2 and 79.9 kDa, respectively. As the monomeric molecular weight of TsaM is 39.6 kDa this analysis suggests that this protein, as expected, is eluting as a trimer. For TsaC, the molecular weight is slightly lower than the expected tetrameric species, but additional experimentation has previously confirmed that TsaC behaves as a tetramer (Tian, Boggs, et al., 2023). (C) Sodium dodecyl sulfate-polyacrylamide gel electrophoresis (SDS-PAGE) is used to assess the purity of the TsaMBCD pathway enzymes. The molecular weight standard used here is a BioRad Precision Plus Protein Dual Color Standard. The lanes correspond to: 1-TsaM (39.6 kDa), 2-TsaC (26.4 kDa), 3-TsaD (51 kDa), and 4-VanB (37 kDa). Of note, this gel shows that TsaM and TsaD contain high levels of impurities. These proteins thus require additional purification steps.

These molecular weight cutoffs were chosen based on the use of the ProtParam tool, which estimates the molecular weight of a TsaC and TsaM monomer to be 26.4 and 39.6 kDa, respectively (Gasteiger et al., 2003).

7. To isolate the protein of interest in the correct oligomeric state, further purify the protein of interest, and eliminate any aggregated protein fraction, the sample should next be applied to a HiPrep 16/60 Sephacryl S200-HR gel filtration column that is pre-equilibrated with buffer C (50 mM 4-(2-hydroxyethyl)-1-piperazineethanesulfonic acid (HEPES) pH 8.0, 200 mM NaCl) for TsaM and buffer D (50 mM HEPES pH 8.0, 200 mM NaCl, 5% glycerol) for TsaC.

8. Confirm fractions that would correspond to the eluted trimeric (TsaM) or tetrameric (TsaC) populations using SDS-PAGE, pool together, and concentrate with respective centrifugal filters (Fig. 2). Of note for this step, a gel filtration standard should be run to estimate the molecular

weight of the purified protein and to facilitate collection of only the protein samples that correspond to the desired oligomer of interest (Fig. 2B). Additional experiments can be used to further verify protein oligomeric state. For example, in the case of TsaC, dynamic light scattering and size-exclusion multiangle light scattering have been used to confirm that TsaC behaves as a tetramer in solution (Tian, Boggs, et al., 2023).

9. Perform quality control experiments as described in Section 5.

10. VanB is purified using the same method as TsaM and TsaC except that the approximate 30–50 mL elution from the HisTrap column should instead be concentrated using a 30 kDa MWCO centrifugal filter, diluted by the addition of 30–50 mL of buffer C, and concentrated again. Relevant to this step, the ProtParam tool calculates the molecular weight of a His_6-tagged VanB monomeric unit to be 37 kDa (Gasteiger et al., 2003).

11. Freeze proteins with liquid nitrogen in 50 μL aliquots and store at $-80\,°C$. The volume chosen for the aliquots should be reflective of the amount that will be needed to perform enzymatic or crystallographic experiments. The goal is to exclude freeze-thaw cycles, and at the same time, not waste purified protein.

4.3 Purification of the NAD$^+$-dependent aldehyde dehydrogenase TsaD

1. Thaw the cell pellet of TsaD-containing *E. coli* cells at room temperature and then resuspend on ice with 5–10 mL of buffer C per gram of pellet.

2. Follow the protocols described in Section 4.2 (steps 3–4) for sonication and centrifugation of the resuspended cell pellet.

3. Decant the clarified solution into a new tube or flask and load onto a pre-equilibrated 5 mL MBP Trap HP (Cytiva) column. Separation of the tagged protein from other cellular contaminants should be carried out using the following method:

   a. As described in Section 4.2 step 5, first wash the column to remove *E. coli* contaminant protein impurities from TsaD. This step should be accomplished with 8 column volumes of buffer C.

   b. Second, isolate TsaD using 5 column volumes of buffer E (50 mM HEPES pH 8.0, 200 mM NaCl, and 10 mM maltose).

4. Identify the TsaD-containing fractions by SDS-PAGE. These fractions should correspond to the fractions that interact with the MBP column.

5. Combine the TsaD-containing fractions and transfer this solution into dialysis tubing. In this step, add 50 mg of TEV protease to the protein solution. Once added, sealed dialysis tubing should be placed into 1 L of pre-chilled buffer C and incubated overnight at 4 °C. Here, TEV protease is typically added in excess due to previous work that suggests tag cleavage in this construct is inefficient (Tian, Boggs, et al., 2023).
6. Load the dialyzed (and tag-free) sample of TsaD onto a pre-equilibrated 5 mL HisTrap HP column (Cytiva) and use the method described in Section 4.2 step 5 to isolate tag-cleaved TsaD. However, in this step, collect the flow-through from the column which contains tag-cleaved TsaD and use buffer F (50 mM HEPES pH 8.0, 200 mM NaCl, 10 mM imidazole) and buffer G (50 mM HEPES pH 8.0, 200 mM NaCl, 200 mM imidazole) for the wash and elution steps, respectively.
7. Confirm that the collected TsaD is cleaved using SDS-PAGE. Here, it is likely for there to be contamination with uncleaved TsaD and free MBP. We hypothesize that the root cause of these impurities involves dimerization of cleaved and uncleaved monomers of TsaD (Fig. 2C). Additional affinity or gel filtration purification steps can be employed, as previously described, to improve protein purity and homogeneity if needed (Tian, Boggs, et al., 2023).
8. Concentrate TsaD with a 30 kDa MWCO centrifugal filter. Again, this molecular weight cutoff was chosen based on the use of the ProtParam tool, which estimates the molecular weight of a TsaD monomer to be 51 kDa (Gasteiger et al., 2003).
9. Perform quality controls as described in Section 5.
10. Freeze proteins with liquid nitrogen in 50 μL aliquots and store at −80 °C.

5. Methods for assessing the quality of the purified TsaMBCD pathway proteins

5.1 Biochemical analysis of purified proteins

To perform *in vitro* assays with Rieske oxygenase pathway proteins, it is important to accurately determine the concentration of the purified proteins. One common practice for calculating protein concentration involves measuring the absorbance at 280 nm (A_{280}). This method necessitates that the protein of interest has Trp or Tyr residues in its primary sequence, and

Rieske oxygenase and NADH pathway proteins

also requires either calculation of an extinction coefficient by hand or by estimation through the use of the ProtParam tool (Gasteiger et al., 2003; Pace, Vajdos, Fee, Grimsley, & Gray, 1995; Simonian, 2002). For metalloproteins, the presence of dithiothreitol and extraneous metal ions that typically accompany purification or reconstitution can impact the accuracy of the A_{280} method (Bradford, 1976; Lanz et al., 2012; Simonian, 2002). Therefore, whereas the concentrations of TsaC and TsaD can be reliably calculated using A_{280} and extinction coefficients of 17,085 and 37,400 M^{-1} cm^{-1}, respectively, the concentrations of TsaM and VanB are measured using the Bradford Assay (Bradford, 1976). This method measures a shift in the maximum absorbance of Coomassie Blue dye from 465 to 595 nm that is induced through interaction with a protein of interest. To correlate this shifted absorbance pattern with protein concentration, this method requires the creation of a standard curve that maps how well dye binds to a standard protein such as bovine serum albumin (Bradford, 1976). Additional quality control experiments that can be performed on these enzymes include circular dichroism and differential scanning fluorimetry, which have been extensively used by our laboratory to probe the folding and stability of wild-type and variant proteins of the TsaMBCD pathway (Tian, Boggs, et al., 2023; Tian, Garcia, et al., 2023; Tian, Liu, et al., 2023).

5.2 Quantification of the iron content in TsaM and VanB

1. To assess the incorporation of iron into TsaM and VanB, a Ferrozine assay should be performed (Fish, 1988; Stookey, 1970). To perform this assessment, concentrated HCl is required and several stock solutions should be prepared, including:
   a. 8 M guanidine HCl (dissolve 7.64 g in 10 mL of ultrapure water).
   b. Ferrozine solution (dissolve 5 mg in 1 mL of ultrapure water). We recommend making this solution fresh or storing the prepared stock in a dark environment.
   c. 10 M Ammonium acetate solution (dissolve 7.64 g in 10 mL of ultrapure water).
   d. 0.04% (w/v) ascorbate solution (dissolve 20 mg in 50 mL of ultrapure water).
2. Once stock solutions are made, the next step is to mix 100 µL of purified TsaM at 20–40 µM with 125 µL of ascorbate solution, 7.5 µL of concentrated HCl, and 12.5 µL of Ferrozine solution. We recommend mixing the final solution using a vortex and performing the assay in triplicate.

3. Add 500 μL of guanidine HCl and 100 μL of ammonium acetate stock solutions to the sample and incubate at room temperature for 30 min.
4. Measure the absorbance at 562 nm using a cuvette and calculate the concentration of iron using an extinction coefficient of 27,900 M^{-1} cm^{-1}. Multiply the resultant value by 8.45 to account for dilution (Fish, 1988; Stookey, 1970).
5. Compare the calculated concentration of iron to the starting concentration of protein. For both VanB and TsaM, there should be 2 or 3 iron ions per monomer, respectively. For VanB, several excellent methods have also been established to quantify whether the needed flavin cofactor is also present in the as-purified protein (Aliverti, Curti, & Vanoni, 1999). In the case of TsaM and other Rieske oxygenases, it is possible that the iron count will be low. For these cases, it is important to be aware of the known lability of the mononuclear iron center, which is likely the culprit of the lower than expected result (Ferraro et al., 2005; Friemann et al., 2009). Depending on the intended use of the purified TsaM sample, supplemental iron can be added to account for the labile nature of this iron center (Liu, Tian, et al., 2022; Lukowski et al., 2018, 2020; Tian, Boggs, et al., 2023; Tian, Garcia, et al., 2023; Tian, Liu, et al., 2023).
6. In the situation that as-purified TsaM protein is also lacking the presence of the Rieske [2Fe-2S] cluster, an Fe-S cluster reconstitution experiment should be performed. The methods used by our laboratory for reconstitution of a [4Fe-4S] cluster have been previously described and can easily be adjusted to account for the [2Fe-2S] content of a Rieske oxygenase (Dill, Li, & Bridwell-Rabb, 2022). To complement the Ferrozine assay, following reconstitution, several additional robust methods for investigating iron or sulfide incorporation into a protein have been previously reported (Dill et al., 2022; Lanz et al., 2012).

6. Enzymatic assays for the TsaMBCD pathway

6.1 Liquid chromatography mass spectrometry (LC-MS) methods for activity assays

LC-MS is a technique that is broadly used for its known ability to separate molecules based on their physical properties (retention time) and their masses (mass-to-charge ratio, m/z) (Pitt, 2009; Thomas, French, Jannetto, Rappold, & Clarke, 2022). However, not all enzyme reaction components

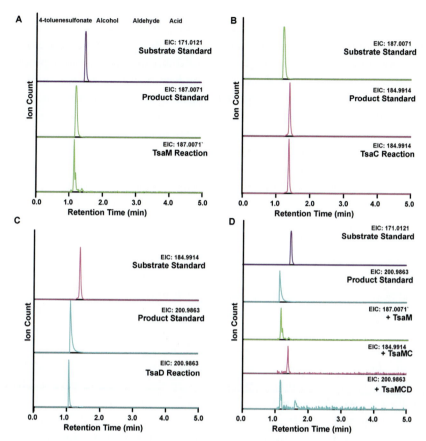

Fig. 3 The substrates and products of the TsaMBCD pathway can be separated by liquid chromatography mass spectrometry (LC–MS). (A) Here, we use the extracted ion chromatograms to show that the substrate (4-methylbenzenesulfonate) and product (4-(hydroxymethyl)benzenesulfonate) standards of the TsaM catalyzed reaction can be separated and that TsaM is active. The retention times of 4-tolunesulfonate and 4-(hydroxymethyl)benzenesulfonate are 1.486 and 1.219 min and the m/z for the extracted masses are 171.0121 and 187.0071, respectively. (B) TsaC catalyzes the formation of 4-benzenesulfonate when 4-(hydroxymethyl)benzenesulfonate is provided as the substrate. The retention time and m/z for 4-formylbenzenesulfonate are 1.400 min and 184.9914, respectively. (C) TsaD catalyzes the formation of 4-sulfobenzoate when 4-formylbenzenesulfonate is provided as the substrate. The retention time and m/z for 4-sulfobenzoate are 1.062 min and 200.9863, respectively. (D) The TsaMBCD pathway can be reconstituted and results in the eventual formation of 4-sulfobenzoate.

are amenable to detection with LC-MS. Therefore, when initially planning an LC-MS experiment, it is important to consider whether the molecules of interest will be observable. To assay enzyme activity using LC-MS, it is often possible to detect either the consumption of a substrate, or the formation of a product. For either case, accurate detection and quantification of these molecules relies on (1) the availability of substrate and product standards, (2) the mass of the substrate and product, (3) the ability of the substrate and product to ionize in a positive or negative ion mode, (4) the separation of reaction components *via* column chromatography, and (5) the identification and availability of an internal standard for the experiment (Pitt, 2009; Thomas et al., 2022).

For the TsaMBCD pathway, enzyme turnover is measured using LC-MS (Fig. 3) (Tian, Boggs, et al., 2023; Tian, Garcia, et al., 2023; Tian, Liu, et al., 2023). This technique, as described above, can be used to study this pathway because (1) all substrates and products of the TsaMBCD pathway enzymes can be purchased commercially. This fact is important because it means that the commercial standards can be used to identify the correct retention times and fragmentation patterns of the reaction components. Additionally, (2) the pathway standards fall into the ideal mass range for small molecule LC-MS (100–1200 Da), and (3) each molecule can be detected when the instrument is run in negative ion mode (Tian, Boggs, et al., 2023; Tian, Garcia, et al., 2023; Tian, Liu, et al., 2023). Furthermore, (4) extensive work has demonstrated that all of the TsaMBCD pathway reaction components can be separated using a reverse-phase Agilent ZORBAX Rapid Resolution HT 3.5 μm, 4.6 × 75 mm SB-CN liquid column (Tian, Boggs, et al., 2023; Tian, Garcia, et al., 2023; Tian, Liu, et al., 2023). And finally, (5) an internal standard (acetaminophen) for this system has been identified that is commercially available, ionizable in negative ion mode, isolatable from the other compounds of interest, and stable in the LC-MS solvents used to separate the pathway intermediates.

6.2 Separation of TsaMBCD pathway intermediates using LC-MS

As described above in Section 6.1, for the LC-MS experiments on the TsaMBCD pathway, a reverse-phase Agilent column has been identified that can be used to separate the different reaction intermediates (Tian, Boggs, et al., 2023; Tian, Garcia, et al., 2023; Tian, Liu, et al., 2023). For these experiments, two solvents should be prepared: solvent A (5% acetonitrile, 95% water, and 20 mM ammonium acetate (pH 5.5)) and solvent B (95% acetonitrile, 5% water, and 20 mM ammonium acetate (pH 5.5)). All LC-MS analysis in our laboratory is performed on an Agilent 5230 time of flight mass spectrometer equipped with a dual AJS ESI source and an

Agilent 1290 Infinity II LC system. Reactions should be analyzed on LC-MS using a flow rate of 0.6 mL/min, a 3 μL sample injection, and the following conditions:

1. An initial 1 min step to equilibrate the column with 10% solvent B.
2. A 3 min gradient step that runs from 10% to 95% solvent B. Whereas the elution in the former step should be sent to waste, in this step, the elution should be sent to the MS for analysis. Of relevance to this step, and to tailoring these methods to other Rieske oxygenase-containing pathways, the best gradient for a system typically separates all molecules of interest and flows from an aqueous solvent to a solvent that has a high percentage of organic solvent.
3. A penultimate 0.5 min isocratic step that washes the column with 95% solvent B.
4. A 1 min re-equilibration of the column with 10% solvent B.

6.3 Identification of the optimal conditions for measuring the activity of TsaM

Before one can begin enzymatic experiments, the reaction conditions must be optimized. For evaluating a Rieske oxygenase catalyzed reaction, as described in Section 2, it is imperative to not only identify an appropriate electron donor, but to also determine the optimal reductase:oxygenase ratio for the system. To accomplish this task, reactions should be assembled as follows:

1. Reactions (50 μL) should be composed of 5 μM TsaM, 2 mM substrate, 500 μM NADH, 100 μM $(NH_4)_2Fe(SO_4)_2{\cdot}6H_2O$, 20 mM Tris-HCl (pH 7.0), and varying concentrations of reductase. For this experiment, we typically employ ratios ranging from 2:1 to 5:1 (reductase:oxygenase). For the TsaM system, as described above, it has been previously determined that the non-native reductase VanB supports higher levels of activity than TsaB (Tian, Boggs, et al., 2023; Tian, Garcia, et al., 2023; Tian, Liu, et al., 2023). In addition, it has also been determined using this trajectory of experiments, that the optimal ratio of reductase:oxygenase is 4:1 (Tian, Boggs, et al., 2023; Tian, Garcia, et al., 2023; Tian, Liu, et al., 2023).
2. Reactions should be initiated by the addition of 5 μM of the oxygenase component (TsaM) and incubated for 3 h at 30 °C.
3. Reactions should be quenched with 100 μL of acetonitrile containing 1 mM of the internal standard (acetaminophen).

4. The quenched reactions should be centrifuged at 12,000 rpm for 15 min.
5. Following centrifugation, 50 μL of the supernatant should be aliquoted for LC-MS analysis.

    The next reaction condition to optimize is the duration of the assay. This reaction condition can be determined using the following trajectory:

6. Assemble reactions as described in steps 1-2 supplemented with the determined amount of needed reductase (20 μM VanB).
7. Quench the reactions at 1, 2, 3, 4, and 6 h time points with 100 μL of acetonitrile containing 1 mM of internal standard (acetaminophen).
8. Centrifuge the quenched reactions and aliquot for analysis as described in steps 4–5. For the TsaM-VanB reaction, it has previously been determined that the optimal time for incubation of the reaction is 3 h (Tian, Boggs, et al., 2023; Tian, Garcia, et al., 2023; Tian, Liu, et al., 2023). This incubation period leads to a maximum amount of product formation and is not increased by longer incubation times.

6.4 Total turnover number (TTN) determination using LC-MS

A TTN describes the number of turnovers an enzyme active site can undergo in its catalytic lifetime (Rogers & Bommarius, 2010). This dimensionless measurement is valuable because it allows the comparison of enzyme variants and also provides a straight-forward method for evaluating the activity of enzymes with a broad range of substrates (Rogers & Bommarius, 2010). There are generally two methods to measure TTN that can be employed using LC-MS. The first method involves calculating the consumption of substrate and the second method involves quantification of the amount of product formed in the reaction. Regardless of the chosen method, both rely on the generation of standard curves. This standard curve should be made by plotting concentration as a function of peak ratio, which is calculated using the equation below:

$$Ratio = \frac{Peak\ area\ of\ product\ or\ peak\ area\ of\ substrate}{Peak\ area\ of\ the\ internal\ standard}$$

These standard curves allow for one to calculate the concentration of substrate consumed or product generated from a least squares regression formula that is fit to the measured data points. It is apt to mention that the prepared standard curve can be used to not only measure TTN, but also to facilitate the measurement of Michaelis-Menten kinetic parameters for a system of interest. This latter type of experiment, however, requires additional steps to ensure that the reactions are being performed under steady state conditions (Brooks et al., 2004). For Rieske oxygenases, accurate measurement of kinetic parameters also includes an investigation into the saturating concentrations of NADH and O_2 (Kincannon et al., 2022; Tian, Liu, et al., 2023). To calculate TTN using substrate consumption the formula is as presented below:

$$TTNs = \frac{(\text{Initial concentration of substrate} - \text{concentration of the substrate left})}{\text{Total concentration of the enzyme used}}$$

Here, the initial concentration of substrate is determined using a control reaction that lacks added enzyme. This step assumes that all of the substrate consumed is generating product in the reaction. Likewise, the calculation of the TTN when product formation is being measured is shown below:

$$TTNs = \frac{(\text{Total concentration of generated product})}{\text{Total concentration of enzyme used}}$$

For the TsaMBCD pathway TTNs, use the following previously described conditions (Tian, Boggs, et al., 2023; Tian, Garcia, et al., 2023; Tian, Liu, et al., 2023):

1. For TsaM reactions: set up 50 µL reactions that contain 20 mM Tris-HCl (pH 7.0), 2.5 µM TsaM, 10 µM VanB, 1 mM substrate, 1 mM NADH, and 100 µM $(NH_4)_2Fe(SO_4)_2\cdot6H_2O$.

2. For TsaC and TsaD catalyzed reactions: set up 50 µL reactions that contain 20 mM Tris-HCl (pH 7.0), 5 µM TsaC or 5 µM TsaD, 2 mM substrate (4-(hydroxymethyl)benzenesulfonate, 4-(hydroxymethyl)benzoate, 4-formylbenzenesulfonate, or 4-formylbenzoate), and 2 mM NAD^+. For this step, it is important to note that TsaC also produces a side product *in vitro* (Tian, Boggs, et al., 2023). This presumably off-pathway product is made through the TsaC-catalyzed oxidation of an aldehyde hydrate intermediate (Tian, Boggs, et al., 2023). The appearance of the resultant carboxylic acid products need to be taken into account as they complicate analysis of TsaC activity. For these proteins, care should be taken, in the form of control experiments, to robustly establish that the observed activity is not due to background contamination from one of the many known *E. coli* dehydrogenases (Sophos & Vasiliou, 2003).

3. All reactions should be centrifuged and quenched as described in Section 6.3 steps 3-4.

4. TTNs should be calculated in triplicate using the appropriate formula.

5. Pathway reactions that allow for measurement of 4-methylbenzene-sulfonate or 4-methylbenzoate consumption and the iterative conversion of these substrates into the corresponding alcohol, aldehyde, and carboxylic acid products over time have been previously performed for the TsaMBCD pathway (Tian, Boggs, et al., 2023). These conditions are reminiscent of those described here in steps 1–3, but instead, the reactions were made to contain 1 mM substrate, 1 mM NADH, 200 μm $(NH_4)_2Fe(SO_4)_2 \cdot 6H_2O$, 5 μM TsaM, 20 μM VanB, and 5 μM of TsaC and TsaD. Samples were taken every 25 min up to three hours and quenched and centrifuged as described in Section 6.3 steps 3-4.

6.5 Spectroscopic assay for analysis of NAD(P)H consumption and production

In the TsaMBCD pathway, NADH supplies electrons to TsaM and NAD^+ accepts electrons, in the form of a hydride, to support the TsaC and TsaD-catalyzed oxidations. As the reduction of the nicotinamide ring bestows an absorbance peak at 340 nm in NAD(P)H, which is not present in NAD $(P)^+$, the spectral change of this transformation affords a mechanism to monitor enzyme activity over time (De Ruyck et al., 2007). For the TsaMBCD pathway, this technique proved to be a valuable tool for evaluating reaction progress using a BioTek Epoch2 microplate reader, and to evaluate both the consumption (TsaM and VanB) and regeneration (TsaC and TsaD) of NADH (Fig. 4) (Tian, Boggs, et al., 2023).

As described in Section 6.3 for measuring TTN, to perform spectroscopic experiments on a Rieske oxygenase or pathway of interest, it is important to optimize the reaction conditions for a chosen system and available instrument. Specifically, the absorbance at 340 nm should be measured for a range of NAD (P)H concentrations (0.01–10 mM) to determine the appropriate scale at which enzymatic activity can be detected. For our purposes, we used this method in a qualitative manner, and as a means to choose a starting concentration for the TsaM-catalyzed conversion of NADH into NAD^+ (Tian, Boggs, et al., 2023). However, a standard curve can also be produced from these initial experiments to determine the linear range of the instrument and to correlate absorbance readings with the amount of NAD(P)H or NAD(P)$^+$ in solution. In addition to determining the optimal NAD(P)H detection range, the components of the enzymatic reaction must also be considered. In particular, the characteristic absorbance spectra of substrates and products should be tested to ensure that they will not interfere or conflate the signal from NAD(P) H (De Ruyck et al., 2007). Likewise, it is important to test whether the

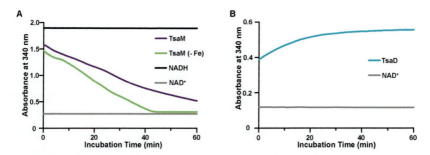

Fig. 4 The consumption or generation of NADH by different enzyme components of the TsaMBCD pathway can be visualized. (A) As indicated by the decrease in absorbance at 340 nm, NADH is consumed by TsaM and VanB. Iron supplementation in this reaction was also tested. The lack of added ferrous ammonium sulfate appears to have an impact on the observed signal from NADH (green). Specifically, this latter data shows that the addition of extra iron into the reaction impacts the measurement of NADH consumption, which is an important consideration for experimental design. (B) As indicated by the increase in absorbance at 340 nm, NADH is generated by TsaD. To emphasize the signal in this panel, 100 μM TsaD was reacted with the 2 mM 4-formylbenzenesulfonate and 2 mM NAD$^+$. In this figure, data were measured in triplicate and presented as the mean value of these measurements. Control reactions are included in both panels.

individual molecules of interest react with NAD(P)H, which can also impact the desired measurements. As previously described, for the TsaMBCD pathway enzymes, NADH consumption and production reactions should be set up in the following manner (Tian, Boggs, et al., 2023):

1. For TsaM: set up 100 μL reactions as described in Section 6.4 step 1.
   a. This volume is the minimum that should be used for accurate absorbance readings in 96-well microplates. In this step, care should be taken to avoid introducing bubbles into the wells.
   b. NADH should be used to initiate the reactions and should be added as the final step to minimize unmeasured data points between the start of the reaction and the first absorbance reading. If preparing several reactions, using a multichannel pipette to add NADH or NAD$^+$ can minimize variability between reactions. To note for this step, both NADH and NAD$^+$ are prone to degradation and should be prepared fresh for all experiments. For stock solutions, both NADH and NAD$^+$ should be stored in a cold and dark environment. Last, for the TsaM-catalyzed reactions, based on the initial tests described above, it has previously been determined that a starting concentration of 1 mM provides a good starting signal with the plate reader (Tian, Boggs, et al., 2023).

2. For TsaC and TsaD: set up 100 μL reactions as described in Section 6.4 step 2.
3. Shake the plate briefly in the plate reader and incubate at 25 °C.
4. Measure the absorbance at 340 nm every 10 s or the minimum time possible for the number of wells based on plate reader limitations (Fig. 4).
5. Pathway reactions that allow for the regeneration of NADH over time have been previously performed using all proteins from the TsaMBCD pathway (Tian, Boggs, et al., 2023). These conditions are reminiscent of the conditions described in steps 1–4, but instead require that all enzyme components be added at different times.

7. Crystallization of the short-chain dehydrogenase/reductase (SDR) enzyme TsaC

As described in Section 1, there is much interest in developing methods to regenerate the reducing equivalents for Rieske oxygenase centered biocatalytic applications. In the TsaMBCD pathway, the annotated SDR enzyme TsaC converts NAD^+ into NADH (Junker et al., 1996; Junker et al., 1997; Locher et al., 1991). This chemistry makes TsaC attractive, not only for its ability to support the chemistry of TsaM, but also for its potential to support other non-native Rieske oxygenases through a coupled reaction (Fig. 1B and C). To broadly use TsaC or other SDR proteins to support Rieske oxygenase chemistry, it is important to identify the design principles that dictate substrate specificity. In this section, we describe how the structure of TsaC was determined (Tian, Boggs, et al., 2023).

1. The protein concentration for screening crystallographic conditions should first be assayed using the Pre-Crystallization Test (Hampton Research). In the case of TsaC, this test revealed that crystallographic conditions for TsaC should be screened at concentrations between 10 and 20 mg/mL (Tian, Boggs, et al., 2023).
2. Following assessment of the correct protein concentration for crystallographic experiments, high-throughput screening should be performed. Typically, for this step, at least 3–4 commercial screens of conditions, or 864–1152 unique combinations of a crystallization solution with different concentrations of the protein of interest should be screened (Rupp, 2010). For TsaC, a Mosquito pipetting robot was used to facilitate this step. Conditions were screened to contain 300 nL of

protein at a concentration of either 10, 15, or 20 mg/mL and 300 nL of reservoir solution. In these sitting drop robot trays, 50 μL of reservoir solution was also included to facilitate vapor diffusion. This screening step identified numerous conditions.
3. To evaluate the crystallization hits, crystals can be harvested from the robot tray directly, or conditions can be scaled-up for optimization. In the case of TsaC, many of the initial conditions and identified crystal forms had high translational non-crystallographic symmetry. This crystal pathology complicated space group assignment and unit cell determination. To overcome this challenge, optimization of the crystallization condition and precise control of crystal growth time and temperature was paramount. Additionally, as many of the optimized wells contained multiple crystal morphologies, the diffraction patterns of many crystals needed to be screened such that clean data sets could be collected (Fig. 5A and B).

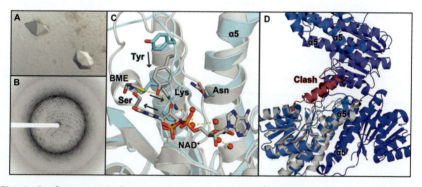

Fig. 5 Conformational changes expected upon NAD$^+$ binding may require different crystallization conditions and alternative crystal packing. (A) Optimized TsaC crystals formed using a vapor diffusion experiment. (B) Representative protein crystal X-ray diffraction frame. (C) A superimposition of the TsaC (cyan) and a *meso*-2,3-butanediol dehydrogenase-NAD$^+$ complex structures (gray) illustrate requisite movements for NAD$^+$ binding. Residues in the catalytic tetrad are labeled, and the direction they must move between the NAD$^+$-free structure and a catalytically competent structure are marked as arrows. (D) In this panel four chains of TsaC are shown in shades of blue. One chain is superimposed with the *meso*-2,3-butanediol dehydrogenase-NAD$^+$ complex (gray, residues 183–224 in red) to demonstrate that ordering of the unmodeled region of TsaC (residues 188–220, equivalent to the ordered red region of *meso*-2,3-butanediol dehydrogenase) would clash with a symmetry related molecule and disrupt the crystal packing. A small clash can also be observed due to a conformational shift of α5. PDB IDs–TsaC: 8SY8 (Tian, Boggs, et al., 2023) and *meso*-2,3-butanediol dehydrogenase-NAD$^+$ complex: 1GEG (Otagiri et al., 2001).

4. Optimized crystals of TsaC can be reproduced by combination of 1 µL of TsaC (10–20 mg/mL) with 1 µL of well solution (0.2 M KCl, 20% polyethylene glycol 3350, and 10% glycerol). Interesting to note here is that our published conditions also include 2 mM NAD^+ (Tian, Boggs, et al., 2023). However, no electron density for this molecule is visible in the final solved structure (Tian, Boggs, et al., 2023). This result suggests that an NAD^+-bound structure likely requires identification of a different crystallization condition. Indeed, an overlay of this NAD^+-free structure with a homolog that has NAD^+ bound in the active site, reveals that the catalytic tetrad residues would need to reorient to point toward the active site upon NAD^+ binding (Fig. 5C). Cosubstrate binding and subsequent rearrangement or ordering of unmodeled residues in this region would likely disrupt the packing interactions of the crystal (Fig. 5D).
5. Crystals of TsaC typically appear within 4 days and can be harvested and frozen in liquid nitrogen following a quick 5–10 s soak in a 20% PEG 3350 cryoprotectant solution.

8. Conclusions

Metalloenzymes catalyze a broad range of complex transformations that typically require the tandem participation of a redox cofactor. One class of metalloenzymes, described in this chapter, the Rieske oxygenases, are regarded for their catalytic prowess and corresponding biocatalytic potential (Ballard et al., 1983; Buckland et al., 1999; Ensley et al., 1983; Fabara & Fraaije, 2020; Newman et al., 2004; Reddy et al., 1999). However, to facilitate their chemistry, Rieske oxygenases, like many other metalloproteins, require the expensive stoichiometric addition of NAD(P) H (Fig. 1A). To circumvent this cost, one appealing attribute of this enzyme family is our emerging understanding that a large number co-occur with accessory NAD(P)H regenerating enzymes (Tian, Boggs, et al., 2023). Therefore, to harness the catalytic power of these co-occurrence systems, we have developed methods to study the Rieske oxygenase TsaM and its NAD(P)H recycling partners found in the TsaMBCD pathway (Fig. 1B). We posit that the methods described here provide a foundation for translation into other enzymatic systems. Furthermore, we anticipate that the TsaC- and TsaD-like enzymes found in this pathway (and others) can be adapted to act as auxiliary systems to continuously supply reducing equivalents (Fig. 1C). Apart from the TsaMBCD pathway, our approaches detailed in this chapter reveal general strategies to explore the wealth of the large, yet underrepresented class of Rieske oxygenases.

Acknowledgments

The work in this publication was supported by the National Institute of General Medical Sciences of the National Institutes of Health under Award Number R35 GM138271 (J.B.R.). The contents of this publication are solely the responsibility of the authors and do not necessarily represent the official views of National Institute of General Medical Sciences or National Institutes of Health.

References

Aliverti, A., Curti, B., & Vanoni, M. A. (1999). Identifying and quantitating FAD and FMN in simple and in iron-sulfur-containing flavoproteins. *Methods in Molecular Biology*, *131*, 9–23.

Ashikawa, Y., Fujimoto, Z., Noguchi, H., Habe, H., Omori, T., Yamane, H., & Nojiri, H. (2006). Electron transfer complex formation between oxygenase and ferredoxin components in Rieske nonheme iron oxygenase system. *Structure (London, England: 1993)*, *14*, 1779–1789.

Ballard, D. G. H., Courtis, A., Shirley, I. M., & Taylor, S. C. (1983). A biotech route to polyphenylene. *Journal of the Chemical Society-Chemical Communications*, *17*, 954–955.

Barry, S. M., & Challis, G. L. (2013). Mechanism and catalytic diversity of Rieske nonheme iron-dependent oxygenases. *ACS Catalysis*, *3*, 2362–2370.

Bradford, M. M. (1976). A rapid and sensitive method for the quantitation of microgram quantities of protein utilizing the principle of protein-dye binding. *Analytical Biochemistry*, *72*, 248–254.

Brimberry, M., Garcia, A. A., Liu, J., Tian, J., & Bridwell-Rabb, J. (2023). Engineering Rieske oxygenase activity one piece at a time. *Current Opinion in Chemical Biology*, *72*, 102227.

Brooks, H. B., Geeganage, S., Kahl, S. D., Montrose, C., Sittampalam, S., Smith, M. C., & Weidner, J. R. (2004). In S. Markossian, A. Grossman, K. Brimacombe, M. Arkin, D. Auld, C. Austin, J. Baell, T. D. Y. Chung, N. P. Coussens, J. L. Dahlin, V. Devanarayan, T. L. Foley, M. Glicksman, K. Gorshkov, J. V. Haas, M. D. Hall, S. Hoare, J. Inglese, P. W. Iversen, S. C. Kales, M. Lal-Nag, Z. Li, J. McGee, O. McManus, T. Riss, P. Saradjian, G. S. Sittampalam, M. Tarselli, O. J. Trask, Y. Wang, J. R. Weidner, M. J. Wildey, K. Wilson, M. Xia, & X. Xu (Eds.). *Basics of enzymatic assays for HTS*. Bethesda, MD: Assay Guidance Manual.

Buckland, B. C., Drew, S. W., Connors, N. C., Chartrain, M. M., Lee, C., Salmon, P. M., ... Greasham, R. L. (1999). Microbial conversion of indene to indandiol: A key intermediate in the synthesis of CRIXIVAN. *Metabolic Engineering*, *1*(1), 63–74.

Dagert, M., & Ehrlich, S. D. (1979). Prolonged incubation in calcium chloride improves the competence of *Escherichia coli* cells. *Gene*, *6*(1), 23–28.

De Ruyck, J., Famerée, M., Wouters, J., Perpète, E. A., Preat, J., & Jacquemin, D. (2007). Towards the understanding of the absorption spectra of NAD(P)H/NAD(P)+ as a common indicator of dehydrogenase enzymatic activity. *Chemical Physics Letters*, *450*(1), 119–122.

Dhindwal, S., Patil, D. N., Mohammadi, M., Sylvestre, M., Tomar, S., & Kumar, P. (2011). Biochemical studies and ligand-bound structures of biphenyl dehydrogenase from *Pandoraea pnomenusa* strain B-356 reveal a basis for broad specificity of the enzyme. *The Journal of Biological Chemistry*, *286*(42), 37011–37022.

Dill, Z., Li, B., & Bridwell-Rabb, J. (2022). Purification and structural elucidation of a cobalamin-dependent radical SAM enzyme. *Methods in Enzymology, 669*, 91–116.

Eaton, R. W., & Chapman, P. J. (1992). Bacterial metabolism of naphthalene: Construction and use of recombinant bacteria to study ring cleavage of 1,2-dihydroxynaphthalene and subsequent reactions. *Journal of Bacteriology, 174*(23), 7542–7554.

Ensley, B. D., Ratzkin, B. J., Osslund, T. D., Simon, M. J., Wackett, L. P., & Gibson, D. T. (1983). Expression of naphthalene oxidation genes in *Escherichia coli* results in the biosynthesis of indigo. *Science (New York, N. Y.), 222*(4620), 167–169.

Fabara, A. N., & Fraaije, M. W. (2020). An overview of microbial indigo-forming enzymes. *Applied Microbiology and Biotechnology, 104*(3), 925–933.

Ferraro, D. J., Gakhar, L., & Ramaswamy, S. (2005). Rieske business: Structure-function of Rieske non-heme oxygenases. *Biochemical and Biophysical Research Communications, 338*(1), 175–190.

Feyza Ozgen, F., Runda, M. E., Burek, B. O., Wied, P., Bloh, J. Z., Kourist, R., & Schmidt, S. (2020). Artificial light-harvesting complexes enable rieske oxygenase catalyzed hydroxylations in non-photosynthetic cells. *Angewandte Chemie (International Ed. in English), 59*(10), 3982–3987.

Fish, W. W. (1988). Rapid colorimetric micromethod for the quantitation of complexed iron in biological samples. *Methods in Enzymology, 158*, 357–364.

Friemann, R., Lee, K., Brown, E. N., Gibson, D. T., Eklund, H., & Ramaswamy, S. (2009). Structures of the multicomponent Rieske non-heme iron toluene 2,3-dioxygenase enzyme system. *Acta Crystallographica. Section D, Biological Crystallography, 65*(Pt 1), 24–33.

Furukawa, K., Hirose, J., Suyama, A., Zaiki, T., & Hayashida, S. (1993). Gene components responsible for discrete substrate specificity in the metabolism of biphenyl (bph operon) and toluene (tod operon). *Journal of Bacteriology, 175*(16), 5224–5232.

Gasteiger, E., Gattiker, A., Hoogland, C., Ivanyi, I., Appel, R. D., & Bairoch, A. (2003). ExPASy: The proteomics server for in-depth protein knowledge and analysis. *Nucleic Acids Research, 31*(13), 3784–3788.

Gibson, B., Wilson, D. J., Feil, E., & Eyre-Walker, A. (2018). The distribution of bacterial doubling times in the wild. *Proceedings. Biological Sciences/the Royal Society, 285*(1880).

Hollmann, F., Opperman, D. J., & Paul, C. E. (2021). Biocatalytic reduction reactions from a chemist's perspective. *Angewandte Chemie-International Edition, 60*(11), 5644–5665.

Hu, W. Y., Li, K., Weitz, A., Wen, A., Kim, H., Murray, J. C., ... Liu, P. (2022). Light-driven oxidative demethylation reaction catalyzed by a Rieske-type non-heme iron enzyme Stc2. *ACS Catalysis, 12*(23), 14559–14570.

Irie, S., Doi, S., Yorifuji, T., Takagi, M., & Yano, K. (1987). Nucleotide sequencing and characterization of the genes encoding benzene oxidation enzymes of *Pseudomonas putida*. *Journal of Bacteriology, 169*(11), 5174–5179.

Junker, F., Kiewitz, R., & Cook, A. M. (1997). Characterization of the *p*-toluenesulfonate operon tsaMBCD and tsaR in *Comamonas testosteroni* T-2. *Journal of Bacteriology, 179*(3), 919–927.

Junker, F., Saller, E., Oppenberg, H. R. S., Kroneck, P. M. H., Leisinger, T., & Cook, A. M. (1996). Degradative pathways for p-toluenecarboxylate and *p*-toluenesulfonate and their multicomponent oxygenases in *Comamonas testosteroni* strains PSB-4 and T-2. *Microbiology-Sgm, 142*, 2419–2427.

Kincannon, W. M., Zahn, M., Clare, R., Lusty Beech, J., Romberg, A., Larson, J., ... DuBois, J. L. (2022). Biochemical and structural characterization of an aromatic ring-hydroxylating dioxygenase for terephthalic acid catabolism. *Proceedings of the National Academy of Sciences of the United States of America, 119*(13), e2121426119.

Knapp, M., Mendoza, J., & Bridwell-Rabb, J. (2021). An aerobic route for C-H bond functionalization: The Rieske non-heme iron oxygenases. In *Encyclopedia of Biological Chemistry*, 3rd edition.

Kovaleva, E. G., & Lipscomb, J. D. (2008). Versatility of biological non-heme Fe(II) centers in oxygen activation reactions. *Nature Chemical Biology, 4*(3), 186–193.

Kweon, O., Kim, S., Baek, S., Chae, J., Adjei, M. D., Baek, D., ... Cerniglia, C. E. (2008). A new classification system for bacterial Rieske non-heme iron aromatic ring-hydroxylating oxygenases. *BMC Biochemistry, 9*(11).

Lanfranchi, E., Trajkovic, M., Barta, K., de Vries, J. G., & Janssen, D. B. (2019). Exploring the selective demethylation of aryl methyl ethers with a *Pseudomonas* Rieske monooxygenase. *Chembiochem: A European Journal of Chemical Biology, 20*(1), 118–125.

Lanz, N. D., Grove, T. L., Gogonea, C. B., Lee, K. H., Krebs, C., & Booker, S. J. (2012). RlmN and AtsB as models for the overproduction and characterization of radical SAM proteins. *Methods in Enzymology, 516*, 125–152.

Lee, J., Simurdiak, M., & Zhao, H. (2005). Reconstitution and characterization of aminopyrrolnitrin oxygenase, a Rieske N-oxygenase that catalyzes unusual arylamine oxidation. *The Journal of Biological Chemistry, 280*(44), 36719–36728.

Lee, J., & Zhao, H. (2006). Mechanistic studies on the conversion of arylamines into arylnitro compounds by aminopyrrolnitrin oxygenase: Identification of intermediates and kinetic studies. *Angewandte Chemie (International Ed. in English), 45*(4), 622–625.

Li, P., Wang, L., & Feng, L. (2013). Characterization of a novel Rieske-type alkane monooxygenase system in *Pusillimonas* sp. strain T7-7. *Journal of Bacteriology, 195*(9), 1892–1901.

Liu, J., Knapp, M., Jo, M., Dill, Z., & Bridwell-Rabb, J. (2022). Rieske oxygenase catalyzed C-H bond functionalization reactions in chlorophyll b biosynthesis. *ACS Central Science, 8*(10), 1393–1403.

Liu, J., Tian, J., Perry, C., Lukowski, A. L., Doukov, T. I., Narayan, A. R. H., & Bridwell-Rabb, J. (2022). Design principles for site-selective hydroxylation by a Rieske oxygenase. *Nature Communications, 13*(1), 255.

Locher, H. H., Leisinger, T., & Cook, A. M. (1991). 4-Toluene sulfonate methylmonooxygenase from *Comamonas testosteroni* T-2: Purification and some properties of the oxygenase component. *Journal of Bacteriology, 173*(12), 3741–3748.

Lukowski, A. L., Ellinwood, D. C., Hinze, M. E., DeLuca, R. J., Du Bois, J., Hall, S., & Narayan, A. R. H. (2018). C-H hydroxylation in paralytic shellfish toxin biosynthesis. *Journal of the American Chemical Society, 140*(37), 11863–11869.

Lukowski, A. L., Liu, J., Bridwell-Rabb, J., & Narayan, A. R. H. (2020). Structural basis for divergent C-H hydroxylation selectivity in two Rieske oxygenases. *Nature Communications, 11*(1), 2991.

Mandel, M., & Higa, A. (1970). Calcium-dependent bacteriophage DNA infection. *Journal of Molecular Biology, 53*(1), 159–162.

Neidle, E. L., Hartnett, C., Ornston, L. N., Bairoch, A., Rekik, M., & Harayama, S. (1992). Cis-diol dehydrogenases encoded by the TOL pWW0 plasmid xylL gene and the *Acinetobacter calcoaceticus* chromosomal benD gene are members of the short-chain alcohol dehydrogenase superfamily. *European Journal of Biochemistry/FEBS, 204*(1), 113–120 (d).

Neidle, E. L., Hartnett, C., Ornston, L. N., Bairoch, A., Rekik, M., & Harayama, S. (1991). Nucleotide sequences of the *Acinetobacter calcoaceticus* benABC genes for benzoate 1,2-dioxygenase reveal evolutionary relationships among multicomponent oxygenases. *Journal of Bacteriology, 173*(17), 5385–5395.

Newman, L. M., Garcia, H., Hudlicky, T., & Selifonov, S. A. (2004). Directed evolution of the dioxygenase complex for the synthesis of furanone flavor compounds. *Tetrahedron, 60*(3), 729–734.

Otagiri, M., Kurisu, G., Ui, S., Takusagawa, Y., Ohkuma, M., Kudo, T., & Kusunoki, M. (2001). Crystal structure of *meso*-2,3-butanediol dehydrogenase in a complex with NAD + and inhibitor mercaptoethanol at 1.7 A resolution for understanding of chiral substrate recognition mechanisms. *Journal of Biochemistry, 129*(2), 205–208.

Pace, C. N., Vajdos, F., Fee, L., Grimsley, G., & Gray, T. (1995). How to measure and predict the molar absorption-coefficient of a protein. *Protein Science, 4*(11), 2411–2423.

Perry, C., de Los Santos, E. L. C., Alkhalaf, L. M., & Challis, G. L. (2018). Rieske non-heme iron-dependent oxygenases catalyse diverse reactions in natural product biosynthesis. *Natural Product Reports, 35*(7), 622–632.

Pitt, J. J. (2009). Principles and applications of liquid chromatography-mass spectrometry in clinical biochemistry. *The Clinical Biochemist. Reviews/Australian Association of Clinical Biochemists, 30*(1), 19–34.

Reddy, J., Lee, C., Neeper, M., Greasham, R., & Zhang, J. (1999). Development of a bioconversion process for production of 1,2-indandiol from indene by recombinant constructs. *Applied Microbiology and Biotechnology, 51*(5), 614–620.

Rogers, T. A., & Bommarius, A. S. (2010). Utilizing simple biochemical measurements to predict lifetime output of biocatalysts in continuous isothermal processes. *Chemical Engineering Science, 65*(6), 2118–2124.

Runda, M. E., de Kok, N. A. W., & Schmidt, S. (2023). Rieske oxygenases and other ferredoxin-dependent enzymes: Electron transfer principles and catalytic capabilities. *Chembiochem: A European Journal of Chemical Biology, 24*(15), e202300078.

Rupp, B. (2010). *Biomolecular crystallography: Principles, practice, and application to structural,* 1–809.

Sgro, M., Chow, N., Olyaei, F., Arentshorst, M., Geoffrion, N., Ram, A. F. J., ... Tsang, A. (2023). Functional analysis of the protocatechuate branch of the beta-ketoadipate pathway in *Aspergillus niger*. *The Journal of Biological Chemistry, 299*(8), 105003.

Simonian, M. H. (2002). Spectrophotometric determination of protein concentration. *Current Protocols in Cell Biology, Appendix, 3* Appendix 3B.

Sophos, N. A., & Vasiliou, V. (2003). Aldehyde dehydrogenase gene superfamily: The 2002 update. *Chemico-Biological Interactions, 143-144,* 5–22.

Stookey, L. L. (1970). Ferrozine - A new spectrophotometric reagent for iron. *Analytical Chemistry, 42*(7), 779.

Thomas, S. N., French, D., Jannetto, P. J., Rappold, B. A., & Clarke, W. A. (2022). Liquid chromatography-tandem mass spectrometry for clinical diagnostics. *Nature Reviews Methods Primers, 2*(1), 96.

Tian, J., Boggs, D. G., Donnan, P. H., Barroso, G. T., Garcia, A. A., Dowling, D. P., ... Bridwell-Rabb, J. (2023). The NADH recycling enzymes TsaC and TsaD regenerate reducing equivalents for Rieske oxygenase chemistry. *The Journal of Biological Chemistry, 299*(10), 105222.

Tian, J., Garcia, A. A., Donnan, P. H., & Bridwell-Rabb, J. (2023). Leveraging a structural blueprint to rationally engineer the Rieske oxygenase TsaM. *Biochemistry, 62*(11), 1807–1822.

Tian, J., Liu, J., Knapp, M., Donnan, P. H., Boggs, D. G., & Bridwell-Rabb, J. (2023). Custom tuning of Rieske oxygenase reactivity. *Nature Communications, 14*(1), 5858.

van der Donk, W. A., & Zhao, H. (2003). Recent developments in pyridine nucleotide regeneration. *Current Opinion in Biotechnology, 14*(4), 421–426.

Wang, X., Saba, T., Yiu, H. H. P., Howe, R. F., Anderson, J. A., & Shi, J. (2017). Cofactor NAD(P)H regeneration inspired by heterogeneous pathways. *Chem, 2*(5), 621–654.

Zylstra, G. J., McCombie, W. R., Gibson, D. T., & Finette, B. A. (1988). Toluene degradation by *Pseudomonas putida* F1: Genetic organization of the tod operon. *Applied and Environmental Microbiology, 54*(6), 1498–1503.

CHAPTER ELEVEN

Whole-cell Rieske non-heme iron biocatalysts

Meredith B. Mock, Shuyuan Zhang, and Ryan M. Summers*

Department of Chemical and Biological Engineering, The University of Alabama, Tuscaloosa, AL, United States
*Corresponding author. e-mail address: rmsummers@eng.ua.edu

Contents

1.	Introduction	244
2.	Before you begin timing: 4–5 days	246
3.	Key resources table	248
4.	Materials and equipment	249
	4.1 Equipment	249
	4.2 Materials and reagents	249
5.	Step-by-step method details	249
	5.1 Cell growth and gene expression	250
	5.2 Resting cell reactions	251
	5.3 Reaction sampling and sample preparation	252
6.	High-performance liquid chromatography (HPLC) analysis	253
7.	Establishing a calibration curve	254
8.	Expected outcomes	255
9.	Quantification and statistical analysis	256
10.	Advantages	258
11.	Limitations	258
12.	Optimization and troubleshooting	258
	12.1 No or low activity detected	258
	12.2 Potential solutions to optimize the procedure	259
13.	Safety considerations and standards	259
14.	Alternative methods/procedures	260
References		260

Abstract

Rieske non-heme iron oxygenases (ROs) possess the ability to catalyze a wide range of reactions. Their ability to degrade aromatic compounds is a unique characteristic and makes ROs interesting for a variety of potential applications. However, purified ROs can be challenging to work with due to low stability and long, complex electron transport chains. Whole cell biocatalysis represents a quick and reliable method for characterizing the activity of ROs and harnessing their metabolic potential. In this

Methods in Enzymology, Volume 703
ISSN 0076-6879, https://doi.org/10.1016/bs.mie.2024.05.008
Copyright © 2024 Elsevier Inc. All rights are reserved, including those for text and data mining, AI training, and similar technologies.

protocol, we outline a step-by-step protocol for the overexpression of ROs for whole cell biocatalysis and characterization. We have utilized a caffeine-degrading, *N*-demethylation system, expressing the RO genes *ndmA* and *ndmD*, as an example of this method.

1. Introduction

Rieske non-heme iron oxygenases (ROs) are notable for their ability to catalyze a wide variety of reactions and are often recognized for the key role they play in the bacterial degradation of aromatic compounds (Perry, De los Santos, Alkhalaf, & Challis, 2018; Wackett, 2002). The first RO enzymes to be characterized were naphthalene dioxygenase (NDO) and toluene dioxygenase, both from *Pseudomonas putida* (Barry & Challis, 2013; Ensley & Gibson, 1983; Wackett, 2002; Yeh, Gibson, & Liu, 1977). Dioxygenases catalyze the incorporation of both atoms of molecular oxygen (O_2) into their products, such as the *cis*-1,2-dihydroxylation of naphthalene to form *cis*-naphthalene dihydrodiol, whereas monooxygenases only incorporate one atom of O_2 into their products (Barry & Challis, 2013; Ferraro, Gakhar, & Ramaswamy, 2005). The *N*-demethylation of caffeine (1,3,7-trimethylxanthine) to theobromine (3,7-dimethylxanthine) by *N*-demethylase A (ndmA) is an example of a monooxygenase catalyzed reaction (Summers, Mohanty, Gopishetty, & Subramanian, 2015). ROs have been shown to possess substrate and reaction versatility or to demonstrate promiscuity in their catalytic behavior. As an example, NDO isolated from *Pseudomonas* sp. NCIB 9816-4 has been reported to oxidize over 75 different substrates (Özgen & Schmidt, 2019; Resnick, Lee, & Gibson, 1996), showing remarkable substrate promiscuity. Toluene dioxygenase from *P. putida* F1 can oxidize more than 200 substrates (Boyd et al., 2001; Özgen & Schmidt, 2019). Additionally, NdmA demonstrates promiscuity in its ability to not only catalyze the N-demethylation of caffeine, but also the *N*-demethylation of theophylline (1,3-dimethylxanthine), paraxanthine (1,7-dimethylxanthine), and 7-methylxanthine (Mock et al., 2021).

Rieske dioxygenases and monooxygenases are believed to function mechanistically in highly similar ways (Wackett, 2002). RO systems use O_2 and NAD(P)H as co-substrates and have been classified into five different types depending on their different components: a reductase, an oxygenase, and sometimes an additional ferredoxin component (Kweon et al., 2008).

Electrons are typically first transferred from NAD(P)H to a flavin and then to an iron-sulfur cluster in the reductase or associated ferredoxin protein. Then the electrons are usually passed to the Rieske iron-sulfur cluster of the oxygenase component. From there, the electrons are shuttled to the non-heme iron center located within the active site, resulting in a ferrous, Fe(II), ion or Fe(III) with the presence of electron rich substrate (Barry & Challis, 2013; Ferraro et al., 2005; Özgen & Schmidt, 2019; Perry et al., 2018; Rogers, Gordon, Rappe, Goodpaster, & Lipscomb, 2022; Wackett, 2002). The ferrous non-heme iron center is then able to bind O_2, to form either an Fe(II)-peroxo or Fe(III)-superoxo intermediate that can subsequently react with a substrate through a wide variety of potential reactions (Ferraro et al., 2005; Rogers et al., 2022; Wackett, 2002). The electron transfer process from NAD(P)H to the non-heme iron center is complex and requires sophisticated electron tunneling architecture to minimize the loss of free energy (Özgen & Schmidt, 2019). The iron-sulfur clusters used in this electron tunneling process are commonly found as either a [2Fe-2S] or a [4Fe-4S] cluster according to the atomic content of iron and sulfide ion (Ferraro et al., 2005; Perry et al., 2018).

Purified ROs are known for being challenging to work with. Their activity relies on a long and complex electron transport chain, which utilizes multiple non-covalently linked components (Özgen & Schmidt, 2019). ROs have also demonstrated low stability once isolated (Barry & Challis, 2013; Özgen & Schmidt, 2019). Specifically, exposed Fe-S clusters can be converted by oxygen species to unstable forms that will quickly decompose (Imlay, 2006). For these reasons, whole cell catalysis is often a preferable method for working with these enzymes (Özgen & Schmidt, 2019). ROs have the potential to be used in a wide range of applications, including the bioremediation of aromatic pollutants via degradation, natural product synthesis of compounds requiring dearomatization and desymmetrization, or in the production of homochiral starting materials (Özgen & Schmidt, 2019).

The sequential *N*-demethylation of caffeine (1,3,7-trimethylxanthine) to xanthine by the encoded proteins of the *NdmABCDE* genes in *P. putida* is an example of a type 1A RO-catalyzed system (Kim et al., 2019; Kweon et al., 2008): a two-component system where the non-heme iron active site and Rieske [2Fe-2S] cluster are both located on the oxygenase, NdmA. The reductase containing the cofactor binding site and the plant-type ferredoxin [2Fe-2S] cluster is located on a separate protein, NdmD

(Summers et al., 2015). Both NdmA and NdmD are required for the demethylation of caffeine to theobromine (3,7-dimethylxanthine). Whole cell catalysis represents a quick and reliable method for exploring the metabolic pathways and product profiles of these ROs in different enzymatic combinations and with various substrates without needing to take the next steps in purifying the enzymes, which may then not retain, or only partially retain their activity. NdmA and NdmD will be used as an example throughout the following protocol. This protocol describes the implementation of whole cell catalysis in a method referred to as a resting cell reaction. The name is based on the idea of the cells being at rest while suspended in a phosphate buffer, which is commonly used in such reactions (Julsing, Kuhn, Schmid, & Bühler, 2012; Panke, Meyer, Huber, Witholt, & Wubbolts, 1999; Pine, 1966; Usvalampi, Kiviharju, Leisola, & Nyyssölä, 2009). The cells are no longer actively growing on a nutrient source, but neither are they dead and, as whole cells containing all necessary cofactors, they will catalyze reactions when supplemented with a substrate. Here, we will describe methods for assaying caffeine-degrading Rieske non-heme enzymes in whole-cell biocatalysts (Algharrawi, Summers, Gopishetty, & Subramanian, 2015; Algharrawi, Summers, & Subramanian, 2017; Mock et al., 2021; Mock et al., 2024; Mock, Cyrus, & Summers, 2022; Mock, Mills, et al., 2022).

2. Before you begin timing: 4–5 days

For this example, we have chosen to use *Escherichia coli* pDdAA (Fig. 1), a derivative of *E. coli* BL21(DE3). This strain contains the *ndmA* and *ndmD* genes under control of the IPTG-inducible T7 promoter and expressed from the pET-28a(+) and pACYCDuet-1 vectors, respectively. This system allows for precise regulation of gene expression so that protein synthesis can be carried out at cold temperatures of around 18 °C. Lower temperatures during expression can promote proper folding during protein synthesis, which may be important in improving enzyme functionality, particularly when using a high-expression system like *E. coli* BL21(DE3). Other expression systems may be more suitable for the RO (s) of interest, such as *P. putida* strain KT2440 (Cook et al., 2018; Elmore, Furches, Wolff, Gorday, & Guss, 2017) for expression of genes derived from *Pseudomonads*.

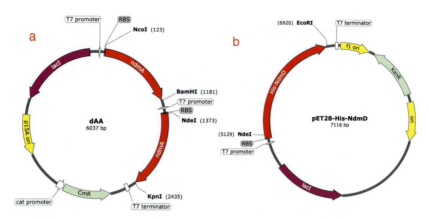

Fig. 1 Plasmid maps of the *E. coli* BL21(DE3) pDdAA strain which contains: (A) pACYCDuet-1 vector with two copies of *ndmA* and (B) pET28a(+) vector with one copy of His-tagged *ndmD*.

The RO(s) of interest can be maintained on a plasmid(s) to allow for the use of antibiotic selective pressure, as well as high expression levels. A genome-integrated system may also be used for improved stability. Because multiple components are required for RO activity, studies of various ratios of one component versus the other(s) may be useful and would benefit from the use of plasmids with varying copy numbers. For example, we have shown that using one copy of *ndmA* and three copies of *ndmD* spread across two plasmids provided optimal activity for the conversion of theophylline to 3-methylxanthine (Algharrawi et al., 2015). Be sure to construct all plasmids of interest and subsequent strains before beginning your experiment. Multiple strains may be tested at one time for comparative purposes. Examples of constructed plasmids and resulting strains used for genetic optimization testing via whole cell catalysis have been established previously (Algharrawi et al., 2015; Mock, Mills, et al., 2022; Summers et al., 2012).

Prepare sufficient Luria-Bertani (LB) medium, appropriate antibiotics, 1 M isopropyl β-d-1-thiogalactopyranoside (IPTG), and 10 mM FeCl$_3$, for the growth and protein expression of all strains to be tested. Prepare sufficient 50 mM potassium phosphate (KP$_i$) buffer (keep refrigerated at 4 °C) for washing and resuspending cells, as well as sufficient buffer and substrate for setting up each reaction. Prepare the appropriate High Performance Liquid Chromatography (HPLC) buffer for the substrate to be used and expected products to be generated. Label sample tubes for all desired time points.

3. Key resources table

Reagent or resource	Source	Identifier
Bacterial strains		
Escherichia coli BL21(DE3)	New England biolabs	CAT# C2527I
Chemicals, peptides, and recombinant proteins		
Tryptone	VWR	CAT# 97063-388
Yeast extract	VWR	CAT# 97063-370
Sodium chloride (NaCl)	VWR	CAT# BDH9286
Agar, bacteriological	VWR	CAT# 97064-336
Potassium phosphate dibasic anhydrous (K_2HPO_4)	VWR	CAT# 97062-234
Potassium phosphate monobasic anhydrous (KH_2PO_4)	AMRESCO	CAT# 97062-350
Ampicillin sulfate (100 ng/mL in water)	AMRESCO	CAT# 470024-776
Kanamycin sulphate (30 ng/mL in water)	VWR	CAT# 75856-684
Chloramphenicol (34 ng/mL in ethanol)	VWR	CAT# IC19032125
Methanol (HPLC grade)	VWR	CAT# 97065-064
Acetic acid (HPLC grade)	VWR	CAT# JT9515-3
Ferric chloride hexahydrate (10 mM $FeCl_3 \cdot 6H_2O$)	VWR	CAT# 97064-992
Isopropyl β-d-1-thiogalactopyranoside (1 M IPTG)	INDOFINE	CAT# 0487-10G
Caffeine (1,3,7-trimethylxanthine)	Sigma-Aldrich	CAT# C0750-100G
Theobromine (3,7-dimethylxanthine)	Sigma-Aldrich	CAT# T4500-100G
Recombinant DNA		
NdmA (rieske oxygenase/ferrodoxin)	GenBank	Accession# JQ061127
NdmD (reductase)	GenBank	Accession# JQ061130
Software and algorithms		
Excel		
LabSolutions (HPLC program)		

4. Materials and equipment
4.1 Equipment
- Bench-top incubator shaker (New Brunswick Scientific, Innova 42R Series; temperature range: 20 °C below ambient to 80 °C)
- Refrigerated high-speed floor centrifuge (Backman Coulter, Avanti JXN-26 Series)
- Refrigerated table-top centrifuge (Eppendorf, Centrifuge 5424 R)
- UV–vis Spectrophotometer (Eppendorf, BioSpectrometer Kinetic)
- Vortex (VWR; Analogue Vortex Mixer)
- ThermoScientific Hypersil BDS C18 HPLC column (4.6 mm inner diameter × 150 mm length)
- Shimadzu LC-20AT HPLC system equipped with a UV photodiode array detector

4.2 Materials and reagents
- LB Medium (10 g/L tryptone, 5 g/L yeast extract, 5 g/L NaCl)
- 1 M Potassium phosphate buffer (KP_i) stock, pH 7.5 (145.26 g/L KH_2PO_4, 22.60 g/L K_2HPO_4)
- 50 mM Potassium phosphate (KP_i) buffer, pH 7.5
- HPLC mobile phase and standard solutions for calibration
  - 15% Methanol (150 mL/L 100% HPLC Grade MeOH, 850 mL/L Ultra-Pure diH_2O, 4.25 mL/L Acetic Acid)
  - High purity caffeine (194.19 g/mol)
  - High purity theobromine (180.16 g/mol)

Alternatives: Terrific Broth (TB) can be used in place of LB medium. Sodium phosphate buffer can be used in place of potassium phosphate buffer.

5. Step-by-step method details

Here, we demonstrate the use of a caffeine-degrading whole-cell system containing two RO components, *ndmA* (Rieske monooxygenase) and *ndmD* (reductase), on two plasmids in *E. coli* BL21(DE3) (Fig. 1). We will detail the cell growth, induction protocol used for optimal protein expression, resting cell assay reaction set-up, and HPLC analysis techniques used to determine the conversion of caffeine to theobromine. Plasmid construction information has been described previously (Algharrawi et al., 2015; Mock, Mills, et al., 2022; Summers et al., 2012).

5.1 Cell growth and gene expression

Timing: 3 days.

1. **Day 1:** Using a sterilized loop or pipette tip, streak a small amount of the desired *E. coli* strain (i.e. pDdAA) from a frozen glycerol stock onto an agar plate with the appropriate antibiotic(s) (i.e. LB + Kanamycin 30 μg/mL + Chloramphenicol 34 μg/mL).

   a) To prepare LB plates, combine 10 g/L tryptone, 5 g/L yeast extract, 5 g/L NaCl, and 16 g/L agar in deionized water (diH$_2$O). Autoclave the mixture and cool to 55 °C before adding the appropriate antibiotic(s) and pouring plates.

   b) Repeat for each desired strain.

2. Incubate the plate(s) for 14–16 h (overnight) at 37 °C.

3. **Day 2:** Select a single well-isolated colony using a sterile inoculation loop or pipette tip and inoculate into 2–3 mL of LB medium with the appropriate antibiotic(s) (i.e. LB + Kanamycin 30 μg/mL + Chloramphenicol 34 μg/mL).

   a) To prepare LB medium, combine 10 g/L tryptone, 5 g/L yeast extract, and 5 g/L NaCl in diH$_2$O. Autoclave the mixture and cool at room temperature prior to adding the appropriate antibiotic(s).

4. Grow the seed culture for 4–6 h at 37 °C with 200 rpm orbital shaking.

5. **Day 3:** Determine the final optical density of the seed culture by use of a spectrophotometer at a wavelength of 600 nm (OD$_{600}$) and a 1 mL cuvette using a 10× dilution of culture in fresh LB medium.

   a) This OD$_{600}$ value is used to determine the volume of seed culture (X mL) required to inoculate 50 mL of fresh LB medium and antibiotic (s) to an OD$_{600}$ of 0.05 by the following calculation: (X mL) = (0.05)(50 mL)/(OD$_{600}$)

   b) Inoculation to a uniform cell density allows for more consistent growth across cultures and simplifies the induction process by making it more predictable.

6. Incubate the freshly inoculated cultures at 37 °C and 200 rpm shaking. When the culture reaches an OD$_{600}$ of 0.5 (approximately 3–3.5 h), supplement with sterile iron chloride (FeCl$_3$) to a final concentration of 10 μM.

7. Shift the culture to 18 °C with 200 rpm shaking.

   a) Supplementation of the iron chloride supports the overexpression of active *N*-demethylases, as the enzymes contain Rieske [2Fe-2S] clusters, as well as a non-heme iron center within the active site.

Whole-cell Rieske non-heme iron biocatalysts   251

b) The temperature shift from 37 °C to 18 °C promotes proper folding of the enzyme and reduces the formation of insoluble inclusion bodies.

8. Once the cultures have reached the desired temperature of 18 °C (~10 min) and reach an OD_{600} of 0.8, add IPTG to a final concentration of 0.1 mM.

9. Grow the cells post-induction for an additional 14–16 h at 18 °C and 200 rpm shaking.

5.2 Resting cell reactions

Timing: 1–3 h.

1. **Day 4:** At the conclusion of the induction period, harvest cells by centrifugation at 10,000 × g for 2.5 min at 4 °C.

   Critical: Keep the cells at 4 °C at all times. Use pre-chilled buffers and equipment (i.e. centrifuge, etc.) from this point forward until you are ready to initiate the reaction.

2. Wash the pelleted cells by resuspension in ice-cold 50 mM potassium phosphate (KP_i) buffer (pH 7.5) using a volume that is 20% of the culture volume.

   a) If the cells were induced overnight in 50 mL of medium, use 10 mL of KP_i buffer to wash the cells.

   b) Sodium phosphate buffer may also be used.

3. Pellet the cells a second time by centrifugation at 10,000 × g for 2.5 min at 4 °C.

4. Resuspend the cells for a second time in ice-cold 50 mM KP_i buffer (pH 7.5) using a volume that is 20% of the initial culture volume.

   Pause Point: Once the cells have been resuspended and placed on ice, the process can be briefly paused; however, it is not recommended to leave the cells unattended for long periods of time, and the reaction should be initiated on the same day that the cells are harvested.

5. Determine the OD_{600} of the resuspended cells ($OD_{600,1}$) using a 100x dilution of cells in KP_i buffer.

   a) The OD_{600} is used to calculate the volume of cells (X mL) required to achieve the desired cell concentration required for the reaction ($OD_{600,2}$) by the following equation: $(X\ mL) = (OD_{600,2})(2\ mL)/(OD_{600,1})$.

   b) For a standard resting cell assay, the reaction volume is usually set at 2 mL, and the cell concentration is set to an OD_{600} of 5. These parameters can be varied for optimization purposes.

6. Initiate the resting cell assay reactions by the addition of substrate (i.e. addition of caffeine to *E. coli* BL21(DE3) cells suspended in 50 mM KP_i buffer).
   a) For a standard resting cell assay, an initial substrate concentration of 1 mM is commonly used. This concentration can be varied based on the nature of the study, the characteristics of the enzyme(s), and the reaction that is catalyzed.
   b) Each reaction should be conducted in three technical replicates.
7. Incubate the reaction mixture at 30 °C with 200 rpm shaking.
   a) The reaction temperature of 30 °C was selected due to the optimal temperature of the enzyme. Adjust the temperature as needed based on optimal enzyme conditions.
   b) The pH of the KP_i buffer used throughout this example is 7.5. This value may also be adjusted as needed for the given reaction system.

Note: It is recommended that the cells + KP_i mixture be incubated at the desired reaction temperature for a couple of minutes prior to reaction initiation to ensure maximum initial activity and enzyme efficiency.

5.3 Reaction sampling and sample preparation

Timing: 5–7 h.
1. Following reaction initiation, take samples periodically for analysis by HPLC.
   a) Typical times points include $t = 1, 15, 30, 45, 60, 90, 120, 180, 240,$ and 300 min. Time points can be modified to most accurately capture the reaction rate.
   b) To sample the reaction, pipette 100 μL of the reaction and add it to 100 μL of 100% methanol to stop the reaction.

   Note: It is recommended that samples be immediately vortexed for approximately 3 s to ensure mixing and to facilitate quenching of the reaction.

   Critical: Store samples at −20 °C before and after sample collection to prevent evaporation of methanol. Evaporation will affect concentration.

   Pause point: Samples can be stored for several days at −20 °C prior to sample preparation and HPLC analysis. However, long-term storage will result in evaporation.
2. To prepare the samples for HPLC analysis, first pellet the cells/debris by centrifugation (14,000 × g for 10 min at 4 °C).
3. Remove the supernatant without disturbing the cell pellet by using a small needle and 1 mL syringe. Save the supernatant and discard the pelleted debris.

4. Filter the supernatant through a 0.2 μm filter to remove any remaining cells or debris.
5. Transfer the processed sample to an HPLC insert/vial.

6. High-performance liquid chromatography (HPLC) analysis

Timing: 45 min (column equilibration); 10–20 min (per sample).

1. Equilibrate the column at a flowrate of 0.5 mL/min for 45 min into the desired mobile phase of either 7.5% or 15% methanol (plus 0.5% acetic acid).
   a) The methanol concentration used is dependent upon the substrate used and products resulting from the reaction. Please refer to Table 1.
   b) An isocratic method can be used to deliver the mobile phase at 0.5 mL/min.

Table 1 An example of the retention times and wavelengths detected for methylxanthine standards generated by a Shimadzu LC-20AT HPLC system (UV PDA detector) equipped with a ThermoScientific Hypersil BDS C18 HPLC column (4.6 mm inner diameter × 150 mm length).

Methylxanthine	Estimated retention times (min)		Compound wavelength (nm)
	15% Methanol	7.5% Methanol	
Caffeine	13.5	–	273
1,3,7-Trimethyluric acid	9.2	–	288
Theophylline	8.2	–	271
Theobromine	5	10.5	272
Paraxanthine	7.5	–	269
7-Methylxanthine	3.5	5.2	268
3-Methylxanthine	3.8	6	271
1-Methylxanthine	4.5	7	267
Xanthine	3	3.5	267

Note: It is recommended that the mobile phase be prepared 2–3 days prior to use or that the mixture be sonicated to reduced air bubble formation that may affect the HPLC.

2. Load 1 μL of sample onto the HPLC column and record the UV spectrum for 10–20 min.
   a) Methylxanthine samples should be analyzed at their maximum absorbance wavelength. An example chromatograph can be seen below in Fig. 2.

7. Establishing a calibration curve

Timing: 2 days.

1. Create solutions covering a range of concentrations from pure compounds of the substrate used (i.e. caffeine) and the expected/known products (i.e. theobromine).
   a) Stocks should be made in ultra-pure water or the buffer used in the assay (KP_i). Filter the solutions with a 0.2 μm filter.
   b) The range of concentrations used should reflect the starting concentration of substrate used in the reaction.

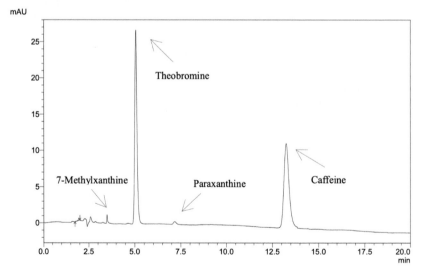

Fig. 2 Example of an HPLC chromatogram showing the conversion of caffeine to theobromine by *E. coli* BL21(DE3) pDdAA whole cell catalysis after 30 min. Minor side products, 7-methylxanthine and paraxanthine, are visible as a result of the promiscuity of ROs. Data was collected at a wavelength of 274 nm.

- Example for caffeine – 0.01, 0.1, 0.25, 0.5, 0.75, 1.0, 2.5, 5, 7.5, and 10 mM.

Note: The solubility of the compound will affect the highest concentration achievable. For example, the maximum theobromine concentration we can reasonably achievable is 1.5–2 mM.

2. To construct the calibration curve, treat the prepared solutions like samples from a resting cell assay before being run on the HPLC.

   a) Add an equal volume of methanol to each solution.

   b) Filter with a 0.2 μm filter.

3. Load 1 μL of sample onto the column and record the UV spectrum for 10–20 min. Run each sample three times to determine variation.

4. Use the HPLC software to determine the area of each peak.

5. Construct a graph comparing the known concentrations of each sample on the y-axis and the averaged peak area on the x-axis.

6. Fit a line-of-best-fit to the data and determine the equation of the line ($y = mx + b$).

   a) Inputting the peak areas from the resting cell data as x will allow for the calculation of the concentration (y).

Note: The R^2 value of the line-of-best-fit should be very close to 1.0. If the variation is greater than 0.90–0.95 it is recommended the that solutions be remade with more care and precision and the HPLC equipment be inspected for parts that may need to be replaced of adjusted such as the guard column, buffer, or the column itself.

8. Expected outcomes

When using a high-expression system such as the T7 promoter in *E. coli* BL21(DE3) to expresses iron-containing ROs, the cell pellets at the conclusion of the induction period should have a red tint (Fig. 3). This coloration is due to the Fe-S clusters found in the enzymes and the sheer quantity of soluble protein present within the cell from overexpression.

The final OD_{600} for one flask containing *E. coli* cells grown in 50 mL of LB is approximately 4. One flask of 50 mL LB medium should generate a sufficient quantity of cells to prepare approximately 20 test tubes of 2 mL reactions at an OD_{600} of 5. This amount is more than sufficient for running reactions in triplicate; however, reactions requiring a higher optical density/greater concentration of cells or higher volume will result in fewer possible reactions.

Fig. 3 Cell pellets of *E. coli* BL21(DE3) strain pDdAA (left) and empty vector strain (right) after a 16-h induction with 0.1 mM IPTG. The pDdAA strain exhibited a noticeable red tint in contrast to the empty vector control.

Five hours is a sufficient amount of time for caffeine degrading ROs to achieve complete conversion or lose activity and plateau in conversion. This time frame may vary based on the efficiency and stability of the RO(s) of interest.

9. Quantification and statistical analysis

When analyzing HPLC data, it is important to use the chromatographs generated by the program properly. For example, concentrations used in the experiments must fall within the range of the established standard curves. Samples from reactions and samples for standard curves must be prepared by the same method (i.e. if reaction samples are added to 100 μL methanol to stop the reaction, standard curve samples should also be added to 100 μL methanol before analyzing by HPLC). Integration of peaks areas must be done consistently. Programs that automatically detect peaks should be monitored and the peaks adjusted when necessary for consistency, especially for complex reaction supernatants where multiple peaks may appear within close proximity. Ensure sufficient separation between all peaks by adjusting the mobile phase accordingly. Gradients can be used to improve separation.

All reactions should be conducted in triplicate as biological replicates to determine the variation in the data. Variation should be low. Significant standard error could be indicative of compound precipitation, HPLC malfunctions, or other unidentified issues. Enough time point samples should be taken to accurately depict the reaction and is subject to the variable of the reaction. For example, sufficient data has been published to confirm that methylxanthine *N*-demethylation reactions at the described conditions conclude within 5 h, and the greatest change in rate occurs within the first third of the reaction (Mock et al., 2021, 2024; Mock, Cyrus, et al., 2022; Mock, Mills, et al., 2022). As a result, six samples (i.e. time = 1, 30, 60, 90 180, and 300 min) are typically sufficient to plot the reaction rate and distinguish differences between varying strains and enzyme mutants. The times and number of samples may vary based on the reaction of interest. An example analysis plot has been constructed based on the *N*-demethylation reaction of caffeine to theobromine by *E. coli* strain pDdAA (Fig. 4). In this example, the reaction is completed in approximately 120 min. As a result, the time points used to analyze the reaction were adjusted to better suit the faster reaction rate (time = 1, 15, 30, 45, 60, 90, and 120 min). Additional samples between 1 and 20 min could be used to better determine the initial reaction rate.

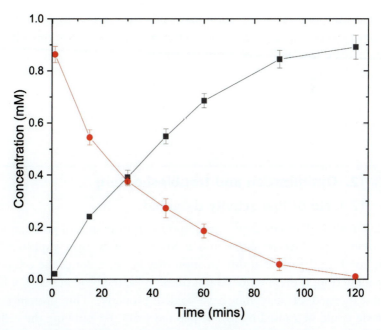

Fig. 4 Conversion of caffeine (red circle) to theobromine (black square) over time by *E. coli* BL21 (DE3) pDdAA. Concentrations reported are shown as means with standard deviations of triplicate results.

10. Advantages

Use of a whole cell method is preferable when working with ROs because stability/solubility and sensitivity issues are common in ROs. The cell provides a better environment for the proteins, promoting stability and colocalizing the enzymes to facilitate contact required to complete the electron transport across multiple protein subunits. Whole cell biocatalysis is a faster, less intensive process than purified enzyme assays, which require cell lysis and purification procedures and may result in substantial loss in enzyme activity.

11. Limitations

One challenge in using whole-cell biocatalysis is the potential for transport limitations of the cell up-taking the substrate. Cell permeabilization may be necessary to improve transport. Further, by using the resting cell method, cells are no longer exposed to the medium's growth factor. The concentration of the co-factor, such as NADH, will become a limiting factor at that point. This limitation may cause the reaction to stop once the co-factor has been consumed. A co-factor recycling system in the cells may help increase availability of required co-factors. For example, we constructed an NADH recycling system using the *frmAB* formaldehyde dehydrogenase genes to convert formaldehyde to formate while generating NADH from NAD^+ during the N-demethylation of caffeine (Mock, Mills, et al., 2022).

12. Optimization and troubleshooting
12.1 No or low activity detected

HPLC analysis does not show consumption of substrate and production of the desired product or shows much lower activity than expected. An indicator as to the cause of this situation may be the color of the cell pellet post-induction. If there is no red tint in the cells, the protein may not have been expressed or may not have folded properly. This protein-based obstacle could be caused by loss of plasmid from not growing the cells in the correct antibiotics, failure to add IPTG to the culture to induce expression, failure to add an exogenous source of iron to facilitate overexpression, use of a non-codon optimized gene, or incubation at too high of an induction temperature. Additional causes of no activity may include failure to add the substrate to the reaction, addition of too much substrate achieving inhibitory concentrations, failure to resuspend cells in a phosphate buffer resulting in cell lysis, poor transport of the substrate across the cell membrane, or use of a non-optimal reaction temperature.

12.2 Potential solutions to optimize the procedure

To trouble shoot an activity issue, start with examining the color of the cell pellet post-induction to determine if the issue is most likely to be upstream (protein expression and folding) or downstream (reaction conditions). Additionally, the cells can be lysed the debris pelleted, and the soluble and insoluble fractions run on a gel to check for expression and protein stability. A solubility tag can be added to improve protein stability, but the location of the tag may impact protein folding or enzyme activity. If the cells do not display a red tint and no protein can be seen on a protein gel, examine your induction procedure and ensure all steps were followed properly. If the procedure is accurate, check the codon-optimization of the protein sequence, try adding a solubility tag to the protein, or try other methods for improving protein stability such as removing flexible regions or altering surface residues.

If the cells do display a red tint but no activity is seen via HPLC, confirm that substrate was added by checking for a peak on the HPLC chromatograph. If substrate has been added, the procedure was followed correctly but no activity is seen, the reaction conditions can be optimized by varying the reaction temperature, the substrate concentration, the length of the reaction, and the concentration of cells used within the reaction.

If transport limitations become an issue, cell permeablization may be necessary. Cell permeabilization can be achieved through the use of organic solvents, such as tween 80, toluene, Triton X-100, chloroform, and sodium dodecyl sulfate (SDS) (Ohnishi et al., 1994; Pathak & Madamwar, 2010; Pope, Chen, & Stewart, 2009).

13. Safety considerations and standards

When working with recombinant bacterial strains, wear appropriate PPE to avoid direct exposure. Sterilize bacterial cells (bleach, autoclave, etc.) after use and before disposal to prevent release of antibiotic resistant organisms into the environment. Methanol is flammable and an irritant. Avoid contact with skin.

14. Alternative methods/procedures

In place of LB medium, TB can be used during the growth and induction phase to increase the final mass of cells produced. TB contains a pH buffer and an additional carbon source (glycerol) to promote growth and protein expression.

A sodium phosphate buffer can be used in place of the potassium phosphate buffer.

A mixed culture method can be employed for more complex reactions or for enzymes that cannot be compatibly co-expressed (Mock & Summers, 2023). In such cases, the cell lines are each grown up individually and proteins expressed as previously described. Once the harvested cells are resuspended, the cell lines can be mixed at varying ratios and assayed.

Some incomplete reactions can be continued by discarding spent cells and adding fresh cells (Mock, Cyrus, et al., 2022).

The reaction of interest may be optimized in a variety of ways. The concentration of cells used per reaction can be increased. The ratio of oxygenase to reductase can be altered by varying genetic properties such as gene copy number, promoter strength, and plasmid copy number. The starting substrate concentration may also be varied to achieve optimal activity as some compounds, especially aromatic compounds, may pose some toxicity to the cells and require a lower initial concentration. (Mock, Mills, et al., 2022).

Other C18 HPLC columns are available and can be used according to the substrates and expected products, including large preparatory columns for processing larger volumes.

References

Algharrawi, K. H., Summers, R. M., Gopishetty, S., & Subramanian, M. (2015). Direct conversion of theophylline to 3-methylxanthine by metabolically engineered *E. coli*. *Microbial Cell Factories, 14*(1), 12. https://doi.org/10.1186/s12934-015-0395-1.

Algharrawi, K. H., Summers, R. M., & Subramanian, M. (2017). Production of theobromine by N-demethylation of caffeine using metabolically engineered *E. coli*. *Biocatalysis and Agricultural Biotechnology, 11*, 153–160.

Barry, S. M., & Challis, G. L. (2013). Mechanism and catalytic diversity of Rieske non-heme iron-dependent oxygenases. *ACS Catalysis, 3*(10), 2362–2370. https://doi.org/10.1021/cs400087p.

Boyd, D. R., Sharma, N. D., Harrison, J. S., Kennedy, M. A., Allen, C. C. R., & Gibson, D. T. (2001). Regio- and stereo-selective dioxygenase-catalysed cis–dihydroxylation of fjord-region polycyclic arenes. *Journal of the Chemical Society, Perkin Transactions 1, 1*(11), 1264–1269. https://doi.org/10.1039/B101833G.

Cook, T. B., Rand, J. M., Nurani, W., Courtney, D. K., Liu, S. A., & Pfleger, B. F. (2018). Genetic tools for reliable gene expression and recombineering in *Pseudomonas putida*. *Journal of Industrial Microbiology and Biotechnology, 45*(7), 517–527. https://doi.org/10.1007/s10295-017-2001-5.

Elmore, J. R., Furches, A., Wolff, G. N., Gorday, K., & Guss, A. M. (2017). Development of a high efficiency integration system and promoter library for rapid modification of *Pseudomonas putida* KT2440. *Metabolic Engineering Communications, 5*, 1–8. https://doi.org/10.1016/j.meteno.2017.04.001.

Ensley, B., & Gibson, D. (1983). Naphthalene dioxygenase: Purification and properties of a terminal oxygenase component. *Journal of Bacteriology, 155*(2), 505–511. https://doi.org/10.1128/jb.155.2.505-511.1983.

Ferraro, D. J., Gakhar, L., & Ramaswamy, S. (2005). Rieske business: Structure–function of Rieske non-heme oxygenases. *Biochemical and Biophysical Research Communications, 338*(1), 175–190. https://doi.org/10.1016/j.bbrc.2005.08.222.

Imlay, J. A. (2006). Iron-sulphur clusters and the problem with oxygen. *Molecular Microbiology, 59*(4), 1073–1082. https://doi.org/10.1111/j.1365-2958.2006.05028.x.

Julsing, M. K., Kuhn, D., Schmid, A., & Bühler, B. (2012). Resting cells of recombinant *E. coli* show high epoxidation yields on energy source and high sensitivity to product inhibition. *Biotechnology and Bioengineering, 109*(5), 1109–1119. https://doi.org/10.1002/bit.24404.

Kim, J. H., Kim, B. H., Brooks, S., Kang, S. Y., Summers, R. M., & Song, H. K. (2019). Structural and mechanistic insights into caffeine degradation by the bacterial N-demethylase complex. *Journal of Molecular Biology, 431*(19), 3647–3661. https://doi.org/10.1016/j.jmb.2019.08.004.

Kweon, O., Kim, S.-J., Baek, S., Chae, J.-C., Adjei, M. D., Baek, D.-H., ... Cerniglia, C. E. (2008). A new classification system for bacterial Rieske non-heme iron aromatic ring-hydroxylating oxygenases. *BMC Biochemistry, 9*(1), 20. https://doi.org/10.1186/1471-2091-9-11.

Mock, M. B., Cyrus, A., & Summers, R. M. (2022). Biocatalytic production of 7-methylxanthine by a caffeine-degrading *Escherichia coli* strain. *Biotechnology and Bioengineering, 119*(11), 3326–3331. https://doi.org/10.1002/bit.28212.

Mock, M. B., Mills, S. B., Cyrus, A., Campo, H., Dreischarf, T., Strock, S., & Summers, R. M. (2022). Biocatalytic production and purification of the high-value biochemical paraxanthine. *Biotechnology and Bioprocess Engineering, 27*(4), 640–651. https://doi.org/10.1007/s12257-021-0301-0.

Mock, M. B., & Summers, R. M. (2023). Mixed culture biocatalytic production of the high-value biochemical 7-methylxanthine. *Journal of Biological Engineering, 17*(1), 2. https://doi.org/10.1186/s13036-022-00316-6.

Mock, M. B., Zhang, S., Pakulski, K., Hutchison, C., Kapperman, M., Dreischarf, T., & Summers, R. M. (2024). Production of 1-methylxanthine via the biodegradation of theophylline by an optimized *Escherichia coli* strain. *Journal of Biotechnology, 379*, 25–32. https://doi.org/10.1016/j.jbiotec.2023.11.005.

Mock, M. B., Zhang, S., Pniak, B., Belt, N., Witherspoon, M., & Summers, R. M. (2021). Substrate promiscuity of the NdmCDE N7-demethylase enzyme complex. *Biotechnology Notes, 2*, 18–25. https://doi.org/10.1016/j.biotno.2021.05.001.

Ohnishi, T., Sled, V., Rudnitzky, N., Meinhardt, S., Yagi, T., Hatefi, Y., ... Daldal, F. (1994). Topographical distribution of redox centres and the Qo site in ubiquinol-cytochrome-c oxidoreductase (complex III) and ligand structure of the Rieske iron-sulphur cluster. *Biochemical Society Transactions, 22*(1), 191–197. https://doi.org/10.1042/bst0220191.

Özgen, F. F., & Schmidt, S. (2019). Rieske non-heme iron dioxygenases: Applications and future perspectives. *Biocatalysis: Enzymatic Basics and Applications*, 57–82. https://doi.org/10.1007/978-3-030-25023-2_4.

Panke, S., Meyer, A., Huber, C. M., Witholt, B., & Wubbolts, M. G. (1999). An alkane-responsive expression system for the production of fine chemicals. *Applied and Environmental Microbiology, 65*(6), 2324–2332. https://doi.org/10.1128/AEM.65.6.2324-2332.1999.

Pathak, H., & Madamwar, D. (2010). Biosynthesis of indigo dye by newly isolated naphthalene-degrading strain *Pseudomonas* sp. HOB1 and its application in dyeing cotton fabric. *Applied Biochemistry and Biotechnology, 160*, 1616–1626. https://doi.org/10.1007/s12010-009-8638-4.

Perry, C., De los Santos, E. L., Alkhalaf, L. M., & Challis, G. L. (2018). Rieske non-heme iron-dependent oxygenases catalyse diverse reactions in natural product biosynthesis. *Natural Product Reports, 35*(7), 622–632. https://doi.org/10.1039/c8np00004b.

Pine, M. J. (1966). Metabolic control of intracellular proteolysis in growing and resting cells of *Escherichia coli*. *Journal of Bacteriology, 92*(4), 847–850. https://doi.org/10.1128/jb.92.4.847-850.1966.

Pope, S. D., Chen, L.-L., & Stewart, V. (2009). Purine utilization by Klebsiella oxytoca M5al: Genes for ring-oxidizing and-opening enzymes. *Journal of Bacteriology, 191*(3), 1006–1017. https://doi.org/10.1128/jb.01281-08.

Resnick, S., Lee, K., & Gibson, D. (1996). Diverse reactions catalyzed by naphthalene dioxygenase from *Pseudomonas* sp. strain NCIB 9816. *Journal of Industrial Microbiology, 17*, 438–457. https://doi.org/10.1007/BF01574775.

Rogers, M. S., Gordon, A. M., Rappe, T. M., Goodpaster, J. D., & Lipscomb, J. D. (2022). Contrasting mechanisms of aromatic and aryl-methyl substituent hydroxylation by the Rieske monooxygenase salicylate 5-hydroxylase. *Biochemistry, 62*(2), 507–523. https://doi.org/10.1021/acs.biochem.2c00610.

Summers, R. M., Louie, T. M., Yu, C.-L., Gakhar, L., Louie, K. C., & Subramanian, M. (2012). Novel, highly specific N-demethylases enable bacteria to live on caffeine and related purine alkaloids. *Journal of Bacteriology, 194*(8), 2041–2049. https://doi.org/10.1128/jb.06637-11.

Summers, R. M., Mohanty, S. K., Gopishetty, S., & Subramanian, M. (2015). Genetic characterization of caffeine degradation by bacteria and its potential applications. *Microbial Biotechnology, 8*(3), 369–378. https://doi.org/10.1111/1751-7915.12262.

Usvalampi, A., Kiviharju, K., Leisola, M., & Nyyssölä, A. (2009). Factors affecting the production of l-xylulose by resting cells of recombinant *Escherichia coli*. *Journal of Industrial Microbiology and Biotechnology, 36*(10), 1323–1330. https://doi.org/10.1007/s10295-009-0616-x.

Wackett, L. P. (2002). Mechanism and applications of Rieske non-heme iron dioxygenases. *Enzyme and Microbial Technology, 31*(5), 577–587. https://doi.org/10.1016/S0141-0229(02)00129-1.

Yeh, W., Gibson, D., & Liu, T.-N. (1977). Toluene dioxygenase: A multicomponent enzyme system. *Biochemical and Biophysical Research Communications, 78*(1), 401–410. https://doi.org/10.1016/0006-291X(77)91268-2.

CHAPTER TWELVE

Photo-reduction facilitated stachydrine oxidative *N*-demethylation reaction: A case study of Rieske non-heme iron oxygenase Stc2 from *Sinorhizobium meliloti*

Tao Zhang, Kelin Li, Yuk Hei Cheung, Mark W. Grinstaff, and Pinghua Liu*

Department of Chemistry, Boston University, Boston, MA, United States
*Corresponding author. e-mail address: pinghua@bu.edu

Contents

1.	Introduction	265
2.	Heterologous expression and purification of Stc2	267
3.	Materials and equipment	269
	3.1 Equipment	269
	3.2 Solutions and consumables	269
4.	Step-by-step procedure details	270
	4.1 *Escherichia coli* starter culture	270
	4.2 Stc2 protein overexpression	270
	4.3 Anaerobic Stc2 protein purification (Daughtry et al., 2012)	271
	4.4 Concentrating the eluted Stc2 in the COY chamber	275
5.	Expected outcomes, advantages, and disadvantages	276
6.	Optimization and troubleshooting	276
7.	Safety considerations and standards	277
8.	Stc2 characterization	277
9.	Materials and equipment	278
	9.1 Equipment	278
	9.2 Solutions and consumables	278
10.	Step-by-step details	279
	10.1 Stc2 protein analysis	279
	10.2 Stc2 iron content quantification (Abbasi, Abbina, Gill, Bhagat, Kizhakkedathu, 2021)	280
	10.3 Quantification of labile sulfur in Stc2 based on the reaction shown in Fig. 6 (Beinert, 1983)	282
	10.4 Optimization and troubleshooting	285

Methods in Enzymology, Volume 703
ISSN 0076-6879, https://doi.org/10.1016/bs.mie.2024.05.002
Copyright © 2024 Elsevier Inc. All rights are reserved, including those for text and data mining, AI training, and similar technologies.

11. Stc2 photo-reduction using eosin Y and Na_2SO_3	285
12. Materials and equipment	286
12.1 Equipment	286
12.2 Solutions and consumables	286
13. Step-by-step method details	286
13.1 Light-driving Stc2 Fe-S cluster reduction using eosin Y and Na_2SO_3 as photosensitizer/sacrificial reagent pair	287
13.2 Light-driving demethylation of stachydrine using eosin Y and Na_2SO_3 under multiple turnover condition (Hu et al., 2022; Xu et al., 2024)	288
13.3 Stc2-catalysis in a flow-setting	290
13.4 Optimization and troubleshooting	292
References	294

Abstract

Rieske-type non-heme iron oxygenases (ROs) are an important family of non-heme iron enzymes. They catalyze a diverse range of transformations in secondary metabolite biosynthesis and xenobiotic bioremediation. ROs typically shuttle electrons from NAD(P)H to the oxygenase component via reductase component(s). This chapter describes our recent biochemical characterization of stachydrine demethylase Stc2 from *Sinorhizobium meliloti*. In this work, the eosin Y/sodium sulfite pair serves as the photoreduction system to replace the NAD(P)H-reductase system. We describe Stc2 protein purification and quality control details as well as a flow-chemistry to separate the photo-reduction half-reaction and the oxidation half-reaction. Our study demonstrates that the eosin Y/sodium sulfite photo-reduction pair is a NAD(P)H-reductase surrogate for Stc2-catalysis in a flow-chemistry setting. Experimental protocols used in this light-driven Stc2 catalysis are likely to be applicable as a photoreduction system for other redox enzymes.

Abbreviations

ROs	Rieske-type non-heme iron oxygenases
Stc2	Rieske oxygenase subunit of stachydrine demethylase
GbcB	glycine-betaine demethylase reductase subunit
NAD(P)H	nicotinamide adenine dinucleotide (phosphate) reduced form
FAD	flavin adenine dinucleotide
NMR	nuclear magnetic resonance
EDTA	ethylenediaminetetraacetic acid
rpm	revolutions per minute
Tris	tris(hydroxymethyl)methylamine
E. coli	*Escherichia coli*
IPTG	isopropyl-β-D-galactopyranoside
LB media	Luria-Bertani media
APS	ammonium persulfate
SDS	sodium dodecyl sulfate
SDS-PAGE	sodium dodecyl sulfate-polyacrylamide gel electrophoresis
TEMED	tetramethylethylenediamine

UV-Vis	ultraviolet-visible spectroscopy
OD	optical density
ppm	parts per million
PS	photo-sensitizer
His	histidine
Cys	cysteine
HABA	2-[4′-hydroxy-benzeneazo] benzoic acid
AHT	anhydrotetracycline hydrochloride
DMPD	N,N-dimethyl-p-phenylenediamine solution
BSA	bovine serum albumin
Ru(bpy)₃	tris (2,2′-bipyridyl) dichlororuthenium (II) hexahydrate
DTT	dithiothreitol

1. Introduction

Rieske-type non-heme iron oxygenases (ROs) are an important family of non-heme iron enzymes. ROs catalyze a diverse range of transformations in the biosynthesis of secondary metabolites and in xenobiotics bioremediation (Barry & Challis, 2013; Perry, de Los Santos, Alkhalaf, & Challis, 2018). ROs showcase *Nature*'s strategy in site-selective C-H bond functionalization reactions (Liu, Knapp, Jo, Dill, & Bridwell-Rabb, 2022), including *cis*-dihydroxylation (Jeffrey et al., 1975; Karlsson et al., 2003; Zhu et al., 2023), N-dealkylation, N-demethylation (Daughtry et al., 2012; Hu et al., 2022; Summers et al., 2012), C-hydroxylation (Fei, Yin, Zhang, & Zabriskie, 2007; Li et al., 2011), C−S bond cleavage (Higgins, Davey, Trickett, Kelly, & Murrell, 1996; Higgins, De Marco, & Murrell, 1997), C−C bond desaturation (Julien, Tian, Reid, & Reeves, 2006), oxidative carbocyclization and heterocyclization (Sydor et al., 2011; Withall, Haynes, & Challis, 2015). In RO-catalyzed oxidation reactions, molecular oxygen (O_2) reduction needs four electrons and two of the electrons come in the form of a reduced nicotinamide adenine dinucleotide cofactor (NAD(P)H). A reductase alone or a reductase/ferredoxin pair is responsible for delivering electrons from NAD(P)H to the catalytic oxygenase component of the enzyme (Runda, de Kok, & Schmidt, 2023). In ROs, a tight coupling between O_2 consumption with substrate oxidation is important. Substrate binding induces changes in the non-heme iron electronic structure, and such alterations play key roles in controlling the iron-center activity (Liu, Tian, et al., 2022; Tian, Garcia, Donnan, & Bridwell-Rabb, 2023).

Stachydrine (**1**, N,N-dimethylproline) can serve as the sole nitrogen and carbon source for the soil bacterium *Sinorhizobium meliloti* 1021

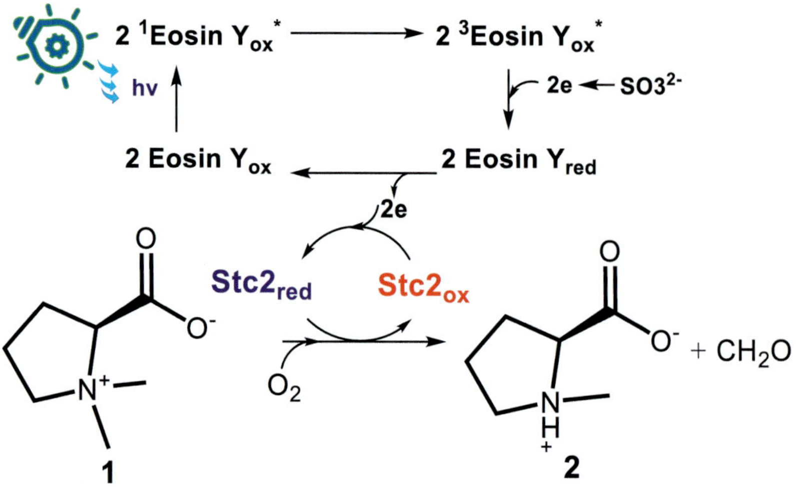

Scheme 1 Stc2-catalyzed *N*-demethylation reaction using a photo-reduction system.

(Phillips et al., 1998). Stachydrine catabolism depends on the stachydrine operon (*stc*), in which *stc2* encodes a Rieske oxygenase and the encoded Stc3/Stc4 proteins serve as the Stc2 reductase pair (Scheme 1) (Burnet et al., 2000; Phillips et al., 1998). In many other Rieske-type oxygenases, the reductase and ferredoxin are fused into one protein (Kweon et al., 2008). Stachydrine demethylase (Stc2) catalyzes the first *N*-demethylation step of stachydrine to produce monomethylproline as the product. Stc2 is one of the structurally characterized ROs (Daughtry et al., 2012; Friemann et al., 2009; Hu et al., 2022), and is a homotrimer, which possesses a conserved mononuclear iron site and a Rieske [2Fe-2S] cluster (Daughtry et al., 2012). This Rieske [2Fe-2S] cluster is part of the electron-transfer chain, which shuttles electrons from NAD(P)H/reductase to the mononuclear iron site for O_2 activation and oxidative *N*-demethylation reaction.

The application of oxygenases in bioremediation or for generating value-added products is attracting increased interest (Dunham & Arnold, 2020; Gibson & Parales, 2000; Hu et al., 2022; Perry et al., 2018; Pyser, Chakrabarty, Romero, & Narayan, 2021; Sun et al., 2023; Wu, Snajdrova, Moore, Baldenius, & Bornscheuer, 2021; Xu et al., 2024; Zwick & Renata, 2020). Most studies on ROs employ an NAD(P)H-dependent reductase pair (Kweon et al., 2008) To apply ROs in biocatalysis, three issues need to be addressed: (1) enzymes must be obtained with a sufficient level of activity, (2) appropriate reductase or reductase pairs must be acquired to achieve an optimal enzymatic performance, and (3) the dependence on the expensive co-substrate NAD(P)H must be bypassed (Hu et al., 2022; Immanuel, Sivasubramanian, Gul, & Dar, 2020). For cytochrome P450 enzymes or flavin-dependent oxygenases, a few methods have been explored, including regenerating NAD(P)H by an enzymatic system (Uppada, Bhaduri, & Noronha, 2014), reducing NADP$^+$ electrochemically or photochemically (Immanuel et al., 2020; Ozgen, Runda, & Schmidt, 2021), using a photo-reduction approach with a photo-sensitizer (PS) to regenerate NAD(P)H (Park et al., 2015), and delivering electrons to reduce iron-sulfur clusters with photo-activated NAD(P)H (Shanmugam, Quareshy, Cameron, Bugg, & Chen, 2021).

In this chapter, we describe the eosin Y/sodium sulfite photo-reduction system as a replacement for the NAD(P)H-reductase pair in the Stc2-catalyzed stachydrine N-demethylation reaction (Scheme 1) (Hu et al., 2022). In addition, using flow-chemistry to separate the reduction half-reaction and oxidation half-reaction, we achieve a performance comparable to that of the Stc2-NAD(P)H-reductase enzymatic system with the Stc2-eosin Y/sodium sulfite photo-biocatalytic system (Hu et al., 2022). The procedures developed in this study are likely be applicable to other oxidoreductase systems.

2. Heterologous expression and purification of Stc2

Key resources table.

Reagent or resource	Source	Identifier
Bacterial strains		
Escherichia coli BL21-pDB1281	Carrying pDB1281 plasmid for iron-sulfur cluster maturation (Daughtry et al., 2012)	
Chemicals		
D-desthiobiotin	Sigma-Aldrich	D1411
Strep-Tactin® Sepharose® resin	IBA life sciences	2-1201-025
Lysozyme	Sigma-Aldrich	L4919-5G
Bradford reagent	Sigma-Aldrich	B6916-500 mL

LB media	Thermo Fisher Scientific	BP1426-2
Tris(hydroxymethyl)methylamine	Thermo Fisher Scientific	10103203
Ammonium bicarbonate (NH_4HCO_3)	Sigma-Aldrich	A6141-500G
Hydrochloric acid (HCl)	Thermo Fisher Scientific	A144-212
Ethylenediaminetetraacetic acid (EDTA)	Thermo Fisher Scientific	A15161.0B
Sodium chloride (NaCl)	Thermo Fisher Scientific	S271-3
Ampicillin sodium	GoldBio	A-301-5
Glycerol	Thermo Fisher Scientific	A 16205.0F
Dimethyl sulfoxide (DMSO)	Thermo Fisher Scientific	J66650. AK
30% (w/v) Acrylamide/bis-acrylamide solution (37.5:1)	Merck	A3699-100 mL
Ammonium persulfate	Thermo Fisher Scientific	17874
Coomassie protein assay reagent	Thermo Fisher Scientific	PI23238
Coomassie brilliant blue R-250	Thermo Fisher Scientific	11876744
Ethanoic acid	Merck	8187552500
2-Mercaptoethanol	Thermo Fisher Scientific	J66742.30
Methanol	Merck	20864.320
3-(N-Morpholino) propanesulfonic acid	Merck	M3183-100G
Tetramethylethylenediamine (TEMED)	Thermo Fisher Scientific	17919
Precision Plus Protein™ all blue prestained protein standards (10−250 kDa)	Bio-Rad	161-0373
2-[4′-Hydroxy-benzeneazo] benzoic acid (HABA)	Sigma-Aldrich	H5126-5G

Anhydrotetracycline hydrochloride (AHT)	Sigma-Aldrich	PHR3175-50MG
Ammonium peroxydisulfate (APS)	Sigma-Aldrich	A3678
Sodium dodecyl sulfate (SDS)	Sigma-Aldrich	L3771
Recombinant DNA		
IBA-pASK5plus plasmid encoding Stc2	This study	None

3. Materials and equipment

3.1 Equipment
- VWR 1525 Digital Incubator.
- C25KC Incubator shaker.
- Beckman Coulter Avanti JXN-26.
- Coy Lab's Vinyl Anaerobic Chamber.
- Fisher Scientific accuSpin Micro 17.
- Fisherbrand™ Model 705 Sonic Dismembrator.
- Agilent Cary 60 UV-Vis Spectrophotometer.
- Electrophoresis Cells, Power supplies for PAGE applications.
- Chemglass Life Sciences Airfree Vacuum System, Welch 8907 Vacuum Pump.
- Mettler-Toledo FiveEasy pH meter F20-Std-Kit.
- Milli-Q® 7000 Ultrapure Water Systems.
- P10, P200, and P1000 Eppendorf adjustable volume pipettes, and corresponding tips.
- Digital thermometer.
- 200 and 50 mL Amicon® Stirred Cells.

3.2 Solutions and consumables
o Miller Luria-Bertani (LB) broth (autoclaved).
o Miller Luria-Bertani (LB) media containing 2% (w/v) agar (autoclaved).
o 100 mg/mL ampicillin (filter sterilized) (1000× stock).
o Anhydrotetracycline hydrochloride in DMSO (filter sterilized) (25 mg/mL).
o Tris-HCl buffer pH 8.0, 1 M (100×).
o Sodium chloride 5 M (100×).
o Lysis buffer 100 mM Tris-HCl (pH 8.0), 50 mM NaCl, 10% glycerol.
o d-Desthiobiotin in lysis buffer (50 mM).
o Ammonium bicarbonate 1 M (100×).

o IBA-pASK5plus plasmid containing Stc2 coding sequence (10 µg/mL) in sterile TE buffer (10 mM Tris–HCl, 1 mM EDTA, pH 8.0).

o Nalgene™ Oak Ridge High-Speed PPCO Centrifuge Tubes.

o 50 mL Falcon tubes, 15 mL tubes (bought sterile).

o 10 and 25 mL serological pipettes (bought sterile).

o P10, P200 and P1000 pipette tips, 1.5 mL Eppendorf tubes (autoclaved).

o pH meter calibration solutions at pH 4.0, 7.0, and 10.0.

4. Step-by-step procedure details

All solutions were prepared using 18.2 MΩ/cm Milli-Q water (Millipore). All consumables were sterilized by autoclaving at 121 °C for 20 min and 0.1 MPa (15 pounds per square inch, PSI) or in a commercial sterilized package. Working stock solutions were prepared freshly each day before use.

4.1 *Escherichia coli* starter culture

Timing: Day 1 to morning of day 2.

1. Using aseptic technique, the *E. coli* BL21–pDB1281 strain carrying IBA-pASK5plus_Stc2 construct was streaked from a frozen glycerol stock (−80 °C) and inoculated on a LB plate supplemented with 100 µg/mL ampicillin and 50 µg/mL kanamycin (10 mL). The plate was incubated in a 37 °C incubator overnight.

2. In the morning of day 2, colonies were picked using a sterile P200 tip to inoculate 100 mL of LB media supplemented with the 100 µg/mL ampicillin and 50 µg/mL kanamycin in a 250 mL Erlenmeyer flask. The culture was incubated at 37 °C and 250 rpm for ∼8 h, serving as a seed culture.

4.2 Stc2 protein overexpression

Timing: Day 2 to 3.

3. The seed culture was diluted in a 1:100 ratio into 4×2 L of LB media, supplemented with 100 µg/mL ampicillin, 50 µg/mL kanamycin, 0.1 mM ferrous ammonium sulfate, and 1 mM cysteine. This step required the use of 4 L Erlenmeyer flasks. The cultures were incubated at 37 °C and 180 rpm. Cell growth was monitored by measuring the optical density at 600 nm (OD600).

4. After ∼2 h, when the OD600 reached ∼0.2−0.4, the temperature of the culture was lowered to room temperature and a few mL of the culture were collected as the pre-induction sample and labeled as [PRE] sample.

The Fe-S cluster maturation operon carried in pDB1281 was induced by L-arabinose at a 0.2 g/L final concentration.

5. After L-arabinose induction, cell growth was continued at 25 °C and 180 rpm for an additional ~2 h to reach OD600 of 0.6~0.8. Stc2 protein overexpression was induced by the addition of AHT to a final concentration of 500 μg/L. After AHT-mediated induction, the culture was left at 25 °C and 180 rpm for an additional 12~14 h. A few mL of cell culture was collected and labeled as [POST] sample.

6. Cells were harvested by centrifugation (Beckman Coulter Avanti JXN-26, JLA 8.1 rotor, 5000 rpm, 20 min, 4 °C) and stored at −80 °C before use. The typical yield is ~3.5 g of wet cells per liter of culture.

7. On this day, all buffers to be used for protein purification were prepared anaerobically and stored in the Coy-chamber. All other needed materials, including columns, tubes, and other consumables were moved to the Coy chamber.

4.3 Anaerobic Stc2 protein purification (Daughtry et al., 2012)

Timing: Day 4.

8. Frozen cell pellets (~10 g) were transferred into the Coy-chamber and left at room temperature for 20 min to be thawed. They were then resuspended using the anaerobic lysis buffer (~40 mL). Lysozyme (40 mg) was then added to a final concentration of 1 mg/mL to lyse the cells. The cell suspension was incubated on ice with gentle shaking for an additional 30 min.

    Items transferred to the Coy-chamber include: *E. coli* cells, lysozyme (50 mg), 50 and 15 mL falcon tubes, Nalgene centrifuge tubes, sonication steel beaker (100 mL), digital thermometer, one bucket of ice. Eppendorf pipette and pipette tips.

9. After incubation, $NH_4Fe(SO_4)_2$ (0.2 mM final concentration), sodium ascorbate (0.4 mM final concentration), and DTT (1 mM final concentration) were added to reconstitute Stc2 with Fe^{2+} under a reducing environment. Cells were then disrupted by sonication (10 cycles: 8 s on/2 min off, amplitude: 80%). The temperature of the cell lysate mixture was carefully monitored to ensure that it is below 10 °C to avoid protein denaturation. The cell disruption process was monitored by measuring the absorbance at 260 nm after the cell lysate supernatant was diluted by 1000-fold. After all cells were disrupted, the resulting cell lysate was clarified by centrifugation (Beckman JA-20 rotor, 19,500 rpm, 45 min, 4 °C) and the supernatant was collected. Approximately 20 μL of the cell lysate were collected and labeled as the [CL] sample for subsequent SDS-PAGE analysis.

Note: To protect the Step-Tactin resin, the supernatant after the first centrifugation step was collected and centrifuged again (Beckman JA-20 rotor, 19,500 rpm, 45 min, 4 °C) to obtain a clear cell lysate solution.

10. To ensure that the O_2 level in the Coy chamber reamins low, when material is transferred into the Coy-chamber, the chamber is re-equilibrated for at least 20 min before conducting any subsequent experiments.

    Transferred items include: tubes after centrifugation, Bradford solution (~2 mL), d-desthiobiotin and NH_4HCO_3 powder, Strep-Tactin® Sepharose® resin (~15 mL), columns, ice, and ice buckets.

11. After the Strep-Tactin® Sepharose® resin was transferred into the Coy chamber, it was pre-conditioned by washing with at least 100 mL of ice-cold anaerobic lysis buffer (~6-fold resin volume). The Strep-Tactin® Sepharose® resin in the anaerobic lysis buffer was kept in the Coy-chamber for at least one more hour before using it in the next step. The supernatant (~20 μL) was collected and labeled as [**SP**] and for subsequent SDS-PAGE analysis. Cell pellets were resuspended in 50 mL lysis buffer and ~20 μL of the resuspension was collected and labeled as [**CP**] and for subsequent SDS-PAGE analysis. The cell lysate supernatant was mixed with 15 mL Strep-tag resin and incubated on ice for 40 min. Every 5 min, the resin was resuspended by gentle agitation to ensure well mixing between the resin and the cell lysate.

12. The supernatant/resin suspension was loaded onto a column. Flow through (~20 μL) was collected and labeled as [**FT**] for subsequent SDS-PAGE analysis. After draining the supernatant, the resin was washed by anaerobic lysis buffer (105 mL, ~7-fold resin volume). The washing process was carefully monitored by collecting the UV–visible spectra of the flow through. The washing step was stopped when the OD_{280} was lower than 0.1. At the end of the washing steps, the first 3 resin volumes of the washes and the last 4 resin volumes of the washing solutions were combined. ~20 μL of each of the combined washing solution was collected and labeled as [**W1**] and [**W2**] separately, for subsequent SDS-PAGE analysis.

    Note: The washing process was stopped when the absorbance at 280 nm reached 0.1, or the ratio between the absorbance at 280 nm ($A_{280\ nm}$) and 260 nm ($A_{260\ nm}$) was higher than 1.5 (Fig. 1).

13. The target Stc2 protein was eluted from the resin using two rounds of elution buffer I (0.5 resin volume of lysis buffer supplemented with

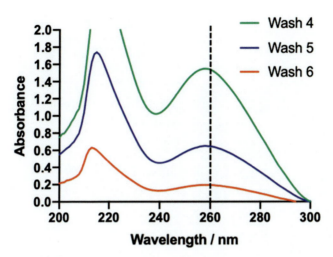

Fig. 1 UV–visible spectra of some of the washing step samples in Stc2 purification. Wash 4: sample from washing the resin with the 4th resin volume buffer. Wash 5: sample from washing the resin with the 5th resin volume buffer. Wash 6: sample from washing the resin with the 6th resin volume buffer.

5 mM d-desthiobiotin each time), then switched to 2–3 rounds of elution buffer II (0.5 resin volume of lysis buffer supplemented with 2.5 mM d-desthiobiotin each time). After elution, all elution fractions were combined. Approximately 20 μL was collected from the combined elution solutions and labeled as [**E**] for SDS-PAGE analysis (Fig. 2). All samples were stored on ice when purification was analyzed by SDS-PAGE analysis.

Note:
  i. Desthiobiotin solution preparation: the solubility of the acidic form of biotin from Sigma is low. To prepare a 50 mM d-desthiobiotin stock solution, d-desthiobiotin (54 mg) from Sigma was dissolved using 0.5 mL of 1 M NH_4HCO_3 solution, and the final volume was adjusted to 5 mL using lysis buffer.
  ii. The first two rounds of elution were carried out using 5 mM d-desthiobiotin-containing elution buffer, and the subsequent elution steps were conducted using the 2.5 mM d-desthiobiotin-containing elution buffer.
  iii. For each round of elution, 0.5 resin volume of ice-cold elution buffer was used to resuspend the resin and the mixture was incubated on ice for 10 min before elution. At the end of each elution, a small aliquot (~10 μL) was collected for the Bradford

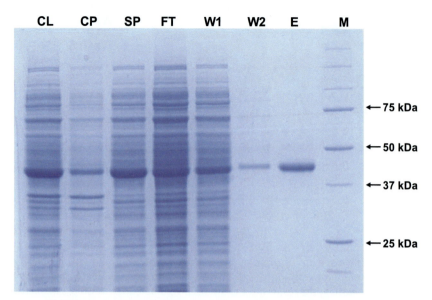

Fig. 2 SDS-PAGE analysis of Stc2 purification samples. *CL*, cell lysate; *CP*, cell pellet suspension; *SP*, supernatant; *FT*, flow through after resin treatment; *W1*, combined wash I solution; *W2*, combined wash II solution; *E*, eluted Stc2 protein by D-desthiobiotin solution; *M*, protein molecular weight marker.

Assay to estimate the protein concentration. This assay provides information on whether the elution process is done. These samples were also analyzed by SDS-PAGE to further assess the degree of purity of the eluted protein.

iv. Strep-Tactin® Sepharose® resin regeneration: Wash the Strep-Tactin® Sepharose® resin from step 13 using the wash buffer (100 mM Tris-HCl buffer, 50 mM sodium chloride, pH 8.0) for 10× resin volume. The Strep-Tactin® Sepharose® resin was washed with 3 resin volumes of the regeneration buffer (100 mM Tris-HCl buffer, pH 8.0, 150 mM sodium chloride, 1 mM HABA, 1 mM EDTA, pH 8.0) first. Then the resin was incubated with 1 resin volume of regeneration buffer (100 mM Tris-HCl buffer, pH 8.0, 150 mM sodium chloride, 1 mM HABA, 1 mM EDTA, pH 8.0) for 15 min. After drainage, the regeneration process was repeated three more times. After resin regeneration by the HABA solution, the excess HABA was removed by washing the resin with washing buffer (100 mM Tris-HCl buffer, 50 mM sodium chloride, pH 8.0). The regenerated resin was stored at 4 °C.

Table 1 Components for casting SDS-PAGE gels.

Reagent	Volume for 12% resolving gel	Volume for 4% stacking gel
40% (w/v) acrylamide stock solution (37.5:1)	1.5 mL	0.24 mL
Deionized water	1.5 mL	1.43 mL
1 M Tris-HCl (pH8.8)	1.88 mL	0
1 M Tris-HCl (pH6.8)	0	0.25 mL
10% (w/v) SDS	50 μL	20 μL
10% (w/v) APS (100 mg/mL)	50 μL	20 μL
TEMED	5 μL	2 μL

4.4 Concentrating the eluted Stc2 in the COY chamber

Timing: Day 4.

14. Samples collected in steps 4–13 were analyzed on a 12% SDS-PAGE gel (Table 1).

    SDS-PAGE samples were prepared in a proper dilution with 5× loading dye (10% SDS, 500 mM DTT, 50% Glycerol, 500 mM Tris-HCl and 0.05% bromophenol blue dye) in a 1.5 mL Eppendorf tube. (CL, CP and SP samples were diluted 10 times; FT and all wash (W) samples were diluted 3-fold; Elution samples (E) samples were diluted 4-fold). 4 μL of each sample (~1–5 μg of Stc2 protein) were used for SDS-PAGE analysis using Tris/glycine/SDS running buffer (Gibbins, 2004). Gels were stained using Coomassie Brilliant Blue R-250 stain (Fig. 2) (Neuhoff, Arold, Taube, & Ehrhardt, 1988).

15. Based on the SDS-PAGE results, proper elution fractions were combined and concentrated using 200 mL Amicon® Stirred Cell with EMD Millipore ultrafiltration disc membranes (10 kDa, 76 mm diameter). When the protein solution volume was ~10 mL, it was transferred to a 50 mL Amicon® Stirred Cell with ultrafiltration disc membranes (10 kDa, 43 mm diameter) and proteins were concentrated to ~100 mg/mL (the final volume is ~2 mL). The concentrated Stc2 was aliquoted and stored in −80 °C.

Note:

In general, if the washing process is carefully monitored, there should be only minor contaminating bands in the elution factions. The elution fractions can be combined directly for biochemical and spectroscopic characterizations.

5. Expected outcomes, advantages, and disadvantages

The host cell *E. coli* BL21-pDB1281-pASK-IBA5plus system was used for Stc2 protein production for a few reasons. First of all, pDB1281 has the iron-sulfur maturation system, which will greatly improve the iron-sulfur cluster content in Stc2 protein (Daughtry et al., 2012; Xiao, Zhao, & Liu, 2008) Secondly, pASK-IBA3plus protein expression system can be induced by anhydrotetracycline (AHT, order no.: PHR3175) at a final concentration of 500 µg/L and it is compatible with the induction system used for iron-sulfur cluster maturation system. Lastly, proteins can be eluted with 2.5 mM of d-desthiobiotin solution from the Strep-Tactin resin because d-desthiobiotin does not interfere with downstream enzymatic reactions. This statement means that the eluted protein can be used for biochemical assays directly, which gives a great chance of obtaining high quality proteins.

The two cofactors in ROs, a Rieske [2Fe-2S] cluster and the mononuclear iron center, are susceptible to oxidative damage (Daughtry et al., 2012; Hu et al., 2022). Due to this reason, all of our proteins were purified under anaerobic conditions in a Coy chamber. Besides preparing anaerobic buffers, all materials were transferred into the Coy chamber ahead of the time and were given enough equilibrating time to ensure that O_2 levels were low. The Coy chamber was maintained in the N_2/H_2 atmosphere with O_2 concentration lower than 5 ppm. In our studies, Stc2 protein was also concentrated in the Coy chamber after purification.

It is also important to monitor every step of the purification process because different proteins may have different expression levels and their binding affinity to the resin might be slightly different.

6. Optimization and troubleshooting

The above methods are robust and work well for several of our non-heme iron enzymes (Cheng et al., 2022; Wang et al., 2023; Zhu et al., 2022).

In some cases, in order to produce the target protein in a high yield, fresh transformation of the protein expression plasmid into the E. coli BL21-pDB1281 competent cell was essential. Stc2 protein has two cofactors: the mononuclear iron site and the [2Fe-2S] cluster. To increase the holo-enzyme yield, besides the incorporation of the Fe-S cluster maturation system, the culture medium was also supplemented with iron salt and cysteine. Because both the [2Fe-2S] cluster and the mononuclear iron site in Stc2 protein can be oxidatively damaged, the purification was conducted under an anaerobic environment. In addition, during the cell lysate process, extra reductants (e.g., DTT and ascorbate) were included to maintain a reducing environment. In our hands, when the purification was conducted under anaerobic conditions, high quality holo-Stc2 was obtained without the need of an additional reconstitution step.

7. Safety considerations and standards

Anything encountering cells, cell lysate, or protein solutions were decontaminated using the appropriate procedures for biohazards. Eye protection and appropriate Personal Protective Equipment (PPE) should be worn (earmuffs, safety glasses, gloves and laboratory coat when handling hazardous chemicals). Chemical waste should be disposed of through approved routes. Acrylamide and bis(acrylamide) are toxic if swallowed, inhaled, or absorbed through the skin and should be handled properly. The Coy-chamber is maintained in 95:5 mixture of N_2/H_2 and the chamber should be placed in a room with a well-functioning ventilation system.

8. Stc2 characterization

Key resources table.

Reagent or resource	Source	Identifier
Chemicals		
Iron standard for AAS	Thermo Fisher Scientific	SI124-100
Nitric acid	Sigma-Aldrich	438073-500 mL
Zinc acetate	Sigma-Aldrich	1724703

Sodium hydroxide (NaOH)	Sigma-Aldrich	S5881-1KG
N, N-dimethyl-p-phenylenediamine	Thermo Fisher Scientific	AAA1417514
Iron (III) chloride, anhydrous	Thermo Fisher Scientific	012357. A1
Hydrochloric acid (12 M)	Thermo Fisher Scientific	A144-212
Tris(hydroxymethyl)methylamine	Thermo Fisher Scientific	10103203
Bovine serum albumin standard (BSA, 2.0 mg/mL)	Thermo Fisher Scientific	23209
Bradford Reagent	Sigma-Aldrich	B6916-500 mL
Software and Algorithms		
Prism	GraphPad	v.8.4.3

9. Materials and equipment
9.1 Equipment
- Coy Lab's Vinyl Anaerobic Chamber.
- Beckman Coulter Microfuge® 22R Centrifuge.
- Agilent 5800 ICP-OES spectrometer.
- Agilent Cary 60 UV-Vis Spectrophotometer.
- Milli-Q® 7000 Ultrapure Water Systems.
- Scientific Industries SI-T236 Vortex-Genie 2T.

9.2 Solutions and consumables
- P10, P200 and P1000 pipette tips, 1.5 mL Eppendorf tubes (autoclaved).
- Iron Reference Standard solution (1000 ppm).
- 2% (w/w) nitric acid.
- 12% (w/v) NaOH solution.
- 0.1% (w/v) N, N-dimethyl-p-phenylenediamine solution (DMPD) in 5 M HCl.
- Deionized water.
- 1% (w/w) zinc acetate solution.
- 23 mM $FeCl_3$ solution in 1.2 M HCl.
- 6 M HCl solution.

- 1.2 M HCl solution.
- 10 mM Tris-HCl buffer, pH 8.0.
- 50 mL, and 15 mL Falcon tubes, 1.5 mL Eppendorf tubes (bought sterile).

10. Step-by-step details

All solutions were prepared using 18.2 MΩ/cm Milli-Q water (Millipore). Unless stated, all Stc2 related operations were conducted in the anaerobic Coy-chamber.

10.1 Stc2 protein analysis

16. SDS-PAGE analysis of Stc2 samples is shown in Fig. 3.
17. Protein standard solutions (0.05–0.5 mg/mL, Table 2) were prepared by serial dilutions of bovine serum albumin (BSA) using a lysis buffer.
18. Stc2 proteins in various dilutions (10×, 50×, and 100×) were prepared from the concentrated Stc2 solution by diluting it using the lysis buffer.
19. Bradford reagents (960 μL) was mixed with 40 μL protein solution (BSA standards or Stc2 samples). After they were thoroughly mixed, the samples were incubated at room temperature for three minutes.

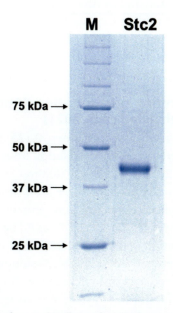

Fig. 3 SDS-PAGE analysis of anaerobically purified Rieske oxygenases Stc2 (final sample).

Table 2 Standard curve of protein quantification.

[BSA] (mg/mL)	$A_{595\,nm}$
0.05	0.1648
0.1	0.2879
0.2	0.4183
0.3	0.5663
0.4	0.7201
0.5	0.8476

20. The mixture samples were transferred to disposable cuvettes. The sample to be used as a blank sample was prepared by mixing 960 µL Bradford reagents with 40 µL of lysis buffer.
21. A standard curve was prepared by correlating the absorbance at 595 nm ($A_{595\,nm}$) with protein concentration (µg/mL) (Fig. 4).
22. Stc2 protein concentration in mg/mL was calculated with the standard curve, a dilution factor, and its A595. The concentration (with the unit µM) was calculated using Stc2 molecular weight of 46 kDa.
    e.g., in one of the samples shown (Fig. 4): Stc2 concentration=C'/ $46,000\times1000\times1000 = [(0.64-0.1167)/1.487\times100]/46,000\times1000\times 1000 = 765$ µM. (C', concentration of protein in mg/mL).

10.2 Stc2 iron content quantification (Abbasi, Abbina, Gill, Bhagat, & Kizhakkedathu, 2021)

23. Iron standards at various concentrations (0.125, 0.25, 0.5, 0.75, 1, 1.5 and 2 ppm) were prepared by serial dilution from an iron stock solution (1000 ppm) using 2% (w/w) HNO_3 solution.
24. Holo-Stc2 protein should have three iron atoms per Stc2 subunit. Based on the concentration of the protein Stc2 stock solution (590 µM), the theoretical iron content should be 98.8 ppm. Therefore, the Stc2 stock solution was diluted by 50-fold using 2% (w/w) nitric acid solution to prepare 1 mL of the Stc2 solution with iron content in the 0.125–2 ppm iron standard curve range. After mixing the protein with nitric acid solution, the mixture was vortexed for 3 min

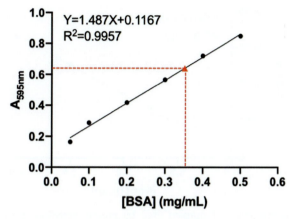

Fig. 4 Bovine serum albumin-based standard curve using Bradford assay for Stc2 protein concentration determination. Stc2 concentration=C′/46,000 × 1000 × 1000 = [(0.64−0.1167)/1.487 × 100]/46,000 × 1000 × 1000 = 765 μM. (C′, concentration of protein in mg/mL).

25. The Stc2 protein sample was then centrifuged at 15,500 rpm for 10 min to remove the protein precipitates and the supernatant was transferred to a new 1.5 mL Eppendorf tube.
26. The iron content in iron standards and the Stc2 samples were then analyzed using Agilent 5800 ICP-OES spectrometer by measuring the emission spectrum at 371.993 nm.

    Note: The sample uptake time was 15 s and stabilization time was 10 s. The emission spectrum at 371.993 nm was recorded for iron concentration calculation. The Agilent January 2021 Edition Microwave Plasma Atomic Emission Spectroscopy (MP-AES) Application eHandbook was used as the guide, especially the procedures of <u>Inductively coupled plasma-atomic emission spectrometry (ICP-AES)</u> outlined in the manual. Data analysis samples are shown in Fig. 5 and Table 3.

Standard curve: $y = 8105x + 522.7$, $R^2 = 0.9980$.
Iron concentration [Fe] = (10,020.13−522.7)/8105 = 1.17 ppm = (1.17/55.845) × 1000 μM = 20.1 μM.
Sample dilution factor D = (18 + 982)/18 = 55.56.
Sample protein concentration [Stc2] = 590 μM.
Iron content of Stc2 protein [Fe]% = [Fe] × D/[Stc2] = (20.1 μM) × 55.56/590 μM = 1.89.

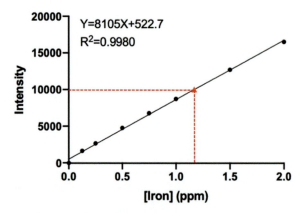

Fig. 5 Iron quantification of purified Stc2 by atomic emission spectroscopy. Iron concentration from Stc2 [Fe] = (10,020.13–522.7)/8105 = 1.17 ppm = (1.17/55.845) × 1000 μM = 20.1 μM. Sample dilution factor D = (18 + 982)/18 = 55.56. Sample protein concentration [Stc2] = 590 μM. Iron content of Stc2 protein [Fe]% = [Fe] × D/[Stc2] = (20.1 μM) × 55.56/590 μM = 1.89.

Note: One molecule of Stc2 should have two irons from the Fe-S cluster and one mononuclear iron. We have tried to adjust the iron and sulfur in the culture medium. In the absence of reconstitution of cofactors in the protein purification step, this iron content is a typical number for Stc2 system Fig. 5.

10.3 Quantification of labile sulfur in Stc2 based on the reaction shown in Fig. 6 (Beinert, 1983)

27. 20 μL of Stc2 protein stock solution was mixed with 180 μL of Tris-HCl buffer (20 mM, pH = 8.0) to make a 200 μL solution containing 90 μM of Stc2.
28. To the above Stc2 protein solution, 600 μL freshly prepared (or freshly diluted) 1% (w/w) zinc acetate solution was added and mixed gently.
29. Immediately to the above mixture, 30 μL 12% (w/w) NaOH solution was added, and the mixture was stirred gently until the schlieren of NaOH disappear. The mixture was left at room temperature in the Coy chamber for about 1 h.
30. Next, 150 μL of 0.1% DMPD solution was added to the mixture and it was stirred gently until only the top 2-mm layer of liquid had undissolved zinc hydroxide. Then, 30 μL of 23 mM FeCl$_3$ solution was added into the mixture under the surface, followed by an immediate acceleration of the stirring bar to make a homogeneous solution quickly.

Table 3 Iron content measurement from AES.

Iron concentration (ppm)	Average intensity	Intensity read 1	Intensity read 2	Intensity read 3	Intensity SD	Intensity %RSD
0	0	11.8	−15.02	3.22	13.69	−
0.125	1645.1	1645.51	1644.99	1644.8	0.37	0.02
0.25	2668.65	2652.23	2676.03	2677.68	14.24	0.53
0.5	4775.29	4747.14	4782.93	4795.81	25.21	0.53
0.75	6776.07	6782.26	6773.09	6772.84	5.37	0.08
1	8716.35	8627.39	8876.61	8645.06	139.07	1.6
1.5	12,725.51	12,661.05	12,778.85	12,736.62	59.68	0.47
2	16,515.91	16,456.06	16,438.24	16,653.42	119.42	0.72
Stc2 sample	10,020.13	10,096.13	10,044.24	9920.01	90.5	0.9

Fig. 6 Reaction scheme for the reaction used for quantification of labile sulfur in purified Stc2.

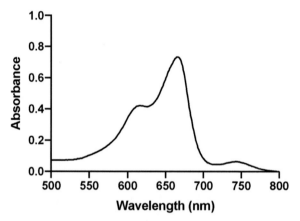

Fig. 7 UV–visible spectrum of one of the Stc2 samples used for labile sulfur quantification. The sulfur concentration [S] = $A_{670\,nm}/(\varepsilon_{670\,nm} \times l)$ = 0.7087/(34.5 $cm^{-1} \cdot mM^{-1}$ × 1 cm) = 20.5 μM. Final Stc2 concentration: $[Stc2]_{start}$/dilution factor = 90 μM/5.05 = 17.8 μM. Labile sulfur content = [S] / [Stc2] = (20.5 μM)/(17.8 μM) = 1.15.

31. After 30 min of color development at room temperature, the mixture was centrifuged at 15,500 rpm for 20 min to precipitate the protein.
32. The supernatant was transferred to a cuvette and the $A_{670\,nm}$, $A_{710\,nm}$, and $A_{750\,nm}$ were recorded. The $A_{670\,nm}$ value was used to calculate the sulfide concentration (Fig. 7) (Beinert, 1983; Hu et al., 2022).
33. Based on the $A_{670\,nm}$ = 0.7087 and $\varepsilon_{670\,nm}$ = 34.5 $cm^{-1} \cdot mM^{-1}$ (Beinert, 1983), the value of sulfide concentration can be calculated and a sample of calculation is shown here:

    the sulfur concentration [S] = $A_{670\,nm}/(\varepsilon_{670\,nm} \times l)$ = 0.7087/(34.5 $cm^{-1} \cdot mM^{-1}$ × 1 cm) = 20.5 μM.
    Final Stc2 concentration: $[Stc2]_{start}$/dilution factor = 90 μM/5.05 = 17.8 μM.
    Labile sulfur content = [S]/[Stc2] = (20.5 μM)/(17.8 μM) = 1.15.
    Because each Rieske iron-sulfur cluster has 2 equivalents of sulfide, the above calculation suggests that our Rieske iron-sulfur content is ∼58%.

Note:

i. Zinc acetate solution (~10%, w/w) is stable for only a few days. It is better to prepare fresh solution.
ii. For the labile iron-sulfur containing proteins, the zinc hydroxide precipitation step takes no longer than 5 min whereas for other stable Fe-S proteins, $l \sim 2$ h might be required.
iii. Standardization may be performed using $Na_2S \cdot 9H_2O$.

10.4 Optimization and troubleshooting

The above protein characterization methods are well-established methods and highly reproducible. For spectroscopic studies, protein concentration determined from amino acid analysis instead of the Bradford Assay might be needed. In addition, the Coy-chamber needs to be maintained well to ensure low O_2 levels. The O_2 detector needs to be calibrated on a regular basis.

11. Stc2 photo-reduction using eosin Y and Na₂SO₃

Reagent table.

Reagent or resource	Source	Identifier
Chemicals		
Tris(hydroxymethyl)methylamine	Thermo Fisher Scientific	10103203
Sodium sulfite (Na₂SO₃)	Sigma-Aldrich	08981
Stachydrine	Biosyn	FS65548
Eosin Y (EY)	Sigma-Aldrich	E4009-5G
Deuterium oxide	Cambridge isotope Laboratories, Inc.	DLM-4-100
Ethyl viologen dibromide	Sigma-Aldrich	384097-1G
Chloroform	Sigma-Aldrich	650498
Hemin	Sigma-Aldrich	51280
Kanamycin	Merck	420311
Isopropyl-β-d-1-thiogalactopyranoside (IPTG)	Merck	I5502
Ammonium sulfate	Sigma-Aldrich	7783-20-2
Glycerol	Thermo Fisher Scientific	A 16205.0F

| L-ascorbate sodium | Sigma-Aldrich | A7631-100G |

Software and algorithms

MestReNova	Mestrelab Research	v.14.1.0
ChemOffice	PerkinElmer	v. 20.0
Prism	GraphPad	v.8.4.3
SpinCount	The Hendrich Metalloprotein Group	

12. Materials and equipment

12.1 Equipment

- Coy Lab's Vinyl Anaerobic Chamber.
- Fisher Scientific accuSpin Micro 17.
- Agilent Cary 60 UV-Vis Spectrophotometer.
- Agilent 500 MHz VNMRS spectrometer.
- Scientific Industries SI-T236 Vortex-Genie 2T.
- Labconco FreeZone 10 L −84 °C Freeze Dryer System.
- Fisher Scientific Isotemp™ digital hotplate stirrer.
- Light Bulb, full spectrum, 24 W (BRIM, #SOL24W)
- Screw-tight quartz cuvette (MES Supplies LLC, LS0703).

12.2 Solutions and consumables

- P10, P200, and P1000 pipette tips, 1.5 mL Eppendorf tubes (autoclaved).
- 10 mM NaOH solution.
- 20 mM Tris-HCl buffer, pH 8.0 (anaerobic preparation).
- 6 M HCl solution (5 mL HCl stock solution in 5 mL deionized water).
- 50 mM Fe(NH$_4$)(SO$_4$) solution.
- 100 mM EDTA solution, pH 8.0.
- 50 mM HEPES solution, pH 8.0.
- Deionized water.
- Glass vial, clear.

13. Step-by-step method details

All solutions were prepared using 18.2 MΩ/cm Milli-Q water (Millipore). All of the experimental operations were conducted under anaerobic conditions in a Coy chamber unless stated otherwise.

13.1 Light-driving Stc2 Fe-S cluster reduction using eosin Y and Na$_2$SO$_3$ as photosensitizer/sacrificial reagent pair

34. At least 20 mL of anaerobic buffer was prepared (20 mM Tris-HCl, pH = 8.0) and transferred into the Coy chamber one-day before the planned experiments.
35. The eosin Y solution was prepared using the anaerobic buffer. In general, the stock solution was prepared at a concentration of ~500 μM. To accurately measure the eosin Y concentration, the stock solution was diluted by 100× and the UV–visible spectrum of the diluted solution was collected (300–800 nm spectrum in Fig. 8 as an example). The eosin Y concentration could be calculated based on its extinction coefficient: $\varepsilon_{525\ nm} = 95\ cm^{-1} \cdot mM^{-1}$.
Example. The eosin Y stock solution's concentration [EY] = $A_{525\ nm} \times D/(\varepsilon \times l)$ = 0.7931 × 100/(95 cm^{-1}•mM^{-1} × 1 cm) = 835 μM ([EY]: EY concentration; D: dilution factor; ε: EY extinction coefficient).
36. The eosin Y stock solution can be aliquoted into small fractions, frozen by liquid nitrogen, and be stored in −20 °C refrigerator in dark (covered by aluminum foil).
37. 1 M of sodium sulfite stock solution was prepared using the anaerobic buffer (20 mM Tris-HCl, pH = 8.0).
38. To monitor Stc2 reduction, the photoreduction reaction was conducted in a screw-tight quartz cuvette directly in the Coy-chamber.

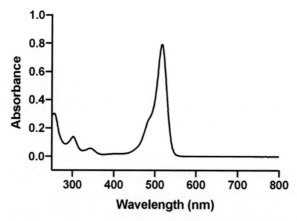

Fig. 8 UV–visible spectrum of an eosin Y for its concentration calculation. The eosin Y stock solution's concentration [EY] = $A_{525nm} \times D / (\varepsilon \times l)$ = 0.7931 × 100/(95 cm^{-1}•mM^{-1} × 1 cm) = 835 μM ([EY], EY concentration; D, dilution factor; ε, EY extinction coefficient).

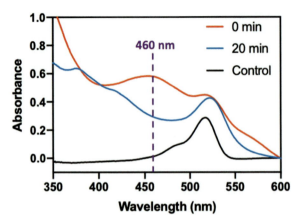

Fig. 9 Time-course of Stc2 reduction monitored by UV–visible spectroscopy. The UV spectra were collected at 0 min (red trace) and 20 min (blue trace). The same mixture without Stc2 is set up as control (black trace, with 1.5 μM of eosin Y) to show the change of UV–visible spectra during the photoreduction process.

A typical 1 mL reaction mixture contained 100 μM Stc2 enzyme, 20 mM sacrificial reagent Na_2SO_3, and 1.5 μM photosensitizer eosin Y in the anaerobic buffer (20 mM Tris-HCl, pH = 8.0).

39. The sealed quartz cuvette was placed on ice and photoreduction was conducted under either white light using mercury lamp or a specific wavelength with LED lamp. When white light was used, to avoid heating up the protein samples, our photo-reduction reaction was conducted on ice.
40. At various time points (0 and 20 min), the UV–visible spectra of the sample were collected to monitor the reducing rate (350–600 nm UV–visible spectra in Fig. 9 as samples).
41. The absorbance at 460 nm, which is related to Stc2 $[2Fe-2S]^{2+}$ cluster, could be used for monitoring the extent of Stc2-reduction reaction.

13.2 Light-driving demethylation of stachydrine using eosin Y and Na_2SO_3 under multiple turnover condition (Hu et al., 2022; Xu et al., 2024)

42. A stachydrine stock solution was prepared by dissolving stachydrine power using the 20 mM Tris-HCl anaerobic buffer. In general, we have an ethyl viologen solution with a known concentration. The stachydrine concentration was determined by using a known concentration of an ethyl viologen with as an internal standard. ^{1}H-NMR

was then used to determine the stachydrine concentration. Fig. 10 shows a sample of stachydrine concentration determination used in this work.

43. Eosin Y concentration calculation has been discussed in prior section (Fig. 8).

44. Sodium sulfite stock solution (1 M) was prepared by dissolving Na_2SO_3 power in the anaerobic buffer (20 mM Tris-HCl, pH = 8.0).

45. To prepare O_2 saturated buffer, the anaerobic buffer (20 mL, 20 mM Tris-HCl, pH = 8.0) in a 100 mL three-neck flask was degassed under vacuum first. An O_2 balloon was then connected to the flask and the mixture was stirred at room temperature for at least 30 min. Buffer degassing and the subsequent O_2 purge were repeated for at least two more times to prepare O_2 saturated buffer.

46. A 250 μL reaction mixture in a 5 mL glass vial (Uline, WI, USA) contained 20 μM Stc2 enzyme, 2 mM stachydrine, 10 μM eosin Y, and 20 mM Na_2SO_3 in the anaerobic buffer (20 mM Tris-HCl, pH = 8.0). Then the mixture was stirred for 30 s. This process should be conducted in the Coy chamber anaerobically.

47. The anaerobic mixture in step 46 was then transferred out of the anaerobic coy chamber and 250 μL oxygenated buffer was immediately mixed with it to initiate the reaction.

48. After reaction initiation, the mixture was stirred in an open vial under white light and aerobic environment for 1 h. The reaction was conducted on ice to avoid protein denaturing due to heating by lamp.

49. The reaction mixture was quenched by 300 μL of chloroform and thoroughly mixed by vortexing for 2 min to inactivate the Stc2 protein.

50. The aqueous and chloroform layers were separated by centrifugation (15,500 rpm, 10 min).

51. The aqueous layer was transferred to a falcon tube, frozen by liquid nitrogen, and lyophilized.

52. The lyophilized sample was re-dissolved by 350 μL deuterated water. After centrifugation (15,500 rpm, 10 min), the supernatant was analyzed by ^{1}H NMR.

53. The turnover numbers were calculated based on the relative intensities of the signals for the N-methyl group in the substrate (stachydrine) vs. the product (mono-methyl proline) in a ^{1}H NMR spectrum (see Fig. 11 as an example).

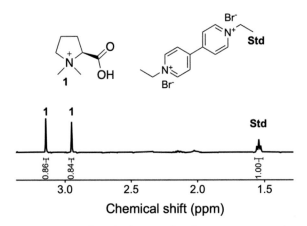

Fig. 10 ^{1}H NMR spectrum of stachydrine with a known concentration of ethyl viologen for stachydrine concentration determination. The NMR samples contain 10 μL of stock stachydrine stock solution, 20 μL of 50 mM ethyl viologen and 300 μL D$_2$O. Each of the two the N-methyl group signals (~3.17 and ~2.97 ppm) from stachydrine contain 3 protons, and the methyl group signal (~1.54 ppm) from the ethyl viologen contains 6 protons. The concentration of the stachydrine stock solution [stachydrine] = (0.86 + 0.84)/2×(6/3)×2 × 50 mM = 170 mM.

13.3 Stc2-catalysis in a flow-setting

54. *E. coli* BL21 competent cells carrying a pET41a-(+)-CLD construct for chlorite dismutase (CLD) was kindly provided by Prof. Yisong Guo. A single colony was inoculated into LB broth supplemented with 50 μg/mL kanamycin (10 mL) and incubated at 37 °C overnight. A 10 mL of seed culture was inoculated into 1 L of LB media supplemented with 90 μM hemin and 50 μg/mL kanamycin. The overexpression of CLD was induced using IPTG at a final concentration of 0.2 mM when OD600 reached 0.6~0.8.

55. The CLD protein was overexpressed at 20 °C for 14 h and the cells were harvested by centrifugation according to the method described previously. *E. coli* BL21-pET41a-CLD cells (8 g) were resuspended in 40 mL of lysis buffer (100 nM Tris-HCl, 50 mM NaCl, pH 8.0) and 40 mg lysozyme was added. Cells were disrupted as outlined in Step 9 except that no iron salt or reductants were included.

56. The supernatant (40 mL) was transferred into a beaker on ice, and 12.04 g ammonium sulfate crystals were added slowly with stirring to reach a level of ~50% ammonium sulfate saturation. This process took about 1 h. After all ammonium sulfate crystals were added to the solution, the mixture was stirred on ice for an additional 30 min

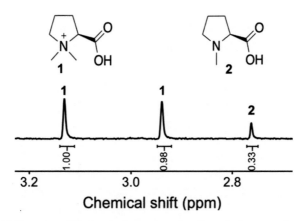

Fig. 11 ¹H NMR assay of Stc2 reaction using the N-methyl group signals from stachydrine and mono-methylproline for quantification.

57. The CLD sample was centrifuged at 15,000 rpm for 20 min. The pellet was collected and resuspended in 5 mL of lysis buffer (100 mM Tris-HCl, 50 mM NaCl, pH 8.0), and incubated on ice for an additional 30 min to fully dissolve CLD. The partially purified CLD protein solution was quickly frozen using liquid nitrogen for future use in Stc2 reactions.
58. A 5 mL reaction mixture that contained 20 μM Stc2, 2 mM stachydrine, 20 mM Na_2SO_3 and 20 μM eosin Y in 20 mM Tris-HCl, pH 8.0 buffer was placed in a reservoir serving as the starting solution. The flow system was set up as shown in Fig. 12.
59. The reaction mixture was circulated between the two halves (reduction half by photocatalysis and oxidation half with O_2 generated from chlorite by CLD) driven by a peristaltic pump. The solution was pumped into a 3 mL PFA tube, which was illuminated by either the LED light or by the white light. The peristaltic pump used in this experiment was operated at a flow rate of 6 mL/min. Therefore, the photoreduction half-reaction time was ~30 s before exiting the PFA tube. The time of the reduction half was determined based on the reduction kinetics determined from the experiments outlined in steps 34–41 in the prior section.
60. O_2 was generated in situ by syringe pump delivery and mixing of two solutions: Solution A was 5 μM CLD in 20 mM Tris-HCl, pH 8.0 buffer and Solution B was 1.25 mM $NaClO_2$ in 20 mM Tris-HCl, pH 8.0 buffer. The CLD/chlorite ratio was 1:1200. The O_2 generating

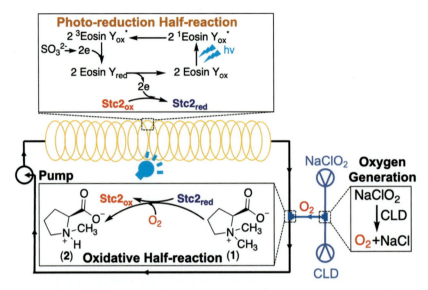

Fig. 12 Schematic diagram of the Stc2-catalysis in flow setting with in situ O₂ generation using CLD-NaClO₂ system.

mixture was then mixed with the Stc2 photo-reduced mixture using a three-way valve as shown in the Fig. 12 setup.

61. The rate of O_2 delivery was achieved by controlling the rate of chlorite delivery and it varied from 7.2 to 2.0 μmol/h to match with the Stc2 circulation rate in the reaction mixture.

62. In some reactions, some additional eosin Y was supplemented to compensate for the loss of eosin Y due to photo-bleaching. At the end of the reaction, the mixture was lyophilized and dissolved in 1 mL D2O and analyzed by ^{1}H NMR (Fig. 13).

13.4 Optimization and troubleshooting

i. When the protocol is applied to other systems, several parameters might be optimized to fit the particular system:

ii. The biocatalyst is temperature dependent, and the lamp generates heat that might denature the enzyme. In our studies, we run the reaction on a water or ice bath to avoid enzyme denaturing by heat.

iii. Light sources used in the photo-reduction might influence the reduction efficiency. According to the absorbance spectrum analysis conducted by EY, it has a pronounced absorbance feature at 525 nm. In our hands

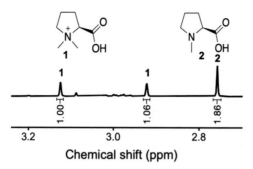

Fig. 13 ¹H NMR assay of Stc2 flow reaction using the N-methyl group signals from stachydrine and mono-methylproline for quantification.

(Fig. 14), the Stc2 batch reaction utilizing green light (525 nm) have an observed turnover that is lower than those of white, blue (467 nm), or purple light (370 nm). When applied in other enzymatic reactions, the light source will likely need to be examined.

iv. In Stc2-catalysis under a flow setting, the rate of the pumping cycle is a crucial factor. Upon doubling the circulation speed of the pump, transitioning from 3 to 6 mL/min, we observed a corresponding doubling in the reaction conversion rate, escalating from approximately 30 to about 65. This phenomenon underscores the significance of pump circulation speed in optimizing reaction efficiency. Given the inherent instability of the protein and the susceptibility of EY to photobleaching under light, modulating the pump's circulation will be needed to achieve an optimal performance.

v. Several O_2 delivery methods have been examined in this Stc2 photocatalytic process. Employing an O_2-saturated buffer as the O_2 source proved to be effective in initiating the reaction. However, this method also posed challenges due to the inherent limitations of O_2 concentration within the buffer. The necessity to deliver a substantial volume of O_2-containing buffer to compensate for this limitation resulted in considerable dilution of the reaction solution. Direct delivery of O_2 gas as slugs used in organic flow-chemistry was also examined (Shimizu, Kato, & Kobayashi, 2023). Unfortunately, this method did not seem to work for Stc2 reaction due to protein denaturation. To resolve the O_2 delivery issue in flow-chemistry setting, we introduced CLD-catalyzed $NaClO_2$ decomposition to O_2 as an in situ O_2 production method. CLD activity is a few orders of magnitude greater than that of Stc2 (Mehboob et al., 2009). To match

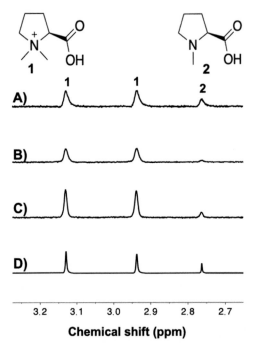

Fig. 14 ¹H NMR analysis of Stc2-catalyzed reaction under representative light source when eosin Y/sulfite serve as the photosensitizer/sacrificial reagent pair. (A) White light lamp, positive control; (B) Greem light, 525 nm wavelength; (C) Blue light, 480 nm wavelength; (D) Purple light, 370 nm wavelength. A typical 500 μL reaction in 20 mM Tirs-HCl buffer (pH 8.0) includes 20 μM Stc2 protein, 2 mM stachydrine, 10 μM eosin Y, 20 mM Na₂SO₃. The reactions were then conducted under aerobic conditions in an open vial under different light source for 1 h before quenching for analysis.

the O_2 delivery rate with that of Stc2-catalysis, co-delivering CLD and NaClO₂ at a proper ratio led to the best Stc2-catalysis performance. When this method is applied in other systems, an optimization of the O_2 delivery rate is needed to fit the particular system of interest.

References

Abbasi, U., Abbina, S., Gill, A., Bhagat, V., & Kizhakkedathu, J. N. (2021). A facile colorimetric method for the quantification of labile iron pool and total iron in cells and tissue specimens. *Scientific Reports, 11*(1), 6008. https://doi.org/10.1038/s41598-021-85387-z.

Barry, S. M., & Challis, G. L. (2013). Mechanism and catalytic diversity of Rieske non-heme iron-dependent oxygenases. *ACS Catalysis, 3*(10), https://doi.org/10.1021/cs400087p.

Beinert, H. (1983). Semi-micro methods for analysis of labile sulfide and of labile sulfide plus sulfane sulfur in unusually stable iron-sulfur proteins. *Analytical Biochemistry, 131*(2), 373–378. https://doi.org/10.1016/0003-2697(83)90186-0.

Burnet, M. W., Goldmann, A., Message, B., Drong, R., El Amrani, A., Loreau, O., ... Tepfer, D. (2000). The stachydrine catabolism region in *Sinorhizobium meliloti* encodes a multi-enzyme complex similar to the xenobiotic degrading systems in other bacteria. *Gene, 244*(1-2), 151–161. https://doi.org/10.1016/s0378-1119(99)00554-5.

Cheng, R., Weitz, A. C., Paris, J., Tang, Y., Zhang, J., Song, H., ... Liu, P. (2022). OvoA$_{Mtht}$ from *Methyloversatilis thermotolerans* ovothiol biosynthesis is a bifunction enzyme: Thiol oxygenase and sulfoxide synthase activities. *Chemical Science, 13*(12), 3589–3598. https://doi.org/10.1039/d1sc05479a.

Daughtry, K. D., Xiao, Y. L., Stoner-Ma, D., Cho, E. S., Orville, A. M., Liu, P. H., & Allen, K. N. (2012). Quaternary ammonium oxidative demethylation: X-ray crystallographic, resonance raman, and UV–visible spectroscopic analysis of a Rieske-type demethylase. *Journal of the American Chemical Society, 134*(5), 2823–2834. https://doi.org/10.1021/ja2111898.

Dunham, N. P., & Arnold, F. H. (2020). Nature's machinery, repurposed: Expanding the repertoire of iron-dependent oxygenases. *ACS Catalysis, 10*(20), 12239–12255. https://doi.org/10.1021/acscatal.0c03606.

Fei, X., Yin, X., Zhang, L., & Zabriskie, T. M. (2007). Roles of VioG and VioQ in the incorporation and modification of the Capreomycidine residue in the peptide antibiotic viomycin. *Journal of Natural Products, 70*(4), 618–622. https://doi.org/10.1021/np060605u.

Friemann, R., Lee, K., Brown, E. N., Gibson, D. T., Eklund, H., & Ramaswamy, S. (2009). Structures of the multicomponent Rieske non-heme iron toluene 2,3-dioxygenase enzyme system. *Acta Crystallographica Section D: Biological Crystallography, 65*(Pt 1), 24–33. https://doi.org/10.1107/S0907444908036524.

Gibbins, J. M. (2004). Techniques for analysis of proteins by SDS-polyacrylamide gel electrophoresis and Western blotting. *Methods in Molecular Biology, 273*, 139–152. https://doi.org/10.1385/1-59259-783-1:139.

Gibson, D. T., & Parales, R. E. (2000). Aromatic hydrocarbon dioxygenases in environmental biotechnology. *Current Opinion in Biotechnology, 11*(3), 236–243. https://doi.org/10.1016/S0958-1669(00)00090-2.

Higgins, T. P., Davey, M., Trickett, J., Kelly, D. P., & Murrell, J. C. (1996). Metabolism of methanesulfonic acid involves a multicomponent monooxygenase enzyme. *Microbiology (Reading), 142*(Pt 2), 251–260. https://doi.org/10.1099/13500872-142-2-251.

Higgins, T. P., De Marco, P., & Murrell, J. C. (1997). Purification and molecular characterization of the electron transfer protein of methanesulfonic acid monooxygenase. *Journal of Bacteriology, 179*(6), 1974–1979. https://doi.org/10.1128/jb.179.6.1974-1979.1997.

Hu, W. Y., Li, K., Weitz, A., Wen, A., Kim, H., Murray, J. C., ... Liu, P. (2022). Light-driven oxidative demethylation reaction catalyzed by a Rieske non-heme iron enzyme Stc2. *ACS Catalysis, 12*(23), 14559–14570. https://doi.org/10.1021/acscatal.2c04232.

Immanuel, S., Sivasubramanian, R., Gul, R., & Dar, M. A. (2020). Recent progress and perspectives on electrochemical regeneration of reduced nicotinamide adenine dinucleotide (NADH). *Chemistry-an Asian Journal, 15*(24), 4256–4270. https://doi.org/10.1002/asia.202001035.

Jeffrey, A. M., Yeh, H. J., Jerina, D. M., Patel, T. R., Davey, J. F., & Gibson, D. T. (1975). Initial reactions in the oxidation of naphthalene by *Pseudomonas putida. Biochemistry, 14*(3), 575–584. https://doi.org/10.1021/bi00674a018.

Julien, B., Tian, Z. Q., Reid, R., & Reeves, C. D. (2006). Analysis of the ambruticin and jerangolid gene clusters of *Sorangium cellulosum* reveals unusual mechanisms of polyketide biosynthesis. *Chemistry & Biology, 13*(12), 1277–1286. https://doi.org/10.1016/j.chembiol.2006.10.004.

Karlsson, A., Parales, J. V., Parales, R. E., Gibson, D. T., Eklund, H., & Ramaswamy, S. (2003). Crystal structure of naphthalene dioxygenase: Side-on binding of dioxygen to iron. *Science (New York, N. Y.), 299*(5609), 1039–1042. https://doi.org/10.1126/science.1078020.

Kweon, O., Kim, S. J., Baek, S., Chae, J. C., Adjei, M. D., Baek, D. H., ... Cerniglia, C. E. (2008). A new classification system for bacterial Rieske non-heme iron aromatic ring-hydroxylating oxygenases. *BMC Biochemistry, 9*, 11. https://doi.org/10.1186/1471-2091-9-11.

Li, W., Estrada-de los Santos, P., Matthijs, S., Xie, G. L., Busson, R., Cornelis, P., ... De Mot, R. (2011). Promysalin, a salicylate-containing *Pseudomonas putida* antibiotic, promotes surface colonization and selectively targets other *Pseudomonas*. *Chemistry & Biology, 18*(10), 1320–1330. https://doi.org/10.1016/j.chembiol.2011.08.006.

Liu, J., Knapp, M., Jo, M., Dill, Z., & Bridwell-Rabb, J. (2022). Rieske oxygenase catalyzed C-H bond functionalization reactions in chlorophyll B biosynthesis. *ACS Central Sciences, 8*(10), 1393–1403. https://doi.org/10.1021/acscentsci.2c00058.

Liu, J., Tian, J., Perry, C., Lukowski, A. L., Doukov, T. I., Narayan, A. R. H., & Bridwell-Rabb, J. (2022). Design principles for site-selective hydroxylation by a Rieske oxygenase. *Nature Communications, 13*(1), 255. https://doi.org/10.1038/s41467-021-27822-3.

Mehboob, F., Wolterink, A. F., Vermeulen, A. J., Jiang, B., Hagedoorn, P. L., Stams, A. J., & Kengen, S. W. (2009). Purification and characterization of a chlorite dismutase from *Pseudomonas chloritidismutans*. *FEMS Microbiology Letters, 293*(1), 115–121. https://doi.org/10.1111/j.1574-6968.2009.01517.x.

Neuhoff, V., Arold, N., Taube, D., & Ehrhardt, W. (1988). Improved staining of proteins in polyacrylamide gels including isoelectric focusing gels with clear background at nanogram sensitivity using Coomassie Brilliant Blue G-250 and R-250. *Electrophoresis, 9*(6), 255–262. https://doi.org/10.1002/elps.1150090603.

Ozgen, F. F., Runda, M. E., & Schmidt, S. (2021). Photo-biocatalytic cascades: Combining chemical and enzymatic transformations fueled by light. *Chembiochem: A European Journal of Chemical Biology, 22*(5), 790–806. https://doi.org/10.1002/cbic.202000587.

Park, J. H., Lee, S. H., Cha, G. S., Choi, D. S., Nam, D. H., Lee, J. H., ... Park, C. B. (2015). Cofactor-free light-driven whole-cell cytochrome P450 catalysis. *Angewandte Chemie-International Edition, 54*(3), 969–973. https://doi.org/10.1002/anie.201410059.

Perry, C., de Los Santos, E. L. C., Alkhalaf, L. M., & Challis, G. L. (2018). Rieske non-heme iron-dependent oxygenases catalyse diverse reactions in natural product biosynthesis. *Natural Product Reports, 35*(7), 622–632. https://doi.org/10.1039/c8np00004b.

Phillips, D. A., Sande, E. S., Vriezen, J. A. C., de Bruijn, F. J., Le Rudulier, D., & Joseph, C. M. (1998). A new genetic locus in *Sinorhizobium meliloti* is involved in stachydrine utilization. *Applied and Environmental Microbiology, 64*(10), 3954–3960. https://doi.org/10.1128/AEM.64.10.3954-3960.1998.

Pyser, J. B., Chakrabarty, S., Romero, E. O., & Narayan, A. R. H. (2021). State-of-the-art biocatalysis. *ACS Central Science, 7*(7), 1105–1116. https://doi.org/10.1021/acscentsci.1c00273.

Runda, M. E., de Kok, N. A. W., & Schmidt, S. (2023). Rieske oxygenases and other ferredoxin-dependent enzymes: electron transfer principles and catalytic capabilities. *Chembiochem: A European Journal of Chemical Biology, 24*(15), e202300078. https://doi.org/10.1002/cbic.202300078.

Shanmugam, M., Quareshy, M., Cameron, A. D., Bugg, T. D. H., & Chen, Y. (2021). Light-activated electron transfer and catalytic mechanism of carnitine oxidation by Rieske-type oxygenase from human microbiota. *Angewandte Chemie-International Edition, 60*(9), 4529–4534. https://doi.org/10.1002/anie.202012381.

Shimizu, K., Kato, K., & Kobayashi, T. (2023). Flow and mixing characteristics of gas-liquid slug flow in a continuous Taylor-Couette flow reactor with narrow gap width. *Chemical Engineering and Processing-Processs Intensification, 183*. https://doi.org/10.1016/j.cep.2022.109226.

Summers, R. M., Louie, T. M., Yu, C. L., Gakhar, L., Louie, K. C., & Subramanian, M. (2012). Novel, highly specific N-demethylases enable bacteria to live on caffeine and related purine alkaloids. *Journal of Bacteriology, 194*(8), 2041–2049. https://doi.org/10.1128/JB.06637-11.

Sun, S., Nicholls, B. T., Bain, D., Qiao, T., Page, C. G., Musser, A. J., & Hyster, T. K. (2023). Enantioselective decarboxylative alkylation using synergistic photoenzymatic catalysis. *Nature Catalysis*. https://doi.org/10.1038/s41929-023-01065-5.

Sydor, P. K., Barry, S. M., Odulate, O. M., Barona-Gomez, F., Haynes, S. W., Corre, C., ... Challis, G. L. (2011). Regio- and stereodivergent antibiotic oxidative carbocyclizations catalysed by Rieske oxygenase-like enzymes. *Nature Chemistry, 3*(5), 388–392. https://doi.org/10.1038/nchem.1024.

Tian, J., Garcia, A. A., Donnan, P. H., & Bridwell-Rabb, J. (2023). Leveraging a structural blueprint to rationally engineer the Rieske oxygenase TsaM. *Biochemistry, 62*(11), 1807–1822. https://doi.org/10.1021/acs.biochem.3c00150.

Uppada, V., Bhaduri, S., & Noronha, S. B. (2014). Cofactor regeneration - An important aspect of biocatalysis. *Current Science, 106*(7), 946–957. http://www.jstor.org/stable/24102378.

Wang, X., Hu, S., Wang, J., Zhang, T., Ye, K., Wen, A., ... Liu, P. (2023). Biochemical and structural characterization of OvoA$_{Th2}$: A mononuclear nonheme iron enzyme from *Hydrogenimonas thermophila* for ovothiol biosynthesis. *ACS Catalysis, 13*(23), 15417–15426. https://doi.org/10.1021/acscatal.3c04026.

Withall, D. M., Haynes, S. W., & Challis, G. L. (2015). Stereochemistry and mechanism of undecylprodigiosin oxidative carbocyclization to streptorubin B by the Rieske oxygenase RedG. *Journal of the American Chemical Society, 137*(24), 7889–7897. https://doi.org/10.1021/jacs.5b03994.

Wu, S. K., Snajdrova, R., Moore, J. C., Baldenius, K., & Bornscheuer, U. T. (2021). Biocatalysis: Enzymatic synthesis for industrial applications. *Angewandte Chemie-International Edition, 60*(1), 88–119. https://doi.org/10.1002/anie.202006648.

Xiao, Y., Zhao, Z. K., & Liu, P. (2008). Mechanistic studies of IspH in the deoxyxylulose phosphate pathway: Heterolytic C-O bond cleavage at C4 position. *Journal of the American Chemical Society, 130*(7), 2164–2165. https://doi.org/10.1021/ja710245d.

Xu, Y., Chen, H., Yu, L., Peng, X., Zhang, J., Xing, Z., ... Huang, X. (2024). A light-driven enzymatic enantioselective radical acylation. *Nature, 625*(7993), 74–78. https://doi.org/10.1038/s41586-023-06822-x.

Zhu, G., Yan, W., Wang, X., Cheng, R., Naowarojna, N., Wang, K., ... Liu, P. (2022). Dissecting the mechanism of the nonheme iron endoperoxidase FtmOx1 using substrate analogues. *JACS Au, 2*(7), 1686–1698. https://doi.org/10.1021/jacsau.2c00248.

Zwick, C. R., & Renata, H. (2020). Harnessing the biocatalytic potential of iron- and alpha-ketoglutarate-dependent dioxygenases in natural product total synthesis. *Natural Product Reports, 37*(8), 1065–1079. https://doi.org/10.1039/c9np00075e.

Zhu, W., Kumar, A., Xiong, J., Abernathy, M. J., Li, X. X., Seo, M. S., ... Nam, W. (2023). Seeing the cis-dihydroxylating intermediate: A mononuclear nonheme iron-peroxo complex in cis-dihydroxylation reactions modeling Rieske dioxygenases. *Journal of the American Chemical Society, 145*(8), 4389–4393. https://doi.org/10.1021/jacs.2c13551.

CHAPTER THIRTEEN

Functional and spectroscopic approaches to determining thermal limitations of Rieske oxygenases

Jessica Lusty Beech[a], Julia Ann Fecko[b], Neela Yennawar[b], and Jennifer L. DuBois[a,*]

[a]Department of Chemistry and Biochemistry, Montana State University, Bozeman, MT, United States
[b]The Huck Institutes of the Life Sciences, The Pennsylvania State University, University Park, PA, United States
*Corresponding author. e-mail address: jennifer.dubois1@montana.edu

Contents

1.	Introduction	300
	1.1 Multimeric mononuclear iron oxygenases and their engineering: Historical perspectives	301
	1.2 Rieske iron oxygenases and their engineering: Efforts towards substrate expansion and functional improvement	302
	1.3 Multimeric Rieske iron oxygenases and their engineering: Outline of unmet potential	304
	1.4 Overview and key parameters derived from each method	305
2.	Expression and purification of a RO system	306
	2.1 Equipment	306
	2.2 Expression and purification of TPA_{DO}	307
	2.3 Expression and purification of TPA_{RED}	309
	2.4 Notes	309
3.	Iron cofactor lability	311
	3.1 Equipment	311
	3.2 Procedure	311
	3.3 Data analysis	312
	3.4 Notes	313
4.	Temperature dependent kinetics	313
	4.1 Equipment	314
	4.2 Procedure	314
	4.3 Data analysis	314
	4.4 Notes	316
5.	Lifetime	317
	5.1 Equipment	317
	5.2 Procedure	318

Methods in Enzymology, Volume 703
ISSN 0076-6879, https://doi.org/10.1016/bs.mie.2024.05.021
Copyright © 2024 Elsevier Inc. All rights are reserved, including those for text and data mining, AI training, and similar technologies.

299

5.3 Data analysis	318
5.4 Notes	320
6. Differential scanning calorimetry (DSC)	320
6.1 Equipment	320
6.2 Procedure	321
6.3 Data analysis	322
6.4 Notes	322
References	323

Abstract

The biotechnological potential of Rieske Oxygenases (ROs) and their cognate reductases remains unmet, in part because these systems can be functionally short-lived. Here, we describe a set of experiments aimed at identifying both the functional and structural stability limitations of ROs, using terephthalate (TPA) dioxygenase (from *Comamonas* strain E6) as a model system. Successful expression and purification of a cofactor-complete, histidine-tagged TPA dioxygenase and reductase protein system requires induction with the *Escherichia coli* host at stationary phase as well as a chaperone inducing cold-shock and supplementation with additional iron, sulfur, and flavin. The relative stability of the Rieske cluster and mononuclear iron center can then be assessed using spectroscopic and functional measurements following dialysis in an iron chelating buffer. These experiments involve measurements of the overall lifetime of the system via total turnover number using both UV-Visible absorbance and HPLC analyses, as well specific activity as a function of temperature. Important methods for assessing the stability of these multi-cofactor, multi-protein dependent systems at multiple levels of structure (secondary to quaternary) include differential scanning calorimetry, circular dichroism, and metallospectroscopy. Results can be rationalized in terms of three-dimensional structures and bioinformatics. The experiments described here provide a roadmap to a detailed characterization of the limitations of ROs. With a few notable exceptions, these issues are not widely addressed in current literature.

1. Introduction

In the golden age of protein discovery, driven by sequencing and bioinformatics, there is a need to have a consistent metric to compare the stability of both structure and function within a class of like proteins. In addition, understanding the detailed limitations of an enzyme of interest allows for engineering efforts of complex multimeric, cofactor–dependent proteins using rational design (Goldenzweig & Fleishman, 2018). Understanding these limitations is especially important as these proteins are often not amenable to directed evolution approaches. Here, we will describe a set of five standard protocols that can be used to provide a metric for the functional and structural stability of complex, multimeric metalloproteins.

We use terephthalic acid dioxygenase (TPA_{DO}), a heteroheximeric Rieske oxygenase (RO) from *Comamonas* sp. E6, as a model system for which we have defined in detail the observed instabilities. The protocols give a multifaceted overview of how and when loss of protein structure leads to functional failure.

1.1 Multimeric mononuclear iron oxygenases and their engineering: Historical perspectives

The first Rieske iron sulfur cluster was described in 1964 via its unique electron paramagnetic resonance (EPR) signal at $g = 1.9$ in a sample of a reduced co-enzyme Q cytochrome C reductase (Complex III) (Rieske, Hansen, & Zaugg, 1964; Rieske, MacIennan, & Colema, 1964). The signal identified in this publication was consistent with an electron transfer mediating, iron-containing complex, but did not match the EPR signals from any existing standard. Four years later in 1968, a benzene dioxygenase was described as the first Rieske cluster containing oxygenase (Gibson, Koch, Schuld, & Kallio, 1968; Gibson, Cardini, Máseles, & Kallio, 1970). Gibson et al. proposed an enzymatic mechanism for O_2 incorporation into chlorobenzene at the expense of nicotinamide adenine dinucleotide (NADH), as observed in *Psuedomonas putida*. At the time, known mammalian hydroxylation events catalyzed by cytochrome P450s were described exclusively as a pair of *trans* monoxygenations; however, Jerina and coworkers rigorously identified the bacterial product of naphthalene-fed *Pseudomonas* sp. to be a *cis*-dihydroxylated, dearomatized 1,2-dihydroxy-1,2-dihydronaphthalene, its production catalyzed by a homologous RO, naphthalene dioxygenase. This observation suggested a new pathway to a novel enzymatic product (Jerina, Daly, Jeffrey, & Gibson, 1971).

ROs have since been identified in the genomes of over 16,000 organisms spanning all kingdoms of life. To date, there are approximately 70,000 identified ROs represented by 21 unique X-ray crystal structures (Bleem et al., 2022; Capyk, D'Angelo, Strynadka, & Eltis, 2009; Daughtry et al., 2012; Dong et al., 2005; Dumitru, Jiang, Weeks, & Wilson, 2009; Friemann et al., 2005; Furusawa et al., 2004; Hou, Guo, Li, & Zhou, 2021; Inoue et al., 2014; Jakoncic, Jouanneau, Meyer, & Stojanoff, 2007; Kauppi et al., 1997; Kim, Kim, Brooks, Kang, & Song, 2019; Kincannon et al., 2022; Kumari, Singh, Ramaswamy, & Ramanathan, 2017; Liu et al., 2022; Lukowski, Liu, Bridwell-Rabb, & Narayan, 2020; Mahto et al., 2022; Martins, Svetlitchnaia, & Dobbek, 2005; Quareshy et al., n.d.). The novel chemistry catalyzed by both mono- and di-oxygenation catalyzing ROs has the potential to fill critical needs in medical, industrial, and environmental applications including the catabolism of various toxic aromatics. RO activity has been identified on pollutants leeched from waste products including toluene, benzene, biphenyl, and phthalate (Axcell & Geary, 1975; Correll, Batiet, Ballou, & Ludwig, 1992; Furusawa et al., 2004; Gibson et al., 1970; Mahto et al., 2021; Martinez-Martinez et al., 2014; Neidle et al., 1991; Parales, Parales, Resnik, & Gibson, 1998). The potential economic and societal impact of these applications provides a driving force to discover, characterize, and improve this class of enzymes.

Structurally, ROs are characterized by the presence of a [2Fe-2S] cluster where one iron center is coordinated to the protein by two cysteine residues and the other by two histidine residues. The iron coordinated by the cysteine residues maintains a ferric state while the other passes through both ferrous and ferric states before and after electron transfer, respectively. ROs also contain a mononuclear non-heme iron center which accepts electrons from the cluster and directly catalyzes the insertion of either one or both oxygens from molecular O_2 into an organic substrate. ROs can have either homo-trimeric α_3 or heteroheximeric $\alpha_3\beta_3$ quaternary protein structures. Each comma-shaped α subunit contains both metallo-cofactors though only half of a complete active site. The head-to-tail orientation of α subunits results in a 12 Å gap between the Rieske cluster in the tail of one subunit and the mononuclear iron in the head of the adjacent subunit. The β subunits are not directly involved in catalysis and likely provide structural support to the α subunits and/or a reductase as it docks with and delivers electrons to the RO. The NAD(P)H derived electrons are delivered to the RO by either a single flavin and plant-type [2Fe-2S] cluster containing reductase, or from separate [2Fe-2S] containing ferredoxin and flavin containing NAD(P)H reductase proteins acting together. The utilization of one or both electron transfer proteins, the RO quaternary structure, and the regiospecificity of oxygenation are used to classify ROs (Fig. 1) (Bopp, Bernet, Kohler, & Hofstetter, 2022; Chakraborty, Suzuki-Minakuchi, Okada, & Nojiri, 2017; Kweon et al., 2008; Pati, Bopp, Kohler, & Hofstetter, 2022).

1.2 Rieske iron oxygenases and their engineering: Efforts towards substrate expansion and functional improvement

The publication of crystal structures from representative ROs such as naphthalene dioxygenase and nitrobenzene dioxygenase led to a variety of site-directed mutagenesis efforts in the late 1990s (Karlsson et al., 2003; Kauppi et al., 1997). The earliest efforts aimed to confirm structure-function

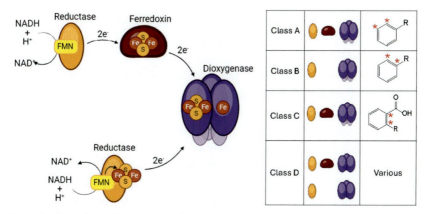

Fig. 1 (Left) Pathway for electron flow from NAD(P)H to the catalytic dioxygenase. Electron transfer proteins mediating the transfer can include a FMN containing reductase, a plant-type or Rieske type iron sulfur cluster ferredoxin, or a combined FMN and iron sulfur cluster containing reductase. (Right) General classification scheme for the four classes of ROs as defined by Chakraborty and coworkers. Classification is based on dioxygenase structure and composition of electron transport proteins.

hypotheses, including mapping electron transfer pathways and catalytically relevant active site residues (Kauppi et al., 1997; Kumari et al., 2017; Massmig et al., 2020; Parales, 2003; Tarasev, Pullela, & Ballou, 2009; Tiwari, Lee, Moon, & Zhao, 2011; Zhu et al., 2014).

As the community's understanding of structure-function relationships developed, the desire to alter existing chemistry arose. Xenobiotic aromatic catabolism is environmentally important, and it makes use of the ROs' unique catalytic capabilities. The regio- and site selectivity and catalytic efficiency of xenobiotic-catabolizing ROs became early targets for engineering, typically through targeted site-directed mutagenesis (Ang, Obbard, & Zhao, 2009; Lessner, Johnson, Parales, Spain, & Gibson, 2002; Liu et al., 2022; Lukowski et al., 2020; Özgen & Schmidt, 2019; Parales, Lee, et al., 2000; Parales, Resnick, et al., 2000). For example, a single point mutation in a biphenyl dioxygenase exhibited an increased k_{cat}/K_M activity on environmentally significant chlorinated biphenyls (Li et al., 2020). Another recent study successfully altered *p*-toluenesulfonate methyl monooxygenase (TsaM) to accept non-native substrate dicamba, a xenobiotic herbicide. Work by this group also expanded the *para*- specific regioselectivity of TsaM to oxygenate at both *meta*- and *ortho*- positions (Tian, Garcia, Donnan, & Bridwell-Rabb, 2023). Additionally, engineering efforts at the subunit interfaces have been undertaken with a view toward understanding communication between cofactor-containing sites involved in O_2 activation and/or electron transfer (Brimberry, Garcia, Liu, Tian, & Bridwell-Rabb, 2023; Tsai et al., 2022).

With notable exceptions, such as documenting the lability of the mononuclear iron cofactor, (Bernhardt & Meisch, 1980; Correll et al., 1992; Suen & Gibson, 1993; Wolfe & Lipscomb, 2003; Wolfe et al., 2002) the overall quaternary stability of ROs remains a largely undiscussed theme. The stabilities of multimeric proteins and multiprotein systems are challenging to quantify because stability is an emergent property that depends on all levels of structure concurrently. An approach to oligomers began with a modified Gibbs free energy equation including a new coefficient, K_{eff}, to describe the equilibrium constant for denaturation, expressed in monomer units (Park & Marqusee, 2004) and accounting for the equilibrium of folded and multiple potential unfolded states. Despite the large number of unfolded states, multimeric proteins are often more stable than monomeric proteins due to the increased rigidity of their globular protein structure (Balcão & Vila, 2015; Carvalho et al., 2012; Chakraborty et al., 2017; Das & Gerstein, 2000; Flores & Ellington, 2002; Gakhar et al., 2005; Goldenzweig & Fleishman, 2018; Jaenicke & Böhm, 1998; Jaenicke, Schurig, Beaucamp, & Ostendorp, 1996; Lisi, Png, & Wilcox, 2014; Park & Marqusee, 2004). Immobilization of the multimer to a solid surface can further increase stability (Arana-Pena et al., 2020; Balcão, Mateo, Fernández-Lafuente, Malcata, & Guisán, 2001; Hobisch et al., 2021; Kumar, 2010; Thompson, Peñafiel, Cosgrove, & Turner, 2018).

1.3 Multimeric Rieske iron oxygenases and their engineering: Outline of unmet potential

Studies characterizing ROs frequently report catalytic efficiency (k_{cat}/K_M) or degrees of uncoupling of O_2 reduction from substrate oxygenation (Ang et al., 2009; Kincannon et al., 2022; Runda, Miao, de Kok, & Schmidt, 2024). However, overall lifetime, turnover number, and structural stability measurements are remarkably absent. The potential use of ROs in bioremediation is a beacon application for this class, and further applications in organic synthesis have been envisioned (Capyk et al., 2009; Friemann et al., 2005; Kumar et al., 2011; Liu et al., 2022; Lukowski et al., 2020; Zhu et al., 2014). One group led by Hudlicky has begun the insertion of ROs into biosynthetic pathways for morphine derivatives and antifungals (Hudlicky & Reed, 2009; Hudlicky, Fan, Luna, Olivo, & Price, 1992; Hudlicky, Gonzalez, & Gibson, 1999; Özgen & Schmidt, 2019), but the utilization of these enzymes is far below their projected potential. Practical application of any enzyme system requires some understanding of its lifetime and efficiency, which may need to be expanded by engineering efforts. Here, we outline an experimental roadmap which can be used to identify several relevant measures of stability for ROs and other multimeric metalloenzymes. While loss of structure can be induced with either heat or chemicals, we have focused on heat as it can be administered progressively and in many experimental contexts. The results from these experiments can be used to develop stability models for the class as a whole and as a foundation on which future stability engineering efforts can be built.

In addition to these methods, we have recently published a paper (Lusty Beech et al., 2024) adding two more structural studies, small-angle X-ray scattering (SAXS) and circular dichroism (CD). SAXS allows users to observe the solution-state globular structure of the protein. When recorded as a function of increasing temperature, SAXS can identify the physical melting curve of a protein and can potentially predict the mechanism through which the protein dissociates. This method, however useful, is a specialized technique not readily available to most users and is consequently left out of this chapter. CD measures the relative content of secondary protein structures in a sample. As protein structure is lost to temperature, the change in composition of α-helices and β-sheets can be used to predict where unfolding occurs, further detailing the mechanism of dissociation. CD is commonly used in studies characterizing protein stability, so we have chosen not to elaborate on that method in this chapter.

1.4 Overview and key parameters derived from each method

Method	Sample type	Sample parameters	Information content
Atomic absorption spectroscopy	Purified dioxygenase, purified reductase	Iron content 1–20 μM	Total iron content, correlated to protein concentration to estimate upper limit on complete active site content.
UV–visible Rieske cluster analysis	Purified dioxygenase	Protein concentration 3–10 μM	Qualitative Rieske cluster content reported as a ratio of Rieske peak absorbance intensity at 450 and 550 nm to protein absorbance at 280 nm.
Steady state activity	Purified dioxygenase, purified reductase	2 μM dioxygenase, 6 μM reductase	Activity following protein incubation at elevated temperatures relative to specific activity at temperature optimum.

Lifetime	Purified dioxygenase, purified reductase	2 μM dioxygenase, 6 μM reductase	Total number of catalytic cycles each enzyme can undergo prior to loss of activity, typically irreversible; Here we outline real-time monitoring of NADH absorbance and discontinuous substrate quantification by HPLC.
DSC	Purified dioxygenase or reductase	20 μM protein	Heat capacity variation due to structural transition event as samples exposed to heat. The result of this method is a quantitative ΔH and mapping of the thermal dissociation pathway for the protein.

2. Expression and purification of a RO system
2.1 Equipment
- Shaker (Innova 44, New Brunswick Scientific)
- 2.8 L Fernbach Flask
- UV-Vis spectrophotometer (Cary 60, Agilent)
- Centrifuge (Beckman)
- Sonicator (Branson)
- Fast protein liquid chromatography (FPLC) (Akta)
- Electrophoresis system (EPS 3500, Pharmacia)
- Amicon centrifugal protein concentration devices (Sigma)
- Desalting column (PD-10, Cytiva)

2.1.1 Buffers, strains, and reagents
- Lemo (DE3) competent *E. coli* cells (NEB) transformed with pETDuet plasmid (Genscript) containing RO α and β (TPA$_{DO}$) (Kincannon et al., 2022) (see note 1)
- Lemo (DE3) competent cells (NEB) transformed with pET 45b(+) plasmid (Genscript) containing RO reductase (TPA$_{RED}$) (Kincannon et al., 2022)
- Terrific Broth (TB) medium
- Ampicillin 50 mg mL^{-1} in water
- Chloramphenicol 34 mg mL^{-1} in ethanol
- Rhamnose 1 M in water
- Isopropyl β-d-1-thiogalactopyranoside (IPTG) 1 M in water
- FeCl$_3$ 1 M in water

- l-cysteine $10\,mg\,mL^{-1}$ in water
- Protease inhibitor phenylmethylsufonyl fluoride
- Chicken egg white lysozyme
- Buffer A (20 mM tris(hydroxymethyl)aminomethane (Tris) pH 8.0, 150 mM NaCl, 10% v:v glycerol)
- Buffer B (20 mM Tris pH 8.0, 150 mM NaCl, 10% v:v glycerol, 250 mM imidazole)
- Buffer C (20 mM Tris pH 8.0, 300 mM NaCl, 15% v:v glycerol)
- Buffer D (20 mM Tris pH 8.0, 300 mM NaCl, 15% v:v glycerol, 250 mM imidazole)
- Ni-NTA agarose resin (McLAB) in FPLC compatible glass column (XK 16/20, Cytiva)

2.2 Expression and purification of TPA$_{DO}$

1. *E. coli* glycerol stocks containing the plasmids with the cloned genes are taken from $-80\,°C$ freezer and kept in a Cryo-Safe box.
2. Use a flame-sterilized steel loop to gently scrape the surface of a glycerol stock. Inoculate an LB agar plate containing $50\,mg\,L^{-1}$ ampicillin and $34\,mg\,L^{-1}$ chloramphenicol. Incubate for $8-16\,h$ at $37\,°C$.
3. The following day, transfer a single colony from the plate to a flask containing 50 mL TB supplemented with $50\,mg\,L^{-1}$ ampicillin and $34\,mg\,L^{-1}$ chloramphenicol (1:1000 dilution from stock). Incubate overnight (16–18 h) at $37\,°C$ with constant shaking at 200 RPM.
4. The next morning, inoculate a 2.8 L Fernbach flask containing 500 mL sterile TB media supplemented with $50\,mg\,L^{-1}$ ampicillin, $34\,mg\,L^{-1}$ chloramphenicol (1:1000 dilution from stock), and 2 mM rhamnose.
5. Grow the cultures at $37\,°C$, 225 RPM until the optical density at 600 nm ($OD_{600\,nm}$) reaches approximately 2.5 (5–6 h) (see note 2).
6. At an $OD_{600\,nm}$ of approximately 2.5, rest cultures in an ice bath for 45 min. Meanwhile, supplement the culture with 0.1 mM $FeCl_3$, $10\,mg\,L^{-1}$ L-cysteine, and 1 mM IPTG (see note 3).
7. Return culture flasks to the incubator and grow at $25\,°C$, 200 RPM overnight (16–18 h) (see note 4).
8. The next morning, harvest cells by centrifugation at $12,000 \times g$ for 15 min, discarding supernatant (see note 5).
9. Resuspend cell pellet in Buffer A, use approximately 10 mL of buffer per gram of cell paste in a sonication safe beaker. Add $0.5\,mg\,mL^{-1}$ (each) chicken egg white lysozyme and phenylmethylsufonyl fluoride.

10. Place in a bath of ice and water. Sonicate for 20 min at 45% amplitude, pulsing 10 s on, 20 s off, while gently stirring.
11. Centrifuge the resulting lysate for 30 min at 40,000 × g at 4 °C. Discard solid cell debris.
12. Attach the glass column containing Ni-NTA resin to the FPLC and wash with 5–10 column volumes (cv) of milli-Q water.
13. Equilibrate the column with 10 cv Buffer A.
14. Using a superloop if necessary, load the clarified supernatant to the column.
15. Wash with 20 cv Buffer A, or until A280 returns to baseline.
16. To remove non-target proteins from the resin, rinse with 10 cv of 30% Buffer B. Discard eluent (see note 6).
17. Elute target protein with an isocratic wash at 100% Buffer B. Collect 2–5 mL fractions.
18. Monitor fractions by sodium dodecyl sulfate–polyacrylamide gel electrophoresis (SDS-PAGE) (Fig. 2).
19. Concentrate pooled pure protein fractions to 5 mL using an Amicon centrifugal filtration device with an appropriate molecular weight cutoff.
20. Remove imidazole from the protein mixture using a PD-10 desalting column (Cytiva) under gravity flow conditions, following the manufacturer's protocol. Here, PD-10 columns were equilibrated using

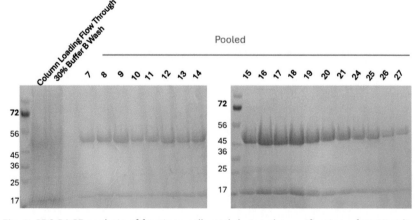

Fig. 2 SDS-PAGE analysis of fractions collected during the purification of 6× Histidine tagged TPA$_{DO}$ seen as bands at 48 and 17 kDa. Fractions (5 mL) were collected during the elution at 100% Buffer B. Fractions 8 through 27 were determined to be of sufficient purity (>95%), pooled together, desalted, concentrated, and stored.

25 mL of Buffer A. Protein was introduced to the equilibrated column in 2.5 mL increments, then was eluted from the column with 3.5 mL Buffer A.

21. Following desalting, protein is concentrated once more using an Amicon centrifugal filtration device until proteins are approximately $10 \, \text{mg} \, \text{mL}^{-1}$. Purified protein is then aliquoted (100 μL), flash frozen in liquid N_2, and stored at $-80\,°C$.

2.3 Expression and purification of TPA$_{RED}$

1. Starter cultures are prepared as for TPA$_{DO}$ and inoculated into 2.8 L Fernbach flasks. *E. coli* cells containing the TPA$_{RED}$ expression plasmid are cultivated and protein expression is induced in the same manner as for TPA$_{DO}$, with the addition of $0.5 \, \text{mg} \, \text{mL}^{-1}$ riboflavin to the cultures before induction. For purification:
2. Attach the glass column containing Ni–NTA resin to the FPLC and wash with 5–10 cv milli-Q water.
3. Equilibrate the column with 10 cv of Buffer C.
4. Using a superloop if necessary, load the clarified supernatant to the column.
5. Wash with 20 cv Buffer C, or as much is necessary such that the absorbance at 280 nm (A$_{280}$) returns to baseline. Discard eluent.
6. Nontarget protein and TPA$_{RED}$ are separated by a linear gradient elution from 0 to 50% Buffer D over 8 cv.
7. Determine purity of fractions through SDS-PAGE and pool only uncontaminated fractions (Fig. 3).
8. Pooled protein fractions are then concentrated to a final volume of 5 mL using an Amicon centrifugal filtration device with an appropriate molecular weight cutoff.
9. Imidazole is removed from the protein mixture using a PD-10 column (Cytiva) under gravity flow conditions, following the manufacturer's protocol.
10. Following desalting, protein is concentrated once more using an Amicon centrifugal filtration device until proteins are approximately $15 \, \text{mg} \, \text{mL}^{-1}$. Purified protein is then aliquoted into 100 μL aliquots, flash frozen in liquid N_2, and stored at $-80\,°C$.

2.4 Notes

1. For TPADO, a TEV-cleavable C-terminal 6× Histidine tag was appended to the α-subunit (TphA2, BAE47077), while the β-subunit

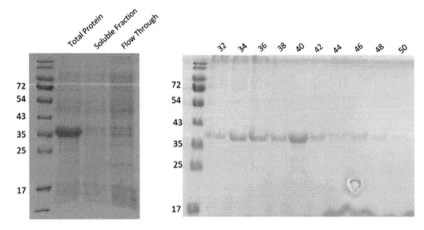

Fig. 3 SDS-PAGE analysis of fractions collected during the purification of 6× Histidine tagged TPA$_{RED}$ seen as a single band at 37 kDa. Fractions (5 mL) were collected during the linear elution 0–50% Buffer D over 8 cv. Fractions 32 through 48 were determined to be of sufficient purity (>95%), pooled together, desalted, concentrated, and stored.

(TphA3, BAE47078) was untagged. For TPA$_{RED}$ (TphA1, BAE47080) a C-terminal 6× Histidine tag was appended (Kincannon et al., 2022).

2. Traditional induction of protein expression occurs at an OD$_{600\,nm}$ of 0.3–0.6; however, induction at a stationary-phase OD$_{600\,nm}$ results in a higher yield of soluble, cofactor complete protein.
3. The process of exposing growth cultures to an ice bath encourages expression of cold-shock chaperones which has been shown to increase the yield of properly folded protein.
4. The Lemo(DE3) system uses rhamnose to encourage the production of a T7 polymerase inhibitor that can be used to adjust the levels of target protein expression. A small-scale preliminary expression test while varying the concentration of rhamnose can be used to determine optimum expression and solubility conditions for your protein.
5. Cell mass can be flash-frozen in liquid N$_2$ and stored at −80 °C for future use. Under these conditions, TPA$_{DO}$ protein typically yields between 20 and 25 g wet cell mass per L culture.
6. The expression system used here includes a 6× Histidine tag available to interact with the affinity column at each of the three α subunits found in TPA$_{DO}$. Other expression systems may require alternative concentrations of Buffer B to remove impurities while allowing the target protein to remain bound to the column.

3. Iron cofactor lability

Metalloproteins rely not only on the preservation of peptide structure, but also on the integrity of the cofactors. Here, we evaluate the relative stability of the coordination of both the Rieske [2Fe-2S] cluster and the mononuclear non-heme iron of TPA_{DO}. By using dialysis against a chelating agent ethylenediaminetetraacetic acid (EDTA), weakly bound iron ions are stripped from the protein. UV-Vis, atomic absorption (AA), and kinetic measurements are used to quantify how much iron was lost, of which type, and to what degree function was compromised.

3.1 Equipment

UV-Visible Spectrophotometer (Cary 60, Agilent).
Atomic Absorption spectrophotometer (SpectrAA 220 Fast Sequential Atomic Absorption Spectrometer, Varian).
Slide-a-lyzer (ThermoScientific).
Buffers and Reagents.
Iron chelating buffer (20 mM Tris, pH 8.0, 150 mM NaCl, 10% v:v glycerol, 5 mM EDTA).
TPA_{DO} storage buffer (20 mM Tris, pH 8.0, 150 mM NaCl, 10% v:v glycerol).
2x Reaction master mix (40 mM (3-(N-morpholino)propanesulfonic acid (MOPS)), pH 7.2, 200 μM NaCl, 20% v-v DMSO, 400 μM NADH).
Terephthalic acid (TPA) (500 μM, in dimethyl sulfoxide (DMSO)).

3.2 Procedure

1. To determine iron content for a newly purified TPA_{DO} sample, first atomic absorption (AA) is used. A series of TPA_{DO} samples at 0.5, 1, and 2 μM TPA_{DO} (approximately 4.5, 9, and 18 μM iron, respectively, assuming each α subunit contains a single [2Fe-2S] cluster and single mononuclear non-heme iron) are prepared. Samples are acidified with 20% nitric acid, incubated at > 90 °C for 10 min, then filtered through a 0.22 μm filter. Absorption measurements for each sample are compared to a standard curve to determine total iron-per-protein concentration.
2. The relative concentrations of Fe-S clusters are observed with UV-Visible spectroscopy. Rieske clusters have a characteristic spectrum with two peaks, one at 450 nm and one at 550 nm. TPA_{DO} is diluted to 3, 6, and 12 μM into storage buffer (20 mM Tris, pH 8.0, 150 mM NaCl, 10% v-v glycerol) in a quartz cuvette and spectra (280–700 nm) are measured (see note 1, Fig. 4A).

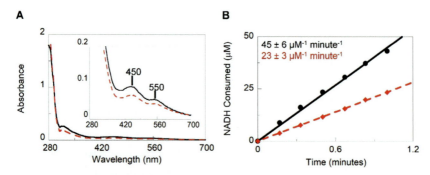

Fig. 4 EDTA removal of labile iron from TPA$_{DO}$. (A) Qualitative UV-visible absorbance spectra of the Rieske cluster consistent with that published in previous literature[2,67,68] indicate two distinct peaks with λ_{max} at 450 and 550 nm (inset). Relative concentrations for the as-isolated (solid black line) and EDTA dialyzed (red dashed line) samples can be referenced to the protein absorbance at 280 nm (B). Specific activity for as-isolated and post-dialysis proteins.

3. Finally, TPA$_{DO}$ activity is measured. To a cuvette containing 150 μM 2× reaction master mix, 50 μL TPA is added. A series of measurements monitoring NADH concentration at 340 nm is measured over time to determine background NADH oxidation. Finally, 50 μL TPA$_{DO}$ (final concentration 2 μM protein) and TPA$_{RED}$ (final concentration 6 μM protein) are added. Absorbance changes at 340 nm were measured every 10 s for 10 min total.
4. To remove highly labile iron, 300 μL TPA$_{DO}$ (10 mg mL^{-1}) are added to a slide-a-lyzer dialysis cassette. The cassette was placed in a 2 L beaker containing iron chelation buffer and gently stirred at 4 °C for 16 h.
5. EDTA is removed by a second dialysis into TPA$_{DO}$ storage buffer.
6. A Bradford assay of the EDTA-treated protein is used to confirm the new concentration (Noble, 2014).
7. Atomic Absorption, UV-Vis, and activity assays are conducted following identical protocols to pre-EDTA treatment.

3.3 Data analysis

1. UV-Vis measurements are a qualitative method for comparing relative concentrations of [2Fe-2S] clusters. After spectra have been obtained for both the as-isolated and post-dialysis samples, A_{450}/A_{280} and A_{550}/A_{280} values can be calculated as an indication of cofactor concentration per concentration of protein. A percentage of signal remaining after dialysis can be reported by dividing the post-dialysis ratios by the as-isolated ratios (Fig. 4A).

Thermal limitations of Rieske oxygenases 313

2. These measurements can be used with the quantitative AA analysis to report if any of the iron lost due to EDTA treatment belongs to the Fe-S cluster. Ideally, the Fe-S clusters, which are more strongly bound than the mononuclear iron, should remain intact while the mononuclear Fe is chelated out of the protein (Fig. 4B).

3. Activity measurements assess whether fully intact active sites (including the mononuclear Fe) remain post-dialysis, since both the mononuclear Fe and adjacent Rieske cluster are necessary for catalysis. Beer's law can be used to convert the absorbance at 340 nm to the concentration of NADH. The change in NADH concentration measured continuously or discontinuously over a short, linear time interval at the initial portion of the reaction gives the activity, which can be referenced to [protein]:

$$Concentration\,(\mu m) = \frac{Absorbance}{Pathlength\,(cm)\cdot\varepsilon\,(\mu m^{-1}\,cm^{-1})} \tag{1}$$

Where concentration is calculated by dividing the absorbance at 340 nm by the multiplication product of the pathlength of the cuvette in centimeters and the extinction coefficient of NADH at 340 nm in units of $\mu m^{-1}\,cm^{-1}$ ($0.00622\,\mu m^{-1}\,cm^{-1}$).

These concentrations can then be plotted as a function of time to give initial velocities. Additionally, the total NADH consumed during the 10 min duration can be calculated using Eq. 2.

$$\Delta NADH = [NADH]_{initial} - [NADH]_n \tag{2}$$

Comparing these calculations for the pre- and post-EDTA treatment can provide a measure to the overall lability of both metallo-cofactors.

3.4 Notes

Use of glycerol in a quartz cuvette can cause wetting of the cuvette surface that is not removed by a standard water and ethanol rinse. An additional soak with a commercially available cuvette specific detergent or 0.1 M NaOH followed by rigorous washes with deionized water can remove the film left on the cuvette surface.

4. Temperature dependent kinetics

Function can be compromised by loss of structure, cofactor, or the dioxygenase-reductase interface as a result of thermally induced destabilization of the dioxygenase/reductase system. Measurements of initial reaction rates are conducted at increasing temperatures and referenced to a normalized temperature that is optimal for the enzyme and/or the host organism from which it derives.

4.1 Equipment

- Plate reader (VarioSkan, Thermo Scienfic; or H1M1, Biotek) (see note 1)
- 96 well plates (Thermo Scientific)
- Thermocycler (Mastercycler EPGradient S, Eppendorf)

4.1.1 Buffers and reagents

- 20 mM MOPS, pH 7.2
- 2X Reaction master mix Blank (40 mM MOPS, pH 7.2, 300 mM NaCl, 20% v:v DMSO)
- 2X Reaction master mix (40 mM MOPS, pH 7.2, 300 mM NaCl, 20% v:v DMSO, 400 μM NADH) (see note 2)
- TPA, 2 mM (see note 3)
- TPA_{DO}, 10 μM active site, in 20 mM MOPS, pH 7.2
- TPA_{RED}, 30 μM active site, in 20 mM MOPS, pH 7.2

4.2 Procedure

1. Divide enzymes into 200 μL aliquots, one for each temperature being evaluated.
2. Place a single aliquot of protein in the thermocycler. Incubate at room temperature for 20 min. (see note 4).
3. Meanwhile, add 125 μL of the 2X reaction master mix to three wells of the 96 well plate.
4. Add 125 μL of the 2X reaction master mix blank and 20 mM MOPS to a fourth well.
5. After the protein has incubated, add 50 μL of each the reductase and RO to each reaction well.
6. Initiate the reaction by addition of 25 μL of TPA. Immediately begin a kinetic absorbance reading at 340 nm, scanning every 10 s for 10 min total (Fig. 5B).
7. Repeat all steps, adjusting the incubation to 10 min at elevated temperatures (ex. 30, 35, 40, 45 °C) followed by a 10 min re-equilibration to room temperature. (see note 5, 6). Fig. 5A shows general schematic for the experiment.

4.3 Data analysis

1. Convert absorbance into concentration using the Beer Lambert law shown in Eq. 1 and the following parameters: extinction coefficient

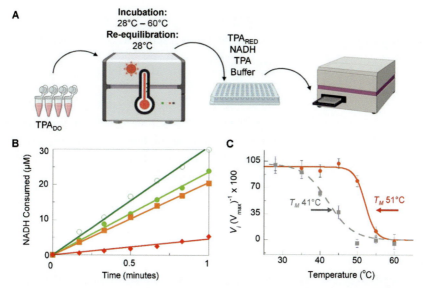

Fig. 5 Determination of functional thermotolerance. (A) Scheme for measuring temperature-dependent function. TPA$_{DO}$ was diluted into a working condition into MOPS buffer then incubated in a thermocycler for 10 min at a series of elevated temperatures. The temperature in the thermocycler was then dropped back down to the reaction temperature (28 °C) for an additional 10 min. The incubated protein was added to reaction wells containing TPA$_{RED}$, substrates NADH and TPA, and a buffer. Reactions were monitored in real-time by the change in NADH $\lambda_{max} = 340$ nm. This process is repeated for each temperature and can be repeated incubating both TPA$_{DO}$ and TPA$_{RED}$. (B) Specific activity plots showing the molar consumption of NADH over time for the first minute of reaction for 28 °C (dark green, hollow circle), 40 °C (light green, filled circle), 50 °C (orange, square), and 55 °C (red, diamond). (C) Relative initial velocity (V_i) for samples when TPA$_{DO}$ is incubated (red, solid line) and when both TPA$_{DO}$ and TPA$_{RED}$ are incubated (gray, dashed line) where the functional melting temperature (T_M) is indicated by arrows.

0.00622 µM^{-1} min^{-1} for NADH at 340 nm, and pathlength of 0.779 cm for a ThermoScientific F-bottom 96 well plate filled with 250 µL solution.

2. The background NADH hydrolysis (uncoupled to TPA dioxygenation) can be subtracted from the measured time course using the value recorded for the blank well. Initial velocity (vi) is defined as the concentration of NADH consumed over time (*t*) within the initial, linear portion of the reaction (see note 7). A value for *vi* is obtained from a line

fitted to the points of the [NADH] versus time plot. Since reaction rates are defined as positive, a negative sign is inserted into Eq. 2 (see note 8):

$$vi = -\frac{\Delta[NADH]}{dt} = -\Delta[NADH]/\Delta t \tag{3}$$

3. The vi at a reference temperature (25 °C) is associated with 100% activity. Relative activities measured at higher temperatures can be calculated using Eq. 3 where Vi_T is the initial velocity at a given temperature and $Vi_{25\,°C}$ is the initial velocity at 25 °C.

$$Relative\ activity\ (\%) = \frac{Vi_T}{Vi_{25°C}} \cdot 100 \tag{4}$$

4. The resulting temperature-response curve can be modeled using Eq. 4, where Max and Min are the maximum and minimum values of the relative activity, x is the temperature, Hill is the fitted Hill coefficient, and k is the value of temperature at which the relatively activity is at 50% (T_M).

$$f(x) = Max - \frac{Max - Min}{1 + (\frac{x}{k})^{Hill}} \tag{5}$$

4.4 Notes

1. For room temperature experiments and those up to 45 °C, the VarioSkan model plate reader or those with similar features will be sufficient. At the time of writing, the BioTek Synergy H1 is the only available plate reader with temperature control up to 70 °C.

2. These concentrations have been optimized through a specific activity screen of buffer type, pH, salt concentration, and DMSO concentration. For a 250 μL reaction, 200 μM final concentration NADH nears the UV-Vis saturation limit.

3. A final reaction concentration of 200 μM for TPA since this concentration is approximately saturating and higher concentrations lead to substrate inhibition.

4. One or both of the RO system proteins (oxygenase/reductase) can be thermally incubated (Fig. 5C).

5. TPA_{DO} has a temperature optimum of 28 °C. Thermal incubation for 10 min at elevated temperatures was followed by re-equilibration to 28 °C prior to activity measurement at 28 °C.

Thermal limitations of Rieske oxygenases

6. The above protocol can additionally be used to identify the temperature optimum of a RO system by performing the kinetic measurements at various temperatures. It is noteworthy that Arrhenius effects will affect the rate of kinetics at elevated temperatures. Additional controls can be used to account for increased background hydrolysis of NADH in the absence of enzymes as well as background hydrolysis due to enzyme activity in the absence of TPA.
7. With the reported conditions, the linear portion of the reaction for TPA_{DO}/TPA_{RED} typically consists of the first 60 s. A plot of [NADH] over the 10 min duration of the measurement will assist in identifying appropriate values to use for specific activity.
8. Specific activity can be derived from this value by referencing the concentration of enzyme active site. Alternatively, values of vi can be compared at a series of temperatures using an equivalent concentration of each enzyme.

5. Lifetime

One of the most useful metrics for applications is enzyme lifetime. Here we describe two methods for determining total turnover number (TTON) for a RO system. First, NADH consumption is monitored by UV-Visible absorbance of NADH at 340 nm. This technique is limited by the saturation limit of the spectrophotometer, so TPA and NADH must be added periodically until RO activity is no longer observed. In the second method, high substrate loading is used to circumvent the need for repeated substrate addition. Here, residual substrate is quantified by HPLC.

5.1 Equipment
- UV-Vis spectrophotometer (Cary 60, Agilent)
- 2 mL cuvettes
- HPLC (LC-300, Shimadzu)
- Hypersil GOLD™ HPLC column (4.6 × 250 mm, 5 μm particle size) (ThermoScientific)

5.1.1 Buffers, strains, and reagents
- MOPS, 20 mM, pH 7.2
- Reaction master mix (24 mM MOPS, pH 7.2, 175 mM NaCl, 20% v:v DMSO)
- NADH, 8 mM (see note 1)

- TPA, 8 mM (see note 1)
- TPA_{DO}, 40 μM active site, diluted in 20 mM MOPS, pH 7.2
- TPA_{RED}, 115 μM active site, diluted in 20 mM MOPS, pH 7.2
- HPLC Buffer A (miliQ water with 0.1% trifluoracetic acid (TFA))
- HPLC Buffer B (acetonitrile with 0.1% TFA)

5.2 Procedure

5.2.1 UV-visible analysis (see note 1)

1. To a 2 mL cuvette, add 1.7 mL of reaction master mix and 100 μL of each protein.
2. Record a full spectrum for background subtraction.
3. Add 50 μL of each NADH and TPA. Begin measuring spectra every 5 min. To maintain oxygenation of the reaction, gently agitate by pipetting (see note 2).
4. When the absorbance at 340 nm reaches the baseline, add an additional 50 μL of NADH and TPA before the next absorbance reading (see note 3, Fig. 6).
5. Periodically add fresh aliquots of NADH and TPA until the absorbance does not change after new addition of NADH and TPA.

5.2.2 HPLC analysis (see note 4)

1. To a 2 mL cuvette, add 1.3 mL of reaction master mix and 100 μL of RO and reductase.
2. Add 250 μL each of NADH and TPA. Quickly mix and remove 20 μL from the reaction for a time = 0 HPLC sample. Quench the sample in 20 μL of ice cold methanol and store at −20 °C.
3. After 5 min has elapsed, remove 20 μL from the reaction for a 5 min HPLC sample. Quench the sample in 20 μL of ice cold methanol and store at −20 °C.
4. Continue this process removing and quenching HPLC samples every 10 min.

5.3 Data analysis

5.3.1 UV-visible analysis

1. Beer's law (Eq. 1) was used to convert the measured absorbance at 340 nm into NADH concentration in μM.
2. The total amount of NADH consumed was calculated by summing the added volumes and multiplying the total by the concentration of the NADH stock.
3. TTON was calculated by dividing the moles of NADH consumed by and the moles of TPA_{DO} active site.

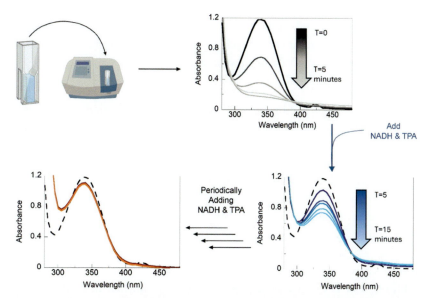

Fig. 6 UV-Visible analysis of enzyme lifetime. Function of the RO system is monitored by a decrease of signal at the $\lambda_{max} = 340$ nm. The reaction containing TPA_{DO}, TPA_{RED}, and buffer was initiated by addition of 100 µM NADH and TPA. After 5 min of continuous measurements, decreased absorbance at 340 nm indicated that the NADH had been consumed, as shown by the black absorbance spectrum. NADH and TPA were then added again in 100 µM increments, as indicated by the blue absorbance spectrum, until no further consumption of NADH was monitored after regeneration (red spectrum).

5.3.2 HPLC analysis

1. The Hypersil GOLD™ HPLC column was pre-equilibrated in 95% buffer A and 5% buffer B. The reaction components were eluted at a rate of 1 mL min^{-1} with the following program: 5% B from 0.0 to 10.0 min, 0–15% B from 10.0 to 18.0 min, 15–75% B from 18.0 to 24.0 min, 75–100% B from 24.0 to 24.1 min, 100% B from 24.1 to 30.1 min, 100–0% B from 30.1 to 30.2 min, and 0% B from 30.2 to 38.0 min. NADH (Alfa Aesar) and TPA (Sigma-Aldrich), were used to generate standard curves from 50 µM to 1 mM. Openlab PostRun software was used to generate standard curves comparing area under the peak to known substrate concentrations.

2. Substrate analysis for each sampled timepoint was reported via integration of peaks monitored at 240 nm and retention times of 16.25 and 23.1 min for NADH and TPA, respectively (Fig. 6). Sample peak areas were then plotted against the standard curve to report molar concentration.

3. As with the UV-Visible measurements, TTON was calculated by converting concentration at each segment to moles and dividing by the total moles of the TPA_{DO} enzyme active site.

5.4 Notes

1. Due to substrate inhibition, this experiment is designed to run under low substrate concentration, augmenting substrate periodically as it is consumed. For spectroscopic analysis, NADH concentration is limited to the saturation limit of the spectrophotometer.

2. Agitation with a stir bar is also acceptable.

3. This procedure assumes close coupling of the consumption of NADH and TPA; consequently, when NADH is depleted, TPA will be as well. By adding NADH and TPA directly prior to the absorbance measurement, we can confirm the new mixture's NADH concentration using Beer's Law.

4. HPLC and UV-Vis measurements can be performed in tandem; however, it is recommended that UV-Vis is performed first to determine approximate rate of substrate consumption and total reaction time.

6. Differential scanning calorimetry (DSC)

While there are several methods for monitoring loss of protein structure, DSC provides detailed quantification of the dissociation landscape (Fig. 7). DSC has long been used to characterize the thermostability of proteins and quantify their folding integrity by reporting variations in heat capacity of a protein sample. Transitions in protein complex dissociation and unfolding are indicated by peaks in heat capacity. The number of peaks, their relative resolution from other peaks, and their areas communicate information about the thermal dissociation profile for a protein sample. Each peak provides a thermal transition midpoint (T_m), and can be used to calculate the apparent enthalpy for the transition. This method can be modified in temperature range as well as the speed at which heat is introduced to the protein so that maximum resolution can be obtained for unique samples.

6.1 Equipment

- Differential scanning calorimeter (VP-capillary DSC, MicroCal)
- Slide-a-lyzer

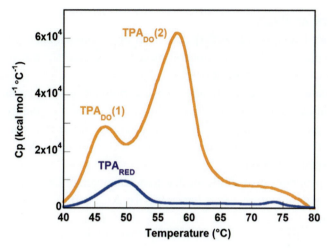

Fig. 7 The DSC plot for the thermal dissociation of TPA$_{DO}$ contains two distinct but overlapping peaks indicating a non-concerted melting pattern. The first, smaller transition peaks at 48 °C while the second, much larger transition has a peak at 58 °C. Origin software is used to deconvolute the overlapping area for the two peaks to report T_M, ΔC_p, and ΔH for each peak.

6.1.1 Buffers and reagents
- VP-Capillary DSC (See note 1)
- TPA$_{DO}$ DSC buffer (20 mM Potassium Phosphate, pH 8.0, 150 mM NaCl, 10% v:v glycerol) (See note 2)
- 400 μL TPA$_{DO}$ protein solution at 0.5 mg mL^{-1}

6.2 Procedure
1. Target protein must be first dialyzed into a DSC appropriate buffer (see note 2). This step can be done using centrifugation buffer exchange or in a small volume dialysis cassette such as a slide-a-lyzer. Following buffer exchange, protein was concentrated to a final concentration 0.5 mg mL^{-1}.
2. Prior to experimentation, both buffer and protein samples must be degassed with stirring under vacuum pressure.
3. Define the parameters for the DSC experiment. Here, we set the starting temperature to 25 °C and final temperature to 100 °C with a pre-scan incubation of 5 min. The scans performed here were taken at a rate of 90 °C h^{-1} (See note 3).
4. Triplicate measurements for the protein sample as well as the DSC buffer were taken. Background thermal changes observed in the buffer sample were subtracted from the protein samples as a reference.

6.3 Data analysis

1. Subtract background (buffer) data from experimental data.
2. Normalize the data for concentration by dividing by the number of moles of the sample in the cell.
3. To create a baseline, a software such as Origin lets you adjust the left and right linear segments and allows you to connect the segments.
4. In Origin, the data is fit by choosing a model, either two state or non-two state. The models use the Levenberg/Marquardt (LM) non-linear least-square method.
5. Some programs provide fit models to auto-integrate DSC data to report melting temperature (T_M), heat capacity ΔC_p and enthalpy (ΔH) (See note 4).
6. The raw data exported from the cap-DSC included temperature (°C), sample weight (mole), and heat flow (kcal s^{-1}). To determine enthalpy, we generated a plot of [Heat flow per weight] on the y-axis and time on the x-axis. The transition state peaks of this plot can then be integrated to report enthalpy (ΔH) in units of kcal mole^{-1}.
7. Heat capacity for the system was calculated by dividing the enthalpy by temperature. Plotting heat capacity as a function of temperature can be used to find specific heat capacity for each transition state at the T_M.

6.4 Notes

1. The VP-Capillary DSC with its autosampling, higher scan rates, and higher temperature capabilities is ideal for protein and therefore we limit our discussions in this section to this instrument, although many considerations apply to the conventional DSC as well.
2. Some buffers, such as Tris, have a low heat of ionization and a high temperature dependent pK_a and are consequently inappropriate choices for temperature dependent experiments. Additionally, the chosen buffer must be suitable for long-term stability of the protein. Here, we conducted an evaluation of activity following buffer exchange. A steady-state assay as described in Section 4 was measured for 10 min every hour for 5 h following buffer exchange. Here, we evaluated HEPES, and various concentrations of potassium phosphate (20–100 mM), using MOPS as a control. For TPA$_{DO}$, 20 mM potassium phosphate was determined to have the most maintained activity for the pH-stable buffer options.
3. The scan-rate parameter will affect the sensitivity and resolution of the DSC data. Higher scan rates (the VP-Capillary DSC can go up to 200 °C h^{-1}, the conventional equivalent has a maximum rate of 90 °C h^{-1}) offer higher

sensitivity and lower noise, but can lead to lost resolution or loss of equilibrium as compared to rates at a slow scan rate ($<60\,^{\circ}C\,h^{-1}$).
4. The DSC chromatograms for TPA_{DO} had significant overlap in the two-phase change peaks. The modeling software was used to deconvolute this overlap and give approximate peak areas for each peak.

References

Ang, E. L., Obbard, J. P., & Zhao, H. (2009). Directed evolution of aniline dioxygenase for enhanced bioremediation of aromatic amines. *Applied Microbiol Biotechnology, 81*(6), 1063–1070. https://doi.org/10.1007/s00253-008-1710-0.

Arana-Pena, S., Carballares, D., Morellon-Sterlling, R., Berenguer-Murcia, A., Alcantara, A. R., Rodrigues, R. C., & Fernandez-Lafuente, R. (2020). Enzyme co-immobilization: Always the biocatalyst designers' choice...or not? *Biotechnology Advances* 107584. https://doi.org/10.1016/j.biotechadv.2020.107584.

Axcell, B. C., & Geary, P. J. (1975). Purification and some properties of a soluble benzene-oxidizing system from a strain of Pseudomonas. *Biochemistry Journal, 146*, 173–183.

Balcão, V. M., Mateo, C., Fernández-Lafuente, R., Malcata, F. X., & Guisán, J. M. (2001). Structural and functional stabilization of L-asparaginase via multisubunit immobilization onto highly activated supports. *Biotechnology Progress, 17*(3), 537–542. https://doi.org/10.1021/bp000163r.

Balcão, V. M., & Vila, M. M. (2015). Structural and functional stabilization of protein entities: State-of-the-art. *Advances in Drug Delivery Reviews, 93*, 25–41. https://doi.org/10.1016/j.addr.2014.10.005.

Bernhardt, F.-H., & Meisch, H.-U. (1980). Reactivation studies on putidamonooxin — The monooxygenase of a 4-methoxybenzoate o-demethylase from *Pseudomonas putida*. *Biochemical and Biophysical Research Communications, 93*(4), 1247–1253. https://doi.org/10.1016/0006-291X(80)90623-3.

Bleem, A., Kuatsjah, E., Presley, G. N., Hinchen, D. J., Zahn, M., Garcia, C. D., ... Michener, J. K. (2022). Discover, characterization, and metabolic engineering of Rieske non-heme iron monooxygenases for guaiacol O-demethylation. *Chem Catalysis, 2*, 1989–2011. https://doi.org/10.1016/j.checat.2022.04.019.

Bopp, C. E., Bernet, N. M., Kohler, H.-P. E., & Hofstetter, T. B. (2022). Elucidating the role of O_2 uncoupling in the oxidative biodegredation of organic contaminants by rieske non-heme iron dioxygenases. *ACS Enviornmental, 2*(5), 428–440.

Brimberry, M., Garcia, A. A., Liu, J., Tian, J., & Bridwell-Rabb, J. (2023). Engineering Rieske oxygenase activity one piece at a time. *Current Opinions in Chemical Biology, 72*, 102227. https://doi.org/10.1016/j.cbpa.2022.102227.

Capyk, J. K., D'Angelo, I., Strynadka, N. C., & Eltis, L. D. (2009). Characterization of 3-ketosteroid 9α-hydroxylase, a Rieske oxygenase in the cholesterol degradation pathway of *Mycobacterium tuberculosis*. *Journal of Biological Chemistry, 284*(15), 9937–9946. https://doi.org/10.1074/jbc.M900719200.

Carvalho, J. W., Santiago, P. S., Batista, T., Salmon, C. E., Barbosa, L. R., Itri, R., & Tabak, M. (2012). On the temperature stability of extracellular hemoglobin of *Glossoscolex paulistus*, at different oxidation states: SAXS and DLS studies. *Biophysical Chemistry, 163-164*, 44–55. https://doi.org/10.1016/j.bpc.2012.02.004.

Chakraborty, J., Suzuki-Minakuchi, C., Okada, K., & Nojiri, H. (2017). Thermophilic bacteria are potential sources of novel Rieske non-heme iron oxygenases. *AMB Express, 7*(1), 17. https://doi.org/10.1186/s13568-016-0318-5.

Correll, C. C., Batiet, C. J., Ballou, D. P., & Ludwig, M. L. (1992). Phthalate dioxygenase reductase: A modular structure for electron transfer from pyridine nucleotides to [2Fe-2S]. *Science (New York, N. Y.), 258*, 1604–1609.

Das, R., & Gerstein, M. (2000). The stability of thermophilic proteins: A study based on comprehensive genome comparison. *Functional Integrated Genomics, 1*(1), 76–88. https://doi.org/10.1007/s101420000003.

Daughtry, K. D., Xiao, Y., Stoner-Ma, D., Cho, E., Orville, A. M., Liu, P., & Allen, K. N. (2012). Quaternary ammonium oxidative demethylation: X-ray crystallographic, resonance raman, and UV–visible spectroscopic analysis of a Rieske-type demethylase. *Journal of the American Chemical Society, 134*(5), 2823–2834. https://doi.org/10.1021/ja2111898.

Dong, X., Fushinobu, S., Fukuda, E., Terada, T., Nakamura, S., Shimizu, K., ... Wakagi, T. (2005). Crystal structure of the terminal oxygenase component of cumene dioxygenase from *Pseudomonas fluorescens* IP01. *Journal of Bacteriology, 187*(7), 2483–2490. https://doi.org/10.1128/JB.187.7.2483-2490.2005.

Dumitru, R., Jiang, W. Z., Weeks, D. P., & Wilson, M. A. (2009). Crystal structure of dicamba monooxygenase: A Rieske nonheme oxygenase that catalyzes oxidative demethylation. *Journal of Molecular Biology, 392*(2), 498–510. https://doi.org/10.1016/j.jmb.2009.07.021.

Flores, H., & Ellington, A. D. (2002). Increasing the thermal stability of an oligomeric protein, beta-glucuronidase. *Journal of Molecular Biology, 315*(3), 325–337. https://doi.org/10.1006/jmbi.2001.5223.

Friemann, R., Ivkovic-Jensen, M. M., Lessner, D. J., Yu, C. L., Gibson, D. T., Parales, R. E., ... Ramaswamy, S. (2005). Structural insight into the dioxygenation of nitroarene compounds: The crystal structure of nitrobenzene dioxygenase. *Journal of Molecular Biology, 348*(5), 1139–1151. https://doi.org/10.1016/j.jmb.2005.03.052.

Furusawa, Y., Nagarajan, V., Tanokura, M., Masai, E., Fukuda, M., & Senda, T. (2004). Crystal structure of the terminal oxygenase component of biphenyl dioxygenase derived from Rhodococcus sp. strain RHA1. *Journal of Molecular Biololgy, 342*(3), 1041–1052. https://doi.org/10.1016/j.jmb.2004.07.062.

Gakhar, L., Malik, Z. A., Allen, C. C., Lipscomb, D. A., Larkin, M. J., & Ramaswamy, S. (2005). Structure and increased thermostability of Rhodococcus sp. naphthalene 1,2-dioxygenase. *Journal of Bacteriology, 187*(21), 7222–7231. https://doi.org/10.1128/JB.187.21.7222-7231.2005.

Gibson, D. T., Cardini, G. E., Máseles, F. C., & Kallio, R. E. (1970). Incorporation of oxygen-18 into benzene by *Pseudomonas putida. Biochemistry, 9*, 1631–1635.

Gibson, D. T., Koch, J. R., Schuld, C. L., & Kallio, R. E. (1968). Oxidative degradation of aromatic hydrocarbons by microorganisms. II. Metabolism of halogenated aromatic hydrocarbons. *Biochemistry, 7*, 3795–3802.

Goldenzweig, A., & Fleishman, S. J. (2018). Principles of protein stability and their application in computational design. *Annual Reviews in Biochemistry, 87*, 105–129. https://doi.org/10.1146/annurev-biochem-062917-012102.

Hobisch, M., Holtmann, D., Gomez de Santos, P., Alcalde, M., Hollmann, F., & Kara, S. (2021). Recent developments in the use of peroxygenases - Exploring their high potential in selective oxyfunctionalisations. *Biotechnology Advances, 51*, 107615. https://doi.org/10.1016/j.biotechadv.2020.107615.

Hou, Y.-J., Guo, Y., Li, D.-F., & Zhou, N.-Y. (2021). Structural and biochemical analysis reveals a distinct catalytic site of salicylate 5-monooxygenase NagGH from Rieske dioxygenases. *Applied and Environmental Microbiology, 87*(6), e01629–01620. https://doi.org/10.1128/AEM.01629-20.

Hudlicky, T., Fan, R., Luna, H., Olivo, H., & Price, J. (1992). Enzymatic hydroxylation of arene and symmetry considerations in efficient synthetic design of oxygenated natural products. *Pure and Applied Chemistry, 64*(8), 1109–1113. https://doi.org/10.1351/pac199264081109 (Pure and Applied Chemistry).

Hudlicky, T., Gonzalez, D., & Gibson, D. (1999). Enzymatic dihydroxylation of aromatics in enantioselective synthesis: Expanding asymmetric methodology. *Aldrichimica acta, 32*, 35–62.

Hudlicky, T., & Reed, J. W. (2009). Applications of biotransformations and biocatalysis to complexity generation in organic synthesis. *Chemical Society Reviews, 38*(11), 3117–3132. https://doi.org/10.1039/B901172M.

Inoue, K., Usami, Y., Ashikawa, Y., Noguchi, H., Umeda, T., Yamagami-Ashikawa, A., ... Nojiri, H. (2014). Structural basis of the divergent oxygenation reactions catalyzed by the Rieske nonheme iron oxygenase carbazole 1,9a-dioxygenase. *Applied and Environmental Microbiology, 80*(9), 2821–2832. https://doi.org/10.1128/AEM.04000-13.

Jaenicke, R., & Böhm, G. (1998). The stability of proteins in extreme environments. *Current Opinions in Structural Biology, 8*(6), 738–748. https://doi.org/10.1016/S0959-440X(98)80094-8.

Jaenicke, R., Schurig, H., Beaucamp, N., & Ostendorp, R. (1996). Structure and stability of hyperstable proteins: Glycolytic enzymes from hyperthermophilic bacterium thermotoga maritima. In F. M. Richards, D. S. Eisenberg, & P. S. Kim (Vol. Eds.), *Advances in protein chemistry: 48*, (pp. 181–269). Academic Press. https://doi.org/10.1016/S0065-3233(08)60363-0.

Jakoncic, J., Jouanneau, Y., Meyer, C., & Stojanoff, V. (2007). The catalytic pocket of the ring-hydroxylating dioxygenase from Sphingomonas CHY-1. *Biochemical and Biophysical Research Communications, 352*(4), 861–866. https://doi.org/10.1016/j.bbrc.2006.11.117.

Jerina, D. M., Daly, J. W., Jeffrey, A. M., & Gibson, D. T. (1971). Cis-1,2-dihydroxy-1,2-dihydronaphthalene: A bacterial metabolite from naphthalene. *Archives of Biochemistry and Biophysics, 142*(1), 394–396.

Karlsson, A., Parales, J. V., Parales, R. E., Gibson, D. T., Eklund, H., & Ramaswamy, S. (2003). Crystal structure of naphthalene dioxygenase: Side-on binding of dioxygen to iron. *Science (New York, N. Y.), 299*.

Kauppi, B., Lee, K., Carredano, E., Parales, R. E., Gibson, D. T., Eklund, H., & Ramaswamy, S. (1997). Structure of an aromatic-ring-hydroxylating dioxygenase—Naphthalene 1,2-dioxygenase. *Structure (London, England: 1993), 6*, 571–586.

Kim, J. H., Kim, B. H., Brooks, S., Kang, S. Y., & Song, H. K. (2019). Structrual and mechanistic insights into caffeine degredatio by the bacterial N-demethylase complex. *Journal of Molecular Biology, 431*, 3647–3661. https://doi.org/10.1016/j.jmb.2019.08.004.

Kincannon, W. M., Zahn, M., Clare, R., Lusty Beech, J., Romberg, A., Larson, J., DuBois, J. L. (2022). Biochemical and structural characterization of an aromatic ring-hydroxylating dioxygenase for terephthalic acid catabolism. *Proceedings of the National Academy of Sciences of the United States of America, 119*(13), e2121426119. https://doi.org/10.1073/pnas.2121426119.

Kumar, S. (2010). Engineering cytochrome P450 biocatalysts for biotechnology, medicine and bioremediation. *Expert Opinions on Drug Metabolism and Toxicology, 6*(2), 115–131. https://doi.org/10.1517/17425250903431040.

Kumar, P., Mohammadi, M., Viger, J. F., Barriault, D., Gomez-Gil, L., Eltis, L. D., ... Sylvestre, M. (2011). Structural insight into the expanded PCB-degrading abilities of a biphenyl dioxygenase obtained by directed evolution. *Journal of Molecular Biology, 405*(2), 531–547. https://doi.org/10.1016/j.jmb.2010.11.009.

Kumari, A., Singh, D., Ramaswamy, S., & Ramanathan, G. (2017). Structural and functional studies of ferredoxin and oxygenase components of 3-nitrotoluene dioxygenase from Diaphorobacter sp strain DS2 (Article) *PLoS One, 12*(4), e0176398. https://doi.org/10.1371/journal.pone.0176398.

Kweon, O., Kim, S. J., Baek, S., Chae, J. C., Adjei, M. D., Baek, D., ... Cerniglia, C. E. (2008). A new classification system for bacterial Rieske non-heme iron aromatic ring-hydroxylating oxygenases. *BMC Biochemistry, 9*, 11. https://doi.org/10.1186/1471-2091-9-11.

Lessner, D. J., Johnson, G. R., Parales, R. E., Spain, J. C., & Gibson, D. T. (2002). Molecular characterization and substrate specificity of nitrobenzene dioxygenase from Comamonas sp. strain JS765. *Applied Environmental Microbiology, 68*(2), 634–641. https://doi.org/10.1128/AEM.68.2.634-641.2002.

Li, J., Min, J., Wang, Y., Chen, W., Kong, Y., Guo, T., Mahto, J. K., Sylvestre, M., & Hu, X. (2020). Engineering *Burkholderia xenovorans* LB400 BphA through site-directed mutagenesis at position 283. *Applied and Environmental Microbiology, 86*(19), e01040-20. https://doi.org/10.1128/AEM.01040.

Lisi, G. P., Png, C. Y. M., & Wilcox, D. E. (2014). Thermodynamic contributions to the stability of the insulin hexamer. *Biochemistry, 53*(22), 3576–3584. https://doi.org/10. 1021/bi401678n.

Liu, J., Tian, J., Perry, C., Lukowski, A. L., Doukov, T. I., Narayan, A. R. H., & Bridwell-Rabb, J. (2022). Design principles for site-selective hydroxylation by a Rieske oxygenase. *Nature Communications, 13*(1), 255. https://doi.org/10.1038/s41467-021-27822-3.

Lukowski, A. L., Liu, J., Bridwell-Rabb, J., & Narayan, A. R. H. (2020). Structural basis for divergent C-H hydroxylation selectivity in two Rieske oxygenases. *Nature Communications, 11*(1), 2991. https://doi.org/10.1038/s41467-020-16729-0.

Lusty Beech, J., Maurya, A. K., da Silva, R., Akpoto, E., Asundi, A., Fecko, J. A., ... DuBois, J. L. (2024). Understanding the stability of a plastic-degrading Rieske iron oxidoreductase system. *Protein Science, 333*(6), e4996. https://doi.org/10.1002/pro.4997.

Mahto, J. K., Neetu, N., Sharma, M., Dubey, M., Vellank, B. P., & Kumar, P. (2022). Structural insights into dihydroxylation of terephthalate, a product of polyethylene terephthalate degradation. *Journal of Bacteriology, 204*(3).

Mahto, J. K., Neetu, N., Waghmode, B., Kuatsjah, E., Sharma, M., Sircar, D., ... Kumar, P. (2021). Molecular insights into substrate recognition and catalysis by phthalate dioxygenase from Comamonas testosteroni. *Journal of Biological Chemistry, 297*(6), 101416. https://doi.org/10.1016/j.jbc.2021.101416.

Martinez-Martinez, M., Lores, I., Pena-Garcia, C., Bargiela, R., Reyes-Duarte, D., Guazzaroni, M. E., ... Ferrer, M. (2014). Biochemical studies on a versatile esterase that is most catalytically active with polyaromatic esters. *Microbial Biotechnology, 7*(2), 184–191. https://doi.org/10.1111/1751-7915.12107.

Martins, B. M., Svetlitchnaia, T., & Dobbek, H. (2005). 2-Oxoquinoline 8-monooxygenase oxygenase component: Active site modulation by Rieske-[2Fe-2S] center oxidation/reduction. *Structure (London, England: 1993), 13*(5), 817–824. https://doi.org/ 10.1016/j.str.2005.03.008.

Massmig, M., Reijerse, E., Krausze, J., Laurich, C., Lubitz, W., Jahn, D., & Moser, J. (2020). Carnitine metabolism in the human gut: Characterization of the two-component carnitine monooxygenase CntAB from *Acinetobacter baumannii. Journal of Biological Chemistry, 295*(37), 13065–13078. https://doi.org/10.1074/jbc.RA120.014266.

Neidle, E. L., Hartnett, C., Ornston, L. N., Bairoch, A., Rekik, M., & Harayama, S. (1991). Nucleotide sequences of the *Acinetobacter calcoaceticus* benABC genes for benzoate 1,2 diozygenase reveal evolutionary relationships among multicomponent oxygenases. *Journal of Bacteriology, 173*(17), 5385–5395. https://doi.org/10.1128/jb.173.17.5385-5395.1991.

Noble, J. E. (2014). Quantification of protein concentration using UV absorbance and Coomassie dyes. *Methods in Enzymology, 536*, 17–26. https://doi.org/10.1016/B978-0-12-420070-8.00002-7.

Özgen, F. F., & Schmidt, S. (2019). Rieske non-heme iron dioxygenases: Applications and future perspectives. *In Biocatalysis*, 57–82. https://doi.org/10.1007/978-3-030-25023-2_4.

Parales, R. E. (2003). The role of active-site residues in naphthalene dioxygenase. *Journal of Industrial Microbiology and Biotechnology, 30*(5), 271–278. https://doi.org/10.1007/ s10295-003-0043-3.

Parales, R. E., Lee, K., Resnick, S. M., Jiang, H., Lessner, D. J., & Gibson, D. T. (2000). Substrate specificity of naphthalene dioxygenase: Effect of specific amino acids at the active site of the enzyme. *Journal of Bacteriology, 82*, 1641–1649.

Parales, J. V., Parales, R. E., Resnik, S. M., & Gibson, D. T. (1998). Enzyme specificity of 2-nitrotoluene 2,3-dioxygenase from Pseudomonas sp. strain JS42 is determined by the C-terminal region of the α subunit of the oxygenase component. *Journal of Bacteriology, 180*, 1194–1199.

Parales, R. E., Resnick, S. M., Sharma, N. D., Yu, C.-L., Boyd, D. R., & Gibson, D. T. (2000). Regioselectivity and enantioselectivity of naphthalene dioxygenase during arene cis-dihydroxylation: Control by phenylalanine 352 in the alpht subunit. *Journal of Bacteriology, 182*, 5495–5504.

Park, C., & Marqusee, S. (2004). Analysis of the stability of multimeric proteins by effective DeltaG and effective m-values. *Protein Science: A Publication of the Protein Society, 13*(9), 2553–2558. https://doi.org/10.1110/ps.04811004.

Pati, S. G., Bopp, C. E., Kohler, H.-P. E., & Hofstetter, T. B. (2022). Substrate-specific coupling of O_2 activation to hydroxylations of aromatic compounds by rieske non-heme iron diozygenases. *ACS Catalysis, 12*, 6444–6456.

Quareshy, M., Shanmugam, M., Townsend, E., Jameson, E., Bugg, T. D. H., Cameron, A. D., & Chen, Y. (n.d.). Structural basis of carnitine monooxygenase CntA substrate specificity, inhibition, and intersubunit electron transfer. *Journal of Biological Chemistry, 296*, 100038. https://doi.org/10.1074/jbc.RA120.016019.

Rieske, J. S., Hansen, R. E., & Zaugg, W. S. (1964). Studies on the electron transfer system. *Journal of Biological Chemistry, 239*(9), 3017–3022. https://doi.org/10.1016/s0021-9258(18)93846-9.

Rieske, J. S., MacIennan, D. H., & Colema, R. (1964). Isolation and properties of an iron-protein from the (reduced coenzyme Q)-cytochrom C reductase complex of the respiration chain. *Biochemical and Biophysical Research Communications, 15*, 338–344.

Runda, M. E., Miao, H., de Kok, N. A. W., & Schmidt, S. (2024). Developing hybrid systems to address oxygen uncoupling in multi-component Rieske oxygenases. *Journal of Biotechnology, 389*, 22–29.

Suen, W.-C., & Gibson, D. T. (1993). Isolation and preliminary characterization of the subunits of the terminal component of naphthalene dioxygenase from *Pseudomonas putida* NCIB 9816-4. *Journal of Bacteriology, 175*.

Tarasev, M., Pullela, S., & Ballou, D. P. (2009). Distal end of 105-125 loop - A putative reductase binding domain of phthalate dioxygenase. *Archives of Biochemistry and Biophysics, 487*(1), 10–18. https://doi.org/10.1016/j.abb.2009.05.008.

Thompson, M. P., Peñafiel, I., Cosgrove, S. C., & Turner, N. J. (2018). Biocatalysis using immobilized enzymes in continuous flow for the synthesis of fine chemicals. *Organic Process Research & Development, 23*(1), 9–18. https://doi.org/10.1021/acs.oprd.8b00305.

Tian, J., Garcia, A. A., Donnan, P. H., & Bridwell-Rabb, J. (2023). Leveraging a structural blueprint to rationally engineer the Rieske oxygenase TsaM. *Biochemistry, 62*(11), 1807–1822. https://doi.org/10.1021/acs.biochem.3c00150.

Tiwari, M. K., Lee, J.-K., Moon, H.-J., & Zhao, H. (2011). Further biochemical studies on aminopyrrolnitrin oxygenase (PrnD). *Bioorganic & Medicinal Chemistry Letters, 21*(10), 2873–2876. https://doi.org/10.1016/j.bmcl.2011.03.087.

Tsai, P.-C., Chakraborty, J., Suzuki-Minakuchi, C., Terada, T., Kotake, T., Matsuzawa, J., ... Nojiri, H. (2022). The a- and b-subunit boundary at the stem of the mushroom like a3 b3-type oxygenase component of Rieske non-heme iron oxygenases is the Rieske-type ferredoxin-binding site. *Applied and Environmental Microbiology, 88*(15).

Wolfe, M. D., Altier, D. J., Stubna, A., Popescu, C. V., Mu¨nck, E., & Lipscomb, J. D. (2002). Benzoate 1,2-dioxygenase from *Pseudomonas putida*: Single turnover kinetics and regulation of a two-component Rieske dioxygenase†. *Biochemistry, 41*, 9611–9626.

Wolfe, M. D., & Lipscomb, J. D. (2003). Hydrogen peroxide-coupled cis-diol formation catalyzed by naphthalene 1,2-dioxygenase. *Journal of Biological Chemistry, 278*(2), 829–835. https://doi.org/10.1074/jbc.M209604200.

Zhu, Y., Jameson, E., Crosatti, M., Schafer, H., Rajakumar, K., Bugg, T. D., & Chen, Y. (2014). Carnitine metabolism to trimethylamine by an unusual Rieske-type oxygenase from human microbiota. *Proceedings of the National Academy of Science of the United States of America, 111*(11), 4268–4273. https://doi.org/10.1073/pnas.1316569111.

Printed in the United States
by Baker & Taylor Publisher Services